High Performance Textiles and Their Applications

The Textile Institute and Woodhead Publishing

The Textile Institute is a unique organisation in textiles, clothing and footwear. Incorporated in England by a Royal Charter granted in 1925, the Institute has individual and corporate members in over 90 countries. The aim of the Institute is to facilitate learning, recognise achievement, reward excellence and disseminate information within the global textiles, clothing and footwear industries.

Historically, The Textile Institute has published books of interest to its members and the textile industry. To maintain this policy, the Institute has entered into partnership with Woodhead Publishing Limited to ensure that Institute members and the textile industry continue to have access to high calibre titles on textile science and technology.

Most Woodhead titles on textiles are now published in collaboration with The Textile Institute. Through this arrangement, the Institute provides an Editorial Board which advises Woodhead on appropriate titles for future publication and suggests possible editors and authors for these books. Each book published under this arrangement carries the Institute's logo.

Woodhead books published in collaboration with The Textile Institute are offered to Textile Institute members at a substantial discount. These books, together with those published by The Textile Institute that are still in print, are offered on the Elsevier website at: http://store.elsevier.com/. Textile Institute books still in print are also available directly from the Institute's website at: www.textileinstitutebooks.com.

A list of Woodhead books on textile science and technology, most of which have been published in collaboration with The Textile Institute, can be found towards the end of the contents pages.

Woodhead Publishing Series in Textiles: Number 153

High Performance Textiles and Their Applications

Edited by

Carl A. Lawrence

The Textile Institute

ELSEVIER

AMSTERDAM • BOSTON • CAMBRIDGE • HEIDELBERG • LONDON
NEW YORK • OXFORD • PARIS • SAN DIEGO
SAN FRANCISCO • SINGAPORE • SYDNEY • TOKYO
Woodhead Publishing is an imprint of Elsevier

WP WOODHEAD PUBLISHING

Published by Woodhead Publishing Limited in association with The Textile Institute
Woodhead Publishing is an imprint of Elsevier
80 High Street, Sawston, Cambridge, CB22 3HJ, UK
225 Wyman Street, Waltham, MA 02451, USA
Langford Lane, Kidlington, OX5 1GB, UK

British Library Cataloguing-in-Publication Data
A catalogue record for this book is available from the British Library

Library of Congress Control Number: 2014940344

ISBN 978-1-84569-180-6 (print)
ISBN 978-0-85709-907-5 (online)

For information on all Woodhead Publishing publications
visit our website at http://store.elsevier.com/

Typeset by Newgen Knowledge Works Pvt Ltd, India

Printed and bound in the United Kingdom

Contents

Contributor contact details

(* = main contact)

Editor and Chapter 8

C. A. Lawrence
School of Design
University of Leeds
Leeds LS2 9JT, UK

E-mail: C.A.Lawrence@leeds.ac.uk

Chapter 1

S. Grishanov
School of Design
University of Leeds
Leeds LS2 9JT, UK

E-mail: S.Grishanov@leeds.ac.uk

Chapter 2

H. M. El-Dessouky
School of Design
University of Leeds
Leeds LS2 9JT, UK

and

Physics Department
Mansoura University
Mansoura, Egypt

E-mail: hassanoptics@yahoo.com; texhed@leeds.ac.uk

Chapter 3

N. Mao
School of Design
University of Leeds
Leeds LS2 9JT, UK

E-mail: n.mao@leeds.ac.uk

Chapter 4

A. R. Horrocks
Institute for Materials Research and Innovation
University of Bolton
Deane Road
Bolton, BL3 5AB, UK

E-mail: a.r.horrocks@bolton.ac.uk

Chapter 5

M. Miraftab
Institute for Materials Research and Innovation
University of Bolton
Deane Road
Bolton, BL3 5AB, UK

E-mail: m.miraftab@bolton.ac.uk

Chapter 6

S. Rajendran* and S. C. Anand
Institute for Materials Research and Innovation

University of Bolton
Deane Road
Bolton, BL3 5AB, UK

E-mail: s.rajendran@bolton.ac.uk

Chapter 7

E. Hardman
formerly of
Madison Filter UK (now Clear Edge Filtration UK)
Suite 1 Festival Way
Festival Park
Stoke-On-Trent, ST1 5SQ, UK

E-mail: ep.hardman@tiscali.co.uk

Chapter 9

Y. Gowayed
Auburn University
Department of Polymer and Fiber Engineering
311 W Magnolia Ave
Auburn, AL 36849, USA

E-mail: gowayya@auburn.edu

Woodhead Publishing Series in Textiles

1 **Watson's textile design and colour Seventh edition**
Edited by Z. Grosicki
2 **Watson's advanced textile design**
Edited by Z. Grosicki
3 **Weaving Second edition**
P. R. Lord and M. H. Mohamed
4 **Handbook of textile fibres Volume 1: Natural fibres**
J. Gordon Cook
5 **Handbook of textile fibres Volume 2: Man-made fibres**
J. Gordon Cook
6 **Recycling textile and plastic waste**
Edited by A. R. Horrocks
7 **New fibers Second edition**
T. Hongu and G. O. Phillips
8 **Atlas of fibre fracture and damage to textiles Second edition**
J. W. S. Hearle, B. Lomas and W. D. Cooke
9 **Ecotextile '98**
Edited by A. R. Horrocks
10 **Physical testing of textiles**
B. P. Saville
11 **Geometric symmetry in patterns and tilings**
C. E. Horne
12 **Handbook of technical textiles**
Edited by A. R. Horrocks and S. C. Anand
13 **Textiles in automotive engineering**
W. Fung and J. M. Hardcastle
14 **Handbook of textile design**
J. Wilson
15 **High-performance fibres**
Edited by J. W. S. Hearle
16 **Knitting technology Third edition**
D. J. Spencer
17 **Medical textiles**
Edited by S. C. Anand

18 **Regenerated cellulose fibres**
Edited by C. Woodings
19 **Silk, mohair, cashmere and other luxury fibres**
Edited by R. R. Franck
20 **Smart fibres, fabrics and clothing**
Edited by X. M. Tao
21 **Yarn texturing technology**
J. W. S. Hearle, L. Hollick and D. K. Wilson
22 **Encyclopedia of textile finishing**
H-K. Rouette
23 **Coated and laminated textiles**
W. Fung
24 **Fancy yarns**
R. H. Gong and R. M. Wright
25 **Wool: Science and technology**
Edited by W. S. Simpson and G. Crawshaw
26 **Dictionary of textile finishing**
H-K. Rouette
27 **Environmental impact of textiles**
K. Slater
28 **Handbook of yarn production**
P. R. Lord
29 **Textile processing with enzymes**
Edited by A. Cavaco-Paulo and G. Gübitz
30 **The China and Hong Kong denim industry**
Y. Li, L. Yao and K. W. Yeung
31 **The World Trade Organization and international denim trading**
Y. Li, Y. Shen, L. Yao and E. Newton
32 **Chemical finishing of textiles**
W. D. Schindler and P. J. Hauser
33 **Clothing appearance and fit**
J. Fan, W. Yu and L. Hunter
34 **Handbook of fibre rope technology**
H. A. McKenna, J. W. S. Hearle and N. O'Hear
35 **Structure and mechanics of woven fabrics**
J. Hu
36 **Synthetic fibres: Nylon, polyester, acrylic, polyolefin**
Edited by J. E. McIntyre
37 **Woollen and worsted woven fabric design**
E. G. Gilligan
38 **Analytical electrochemistry in textiles**
P. Westbroek, G. Priniotakis and P. Kiekens

39 **Bast and other plant fibres**
R. R. Franck
40 **Chemical testing of textiles**
Edited by Q. Fan
41 **Design and manufacture of textile composites**
Edited by A. C. Long
42 **Effect of mechanical and physical properties on fabric hand**
Edited by H. M. Behery
43 **New millennium fibers**
T. Hongu, M. Takigami and G. O. Phillips
44 **Textiles for protection**
Edited by R. A. Scott
45 **Textiles in sport**
Edited by R. Shishoo
46 **Wearable electronics and photonics**
Edited by X. M. Tao
47 **Biodegradable and sustainable fibres**
Edited by R. S. Blackburn
48 **Medical textiles and biomaterials for healthcare**
Edited by S. C. Anand, M. Miraftab, S. Rajendran and J. F. Kennedy
49 **Total colour management in textiles**
Edited by J. Xin
50 **Recycling in textiles**
Edited by Y. Wang
51 **Clothing biosensory engineering**
Y. Li and A. S. W. Wong
52 **Biomechanical engineering of textiles and clothing**
Edited by Y. Li and D. X-Q. Dai
53 **Digital printing of textiles**
Edited by H. Ujiie
54 **Intelligent textiles and clothing**
Edited by H. R. Mattila
55 **Innovation and technology of women's intimate apparel**
W. Yu, J. Fan, S. C. Harlock and S. P. Ng
56 **Thermal and moisture transport in fibrous materials**
Edited by N. Pan and P. Gibson
57 **Geosynthetics in civil engineering**
Edited by R. W. Sarsby
58 **Handbook of nonwovens**
Edited by S. Russell
59 **Cotton: Science and technology**
Edited by S. Gordon and Y-L. Hsieh

60 **Ecotextiles**
Edited by M. Miraftab and A. R. Horrocks
61 **Composite forming technologies**
Edited by A. C. Long
62 **Plasma technology for textiles**
Edited by R. Shishoo
63 **Smart textiles for medicine and healthcare**
Edited by L. Van Langenhove
64 **Sizing in clothing**
Edited by S. Ashdown
65 **Shape memory polymers and textiles**
J. Hu
66 **Environmental aspects of textile dyeing**
Edited by R. Christie
67 **Nanofibers and nanotechnology in textiles**
Edited by P. Brown and K. Stevens
68 **Physical properties of textile fibres Fourth edition**
W. E. Morton and J. W. S. Hearle
69 **Advances in apparel production**
Edited by C. Fairhurst
70 **Advances in fire retardant materials**
Edited by A. R. Horrocks and D. Price
71 **Polyesters and polyamides**
Edited by B. L. Deopura, R. Alagirusamy, M. Joshi and B. S. Gupta
72 **Advances in wool technology**
Edited by N. A. G. Johnson and I. Russell
73 **Military textiles**
Edited by E. Wilusz
74 **3D fibrous assemblies: Properties, applications and modelling of three-dimensional textile structures**
J. Hu
75 **Medical and healthcare textiles**
Edited by S. C. Anand, J. F. Kennedy, M. Miraftab and S. Rajendran
76 **Fabric testing**
Edited by J. Hu
77 **Biologically inspired textiles**
Edited by A. Abbott and M. Ellison
78 **Friction in textile materials**
Edited by B. S. Gupta
79 **Textile advances in the automotive industry**
Edited by R. Shishoo
80 **Structure and mechanics of textile fibre assemblies**
Edited by P. Schwartz

81 **Engineering textiles: Integrating the design and manufacture of textile products**
Edited by Y. E. El-Mogahzy
82 **Polyolefin fibres: Industrial and medical applications**
Edited by S. C. O. Ugbolue
83 **Smart clothes and wearable technology**
Edited by J. McCann and D. Bryson
84 **Identification of textile fibres**
Edited by M. Houck
85 **Advanced textiles for wound care**
Edited by S. Rajendran
86 **Fatigue failure of textile fibres**
Edited by M. Miraftab
87 **Advances in carpet technology**
Edited by K. Goswami
88 **Handbook of textile fibre structure Volume 1 and Volume 2**
Edited by S. J. Eichhorn, J. W. S. Hearle, M. Jaffe and T. Kikutani
89 **Advances in knitting technology**
Edited by K-F. Au
90 **Smart textile coatings and laminates**
Edited by W. C. Smith
91 **Handbook of tensile properties of textile and technical fibres**
Edited by A. R. Bunsell
92 **Interior textiles: Design and developments**
Edited by T. Rowe
93 **Textiles for cold weather apparel**
Edited by J. T. Williams
94 **Modelling and predicting textile behaviour**
Edited by X. Chen
95 **Textiles, polymers and composites for buildings**
Edited by G. Pohl
96 **Engineering apparel fabrics and garments**
J. Fan and L. Hunter
97 **Surface modification of textiles**
Edited by Q. Wei
98 **Sustainable textiles**
Edited by R. S. Blackburn
99 **Advances in yarn spinning technology**
Edited by C. A. Lawrence
100 **Handbook of medical textiles**
Edited by V. T. Bartels
101 **Technical textile yarns**
Edited by R. Alagirusamy and A. Das

102 **Applications of nonwovens in technical textiles**
Edited by R. A. Chapman
103 **Colour measurement: Principles, advances and industrial applications**
Edited by M. L. Gulrajani
104 **Fibrous and composite materials for civil engineering applications**
Edited by R. Fangueiro
105 **New product development in textiles: Innovation and production**
Edited by L.Horne
106 **Improving comfort in clothing**
Edited by G. Song
107 **Advances in textile biotechnology**
Edited by V. A. Nierstrasz and A. Cavaco-Paulo
108 **Textiles for hygiene and infection control**
Edited by B. McCarthy
109 **Nanofunctional textiles**
Edited by Y. Li
110 **Joining textiles: Principles and applications**
Edited by I. Jones and G. Stylios
111 **Soft computing in textile engineering**
Edited by A. Majumdar
112 **Textile design**
Edited by A. Briggs-Goode and K. Townsend
113 **Biotextiles as medical implants**
Edited by M. W. King, B. S. Gupta and R. Guidoin
114 **Textile thermal bioengineering**
Edited by Y. Li
115 **Woven textile structure**
B. K. Behera and P. K. Hari
116 **Handbook of textile and industrial dyeing. Volume 1: Principles, processes and types of dyes**
Edited by M. Clark
117 **Handbook of textile and industrial dyeing. Volume 2: Applications of dyes**
Edited by M. Clark
118 **Handbook of natural fibres. Volume 1: Types, properties and factors affecting breeding and cultivation**
Edited by R. Kozłowski
119 **Handbook of natural fibres. Volume 2: Processing and applications**
Edited by R. Kozłowski
120 **Functional textiles for improved performance, protection and health**
Edited by N. Pan and G. Sun
121 **Computer technology for textiles and apparel**
Edited by J. Hu

122 **Advances in military textiles and personal equipment**
Edited by E. Sparks
123 **Specialist yarn and fabric structures**
Edited by R. H. Gong
124 **Handbook of sustainable textile production**
M. I. Tobler-Rohr
125 **Woven textiles: Principles, developments and applications**
Edited by K. Gandhi
126 **Textiles and fashion: Materials design and technology**
Edited by R. Sinclair
127 **Industrial cutting of textile materials**
I. Viļumsone-Nemes
128 **Colour design: Theories and applications**
Edited by J. Best
129 **False twist textured yarns**
C. Atkinson
130 **Modelling, simulation and control of the dyeing process**
R. Shamey and X. Zhao
131 **Process control in textile manufacturing**
Edited by A. Majumdar, A. Das, R. Alagirusamy and V. K. Kothari
132 **Understanding and improving the durability of textiles**
Edited by P. A. Annis
133 **Smart textiles for protection**
Edited by R. A. Chapman
134 **Functional nanofibers and applications**
Edited by Q. Wei
135 **The global textile and clothing industry: Technological advances and future challenges**
Edited by R. Shishoo
136 **Simulation in textile technology: Theory and applications**
Edited by D. Veit
137 **Pattern cutting for clothing using CAD: How to use Lectra Modaris pattern cutting software**
M. Stott
138 **Advances in the dyeing and finishing of technical textiles**
M. L. Gulrajani
139 **Multidisciplinary know-how for smart textiles developers**
Edited by T. Kirstein
140 **Handbook of fire resistant textiles**
Edited by F. Selcen Kilinc
141 **Handbook of footwear design and manufacture**
Edited by A. Luximon

142 **Textile-led design for the active ageing population**
Edited by J. McCann and D. Bryson
143 **Optimizing decision making in the apparel supply chain using artificial intelligence (AI): From production to retail**
Edited by W. K. Wong, Z. X. Guo and S. Y. S. Leung
144 **Mechanisms of flat weaving technology**
V. V. Choogin, P. Bandara and E. V. Chepelyuk
145 **Innovative jacquard textile design using digital technologies**
F. Ng and J. Zhou
146 **Advances in shape memory polymers**
J. Hu
147 **Design of clothing manufacturing processes: A systematic approach to planning, scheduling and control**
J. Gersak
148 **Anthropometry, apparel sizing and design**
D. Gupta and N. Zakaria
149 **Silk: Processing, properties and applications**
Edited by K. Murugesh Babu
150 **Advances in filament yarn spinning of textiles and polymers**
Edited by D. Zhang
151 **Designing apparel for consumers: The impact of body shape and size**
Edited by M.-E. Faust and S. Carrier
152 **Fashion supply chain management using radio frequency identification (RFID) technologies**
Edited by W. K. Wong and Z. X. Guo
153 **High performance textiles and their applications**
Edited by C. A. Lawrence
154 **Protective clothing: Managing thermal stress**
Edited by F. Wang and C. Gao
155 **Composite nonwoven materials: Improving comfort, performance and protection**
Edited by D. Das and B. Pourdeyhimi
156 **Functional finishes for textiles: Improving comfort, performance and protection**
Edited by R. Paul
157 **Assessing the environmental impact of textiles and the clothing supply chain**
S. S. Muthu
158 **Braiding technology for textiles**
Y. Kyosev

159 **Principles of colour appearance and measurement Volume 1: Object appearance, colour perception and instrumental measurement**
A. K. R. Choudhury

160 **Principles of colour appearance and measurement Volume 2: Visual measurement of colour, colour comparison and management**
A. K. R. Choudhury

1

Modelling structure–property relationships in advanced textile materials

S. GRISHANOV, University of Leeds, UK

DOI: 10.1533/9780857099075.1

Abstract: This chapter provides an overview of various methods used for modelling the geometry, structure, and properties of a wide range of textile materials. Methods of experimental design and data analysis leading to regression models are briefly discussed. Models of internal geometry and structural models based on knot theory are described. Models of mechanical and physical properties are considered next, highlighting the link between structure and properties. Examples of continuous and discrete models of textile processes are presented. This chapter concludes with discussion of linear and non-linear optimization methods in application to the performance characteristics of materials and processes.

Key words: fibre, yarn, woven fabric, knitted fabric, non-wovens, design of experiment, hypothesis testing, analysis of variance (ANOVA), regression analysis, geometrical models, structural models, fibre migration, unit cell, knot invariants, textile mechanics, physical properties of textiles, homogenization, optimization.

1.1 Introduction to textile modelling

New applications of textile fibrous assemblies for automotive, aerospace, construction, agriculture, geo-engineering, maritime industries, medical, smart and sensory materials, protective clothing, wearable electronics, pervasive computing, and other areas require an engineering approach towards the design of materials for specified end-use and performance characteristics. The performance characteristics of textiles depend on two major factors, which are the properties of the constituting elements and their structural arrangement. The relationship between these two sets of factors must be known, preferably in a mathematical form, before the engineering methods of predictive modelling and optimization of textile structures can be effectively implemented.

The latest developments in computer technology and textile machinery have made it desirable to apply mathematical methods to the computer-aided design and predictive modelling of textiles. Modelling can be considered as a research and engineering design method aimed at time reduction, increase in quality, and cost reduction in the development of products and processes by generating a comprehensive understanding of the relationships and factors that affect their performance characteristics. Modelling enables the products and processes to be designed, simulated, tested, and optimized without physical production of prototypes. The application of modelling, which started in the aerospace industry, now finds its way into other branches of industry, including textiles. Market analysis of the reasons for the wide-spread use of modelling methods by the Aberdeen Group (ANSYSa, ANSYSb) has shown that the early application of these methods helps in preventing very costly mistakes which are difficult to correct at the later stages of product development. Companies that have integrated modelling and simulation methods into their practise have managed to maintain consistently high performance and have achieved highest improvements in their new products, at the same time lowering the design and production costs.

Mainstream research in textiles has been, and still is, focused on the investigation and modelling of the relationship between physical properties of fibres and yarns, and the internal geometry and physical properties of textile materials. This focus is explained by the lack of a universal numerical parameter that would be able to characterize the structure of fibrous assemblies. In this respect it is necessary to make a distinction between the terms 'structure' and 'geometry', which often, but not quite correctly, are used interchangeably.

Structural characteristics of textiles can be defined by the mutual positioning of the constitutive elements. Depending on the scale at which the structure is considered, these elements can be molecular chains in the fibre, fibres in yarns and non-wovens, and threads in woven, knitted, and braided materials. The position of elements, together with their geometry, affects the mode of their interaction, e.g. strength of intermolecular bonds or friction at fibre-to-fibre contact points. The position of elements can be described by the above–below relationship between structural elements on an arbitrary chosen projection of a fibre assembly. For example, warp and weft threads in a plain woven fabric are alternatively above or below each other. These threads can be different in diameter, can have different local curvatures, and can be located at different distances from each other, but all these changes in the geometry of the structure do not change their mutual position, i.e. the structure itself. Once the position of the elements has been set and their ends fixed to limit their mutual movement, the action of physical principles, such as the minimum of potential energy of deformation or energy of intermolecular interaction, defines the final geometry of the structure. It is therefore possible to say that for regular textiles and, to a lesser extent, for the random fibre assemblies, the geometry is a function of position. The position of the elements is defined by the processing technology which follows the route from fibres to yarns and then to fabrics and composites. This explains the dual nature of textiles, because they are continuous materials produced from discrete sets of fibres.

According to their fibrous structure the textile materials can be classified into semi-regular, regular, and random. Semi-regular textiles, such as yarns, can be characterized by a high degree of fibre orientation; they are produced directly from fibres by various spinning methods. Many non-woven materials can be considered as semi-regular, in particular those that are sheet materials made from directionally or randomly orientated fibres which are then bonded by regular actions. Needle-punching, stitching, patterned thermal bonding, and hydro-entanglement are examples of the methods used for their production. The structure of regular textiles can be defined by a unit cell repeated in a periodic manner in one, two, or three directions across the fabric so that they can be considered as a particular case of periodic interlaced structures in three dimensions. These are produced from yarns and threads by weaving, knitting, and braiding as well as combination of weaving and knitting, where Malimo knitting-through technique is one example (Zeisberg, 1990). Random textile structures, where many of them are known as non-woven materials, are mainly made directly from fibres and then bonded by cohesion and/or adhesion. The production processes, however, can change the properties of fibres on their way through the processing line by shortening the fibres, or by changing or damaging fibre surface or fibre internal structure. It can be said that the final products are made of fibres which are different in their properties from those that enter the production process.

The relationships between the production process, fibre properties, material structure, and the properties of the final product can be summarized in two diagrams. Figure 1.1 shows the general production route of textile materials from fibres to composites, and at the same time explains the structural hierarchy and inheritance of properties from parent products to those that result from further processing. Figure 1.2, in a simple form, encapsulates the relationships between the properties of fibres, processing technology, material structure, and the material's properties which are affected by temperature, humidity, and other environmental parameters.

The above-mentioned complexity of the relationships between the structure and properties of textiles is further complicated by the non-linear mechanical properties of individual fibres caused by their visco-elastic behaviour, friction between fibres and threads, anisotropy, and statistical distribution of all properties. Modelling such complex materials requires application of a combination of experimental, analytical, and numerical methods, which will be considered in this chapter.

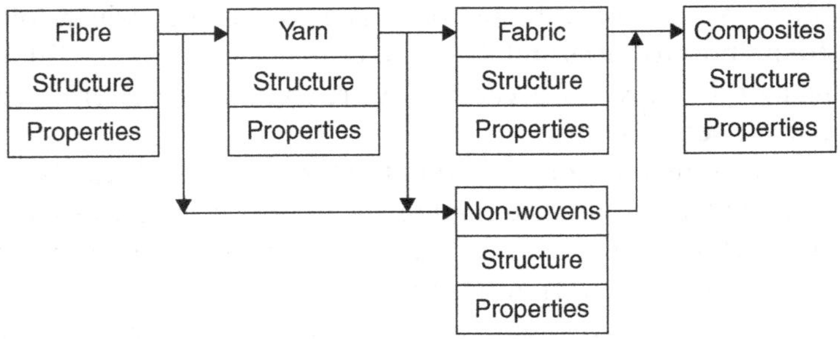

1.1 General production routes of textile materials.

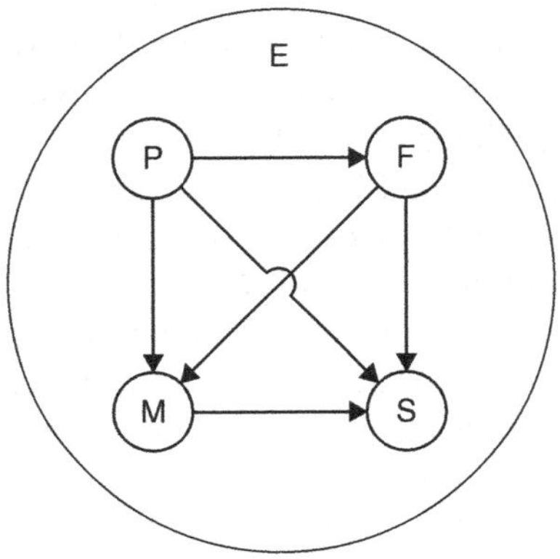

1.2 Process–structure–property relationships in textiles. E – environment parameters, P – process, F – fibre properties, S – material structure, M – material properties.

1.2 Approaches to modelling geometry, structure, properties and processes

It follows from the previous section that there can be several approaches to modelling of textiles. The model can be:

- physical prototype – this replicates the actual process or material on a smaller scale or with a limited range of parameters;
- empirical – based on observations or data measured in an experiment;
- analytical – based on consideration of physical laws expressed as mathematical equations;
- numerical – provides solutions to the mathematical model or simulates the functionality of the process by an algorithm;
- compound – a combination of the above four approaches.

Depending on the degree of idealization assumed in the model it can be:

- Discrete – where the physical objects are considered as consisting of a number of individual elements, and the processes as a number of distinct stages. For example, yarn can be considered as formed from individual fibres, and knitted fabric can be thought of as consisting of a set of threads or a set of unit cells, i.e. loops and their derivatives, repeating in two directions.
- Continuous – in which no distinction is made with regards to structural elements of the physical objects or individual stages of the processes. For example, yarn can be considered as a solid cylindrical body whereas fabric can be considered as a solid sheet material. Following this approach, the movement of a fibrous web through the carding machine can be modelled as a flow of a 'fluid' where the interaction of the machine's working parts with individual fibres and fibres with one another is not taken into account.

Analytical models can be classified into deterministic and stochastic. The former formulates the relationship between the known and unknown factors in the form of equations, the solution of which often requires application of numerical methods. By following prescribed rules the same result can always be obtained from the same starting conditions and initial values of known factors. In the latter, the model contains a degree of uncertainty caused by random events or variations in the values of factors, thus leading to potentially different results even when starting from the same initial conditions.

Models can consider materials and processes in their static condition or describe the changes in parameters against time, in which case the model is usually called 'dynamic'. The parameters of dynamic models can continuously develop with time or converge to a dynamic equilibrium.

The modelling equations can be:

- linear algebraic equations in one or many variables;
- polynomial or transcendental algebraic equations in one or many variables;
- ordinary differential equations;
- partial differential equations;
- integral equations.

Modelling is useful only if it can provide information which is not readily available, or difficult to obtain. This information can usually be obtained from the solution of the modelling equations or from the specific points which characterize these equations. It is worth remembering that the solutions of algebraic equations are numbers, whereas the solutions of differential and integral equations are functions. These specific points can be:

- points of minimum or maximum;
- inflection points;
- limiting boundaries.

For the majority of real-life cases, finding the solutions of modelling equations requires application of numerical methods.

1.3 Experimental models

Experimental investigation is the most common method of studying the properties of objects and processes in modern engineering and science. For the experiment to be carried out effectively, resulting in a valid data that enables meaningful conclusions to be drawn, it is necessary to use statistical principles in the experimental design. The way in which the experiment is designed determines the method of data processing and the results that will be obtained. That is why it is very important that the whole problem should receive comprehensive consideration and be very well thought out before starting actual tests.

Before starting any experiment, as much background information as is available on the problem should be analysed to establish general understanding of the relationships that may or may not exist. On the basis of this, a clear aim for the experiment should be set by determining what exactly is to be found.

The experiment is essentially a series of tests, with a product or a process where controllable changes are made to the independent variables and the effect of these on the dependent variable is measured. It is thus assumed that the dependent variable, y, is an (unknown) function of the independent variables, $x_1, x_2, \ldots, x_m$, as follows:

$$y = f(x_1, x_2, \ldots, x_m) \qquad [1.1]$$

The term 'independent' highlights the fact that it should be possible to change only one of the variables without changing all others.

There are two basic forms of the experimental studies, as follows:

1. Descriptive studies concerned with the characterization of average properties and variation of properties of products and processes. They serve to establish reference points for other types of investigations, and help to avoid obvious mistakes. They may also be focused on the investigation of the opinions of general public and specialists in the relevant area on the quality of products and services, and on specific social, political, and economic problems.
2. Relational experiments are aimed at uncovering the nature and the pattern of the relationships between the variables in question. These experiments

use the results of descriptive studies in order to generate model equations. This approach is most relevant to the multi-variable cases where there are many factors which are difficult to model from first principles.

The quantitative experiment results in numerical data consisting of measurements related to the characteristics of product or process under consideration. The variables associated with these characteristics are quantitative variables. The measured property may be continuous, which may take any value limited only by the precision of the measuring device, or discrete, in which case it takes only integer values resulting from counting.

There may be many independent variables that affect the dependent variable in question, but only the factors that do not correlate with each other must be taken into consideration.

In a controllable (or active) experiment the independent variables are set and maintained on certain levels. The levels should be chosen in such a way that they are practically achievable, and can be reliably measured and maintained; the change between the levels should produce a measurable effect on the dependent variable.

Sometimes the experiment is conducted such that the independent variables are not, or cannot be, controlled. This usually happens in unplanned or passive experiments where samples or objects are measured as they come; this type of the experiment may be easier or less expensive to run, but it requires a greater number of measurements to be taken.

Depending on the number of variables that have been chosen, there may be single- or multi-factor experiments. In the case of a single-factor experiment, generally there are two options for the number of levels. The first is a simple comparison experiment at two levels. The experimental results in this case are processed using hypothesis testing. The second option is concerned with the generation of a mathematical model in the form of Equation [1.1] that describes the relationship between the independent and the dependent variables. The process of fitting the best-suited mathematical model to the experimental data is called regression analysis, which is based on the least squares method. The model itself is called a regression equation, for which polynomial functions are widely used. In the case of one independent variable, it is recommended to have at least five different levels of the independent variable, even if a linear model is to be found. In any case, the number of levels should at least be equal to the number of unknown coefficients plus one.

1.3.1 Design of the experiment

In traditional designs, the effect of independent variables on the dependent variable is studied in a series of tests in which only one factor changes its value while all the others remain constant. In factorial designs, the effect of

Table 1.1 Levels of variables in the 2^2 factorial design

Test no.	Level of variable 1 x_1	Level of variable 2 x_2
1	–1	–1
2	–1	+1
3	+1	–1
4	+1	+1

all possible combinations of levels of all dependent variables is investigated. The advantage of the factorial designs in comparison with the traditional designs is that they require fewer experiments. This advantage increases with the increase in the number of the variables. Factorial designs can be used to study the joint effect (or interaction) of the dependent variables on a response. If there is an interaction between the independent variables, then the effect produced by one of the independent variables at different levels is not the same at all levels of the other factors.

In the simplest factorial design, called a 2^2 design, there are only two independent variables, x_1 and x_2, which can be set at two different levels, termed as 'low' and 'high'; they are usually coded as –1 and +1, as shown in Table 1.1. The link between the natural and the coded values can be expressed by simple equations as follows:

$$x^c_{+1i} = \frac{x_{\max i} - x_{0i}}{\delta_i};\ x^c_{-1i} = \frac{x_{\min i} - x_{0i}}{\delta_i};\ x^c_{0i} = 0 \qquad [1.2]$$

where x^c_{+1i}, x^c_{0i}, and x^c_{-1i} are the coded values for the highest, the middle, and the lowest level of variable i, respectively; $x_{\max i}$, x_{0i}, and $x_{\min i}$ are the natural values for the highest, the middle, and the lowest level of variable i, respectively; and δ_i is the interval between the levels of variable i.

The model equation that can be derived from this experiment is as follows:

$$y = a_0 + a_1x_1 + a_2x_2 + a_{12}x_1x_2 \qquad [1.3]$$

where a_0, a_1, a_2, and a_{12} are unknown coefficients.

The number of possible combinations of factors is $2^2 = 4$, and thus is the minimum number of tests to be done. The tests can be repeated at the same conditions several times in order to increase the accuracy in estimating the value of the dependent variable.

In more complex cases, for example, if there are more than two variables, or if it is necessary to study the effects at three levels of variables, 2^k and 3^k

factorial designs can be used. In general, the factorial designs are denoted as m^k, where m is the number of levels and k is the number of variables.

Specific factorial designs can be considered if the resources are insufficient to run all tests prescribed by the complete factorial design (see for example, Feng *et al.*, 2013). This, in particular, is true for the experiments with the number of factors $k \geq 3$. These designs include fractional factorial designs such as 2^{k-1} and 2^{k-2}, which are called one-half and one-quarter designs, respectively, because they require respectively a half or a quarter of the total number of tests of the full factor experiment.

Response surface designs are particularly useful if:

- there are several independent variables that affect the dependent variable;
- the relationship between the variables is expected to be non-linear where the second degree polynomial is adequate in most cases; and
- the aim of the experiment is to optimize the dependent variable.

A rotatable central composite design is often used in this case. This design consists of 2^k factorial design combined with a number of repeated tests in the central (zero) point plus $2k$ star points. In the central composite factorial design, the independent variables are varied at five levels that are called a zero level, +1 and −1 levels, and two star levels. The +1 and −1 levels provide an increase or decrease in the level of the factor x_i from its zero level, x_{0i}, by one increment, δ_i. Star levels provide an increase or decrease in the level of the factor x_i from its zero level by one increment multiplied by a coefficient, α, which depends on the number of variables in the design. The value of α can be calculated as $\alpha = 2^{k/4}$ (Cochran and Cox, 1957). An example of the two-variable rotatable central composite design is shown in Table 1.2.

The model equation that can be fitted to the experimental data is as follows:

$$y = a_0 + \sum_{i=1}^{k} a_i x_i + \sum_{i=1}^{k} a_{ii} x_i^2 + \sum_{i=1}^{k-1} \sum_{j=i+1}^{k} a_{ij} x_i x_j \qquad [1.4]$$

The designs that serve to reduce the variability due to the known and controllable extraneous variables are:

- randomized complete block design;
- the Latin square design;
- the Greco-Latin square design.

The randomized complete block designs are used when there is one extraneous variable that may affect the result. For example, in testing the fabric

Table 1.2 Levels of variables in the two-variable rotatable central composite design

Test no.	Level of variable 1 x_1	Level of variable 2 x_2	Note
1	−1	−1	2^2 factorial design
2	−1	+1	
3	+1	−1	
4	+1	+1	
5	−1.4142	0	Star points
6	+1.4142	0	
7	0	−1.4142	
8	0	+1.4142	
9	0	0	Zero points
10	0	0	
11	0	0	
12	0	0	
13	0	0	

abrasion resistance on two testers, the result may be affected by variation in operating characteristics of these two testers. In a similar way, Latin square and Greco-Latin square designs are used if it is necessary to take into account two and three extraneous variables, respectively.

1.3.2 Data analysis and interpretation of results

Analysis of data obtained in experiments usually starts with the estimation of statistical measures that characterize the range, the mean value, the variance of the data, and their confidence intervals. Sometimes, when the experiment concerns the identification of changes in the distribution of the dependent factor, such as fibre length or fibre diameter distribution, the analysis continues with the estimation of the skewness and kurtosis, which are measures of the distribution symmetry and sharpness, respectively. Table 1.3 summarizes equations for the calculation of statistical measures. In this table $x_1, x_2, \ldots, x_i, \ldots, x_n$ are individual measurements or observations for a sample of n measurements.

Important property of the variances is that they are additive. In the general case, if two dependent random variables x_1 and x_2 produce the result y, then:

$$y = x_1 \pm x_2;$$

$$\bar{y} = \bar{x}_1 \pm \bar{x}_2;$$

$$s_y^2 = s_1^2 + s_2^2 \pm 2\,\mathrm{cov}(x_1, x_2) \qquad [1.5]$$

Table 1.3 Basic statistical measures

Statistical measure	Equation	Note
Sample mean value	$\bar{x} = \sum_{i=1}^{n} x_i \Big/ n$	A measure of central tendency
Sample variance	$s^2 = \sum_{i=1}^{n} (x_i - \bar{x})^2 \Big/ n-1$	Absolute measure of dispersion
Sample standard deviation	$s = \sqrt{\sum_{i=1}^{n} (x_i - \bar{x})^2 \Big/ n-1}$	
Variation coefficient	$CV = s/\bar{x}$	Relative measure of dispersion
Limits of the confidence interval for the mean	$x_U = \bar{x} + t\frac{s}{\sqrt{n}} \qquad x_L = \bar{x} - t\frac{s}{\sqrt{n}}$	x_U and x_L are upper and lower limits. t is Student's statistics for (n–1) degrees of freedom and confidence probability (1–α)
Limits of the confidence interval for the standard deviation	$s_U = s\sqrt{\frac{n-1}{\chi^2_{\alpha/2,n-1}}} \qquad s_L = s\sqrt{\frac{n-1}{\chi^2_{1-\alpha/2,n-1}}}$	s_U and s_L are upper and lower limits. $\chi^2_{\alpha/2,n-1}$ and $\chi^2_{1-\alpha/2,n-1}$ are Chi-squared statistics for $\alpha/2$ and (1– $\alpha/2$) confidence levels and (n–1) degrees of freedom
Skewness	$g_1 = \frac{\sqrt{n(n-1)}}{n-2}\frac{\frac{1}{n}\sum_{i=1}^{n}(x_i - \bar{x})^3}{\sqrt{(\frac{1}{n}\sum_{i=1}^{n}(x_i - \bar{x})^2)^3}}$	Skewness is positive if in comparison to the normal distribution the peak of the distribution is shifted to the left with a low tail extending to the right, and negative if the peak is shifted to the right and a long low tail extends to the left towards smaller values
Kurtosis	$g_2 = \frac{n(n-1)(n+1)}{(n-2)(n-3)}\frac{\sum_{i=1}^{n}(x_i - \bar{x})^4}{(\sum_{i=1}^{n}(x_i - \bar{x})^2)^2} - 3\frac{(n-1)^2}{(n-2)(n-3)}$	Kurtosis is positive if in comparison to the normal distribution there is a more acute peak around the mean and thicker tails, and negative if distribution has a lower, wider peak and thinner tails

where $\bar{y}$, $\bar{x}_1$, and $\bar{x}_2$ are mean values of y, x_1, and x_2, respectively; s_y^2 is the total variance; s_1^2 and s_2^2 are the variances of factors 1 and 2, respectively; $\text{cov}(x_1, x_2)$ is covariance of variables x_1 and x_2.

This property of the variance is used in the data analysis technique called analysis of variance (ANOVA).

Hypothesis testing is used in the experimental research if the aim of the experiment is to determine whether the difference between two characteristics, such as two means or two standard deviations, is caused by controlled changes in independent variables or to examine the significance of correlation between two sets of data. In statistical terms, hypothesis is a statement about the relationship between two statistical parameters. It includes a null hypothesis, usually stating that two parameters are equal, which is tested against an alternative hypothesis that they are not. Table 1.4 summarizes the equations and the rules for making the decision in favour of one of the hypotheses.

Experimental results obtained from the tests may demonstrate that manipulations with independent variables in some way affect the dependent variable. The question, however, is whether this is caused by the controlled change in the levels of the independent variables or is purely due to the experimental errors. The idea of ANOVA is to split the total variation observed into its components and then to estimate the significance of each component. This results in a relatively simple procedure whereby individual variances can be estimated using sum of squares of the differences between the measurements or sums of measurements and corresponding averages (Montgomery, 1991).

For example, in the case of a one-factor experiment, the total variability of the data, SS_T, is the sum of the variability due to changes in the independent variable, SS_1, and the variability due to uncontrolled factors (experimental errors), SS_E:

$$SS_T = SS_1 + SS_E \qquad [1.6]$$

In the case of a 2^2 factorial design, the total variability includes the variability due to two factors, SS_1 and SS_2, their interaction, SS_{12}, and experimental errors, SS_E:

$$SS_T = SS_1 + SS_2 + SS_{12} + SS_E \qquad [1.7]$$

The total variability of the data and all its components, except for the experimental error, can be estimated through the data; the latter can then be calculated by subtraction. The test of the significance of the component can be done using the ratio of the mean sum of squares related to the

Table 1.4 Hypothesis testing

Hypothesis	Equation		Note
Null hypothesis H_0: $s_1^2 = s_2^2$ Alternative H_1: $s_1^2 \neq s_2^2$	$F_c = \frac{(n_1/n_1-1)s_1^2}{(n_2/n_2-1)s_2^2}$; $df_1 = n_1 - 1$; $df_2 = n_2 - 1$		Fisher's test for the equality of the variances Accept H_0 if $F_c < F_t$ for df_1 degrees of freedom for the numerator and df_2 degrees of freedom for the denominator and significance level α
Student's test for the means			
Null hypothesis H_0: $\bar{x}_1 = \bar{x}_2$ Alternative H_1: $\bar{x}_1 \neq \bar{x}_2$	Variances are equal $s^2 = \frac{(n_1-1)s_1^2 + (n_2-1)s_2^2}{n_1+n_2-2}$	Variances are not equal $c = \frac{s_1^2/n_1}{\frac{s_1^2}{n_1}+\frac{s_2^2}{n_2}}$	Two-sided test. H_1 can be true in two cases: $\bar{x}_1 < \bar{x}_2$ or $\bar{x}_1 > \bar{x}_2$. Accept H_0 if $t_c < t_t$ for df degrees of freedom and level of significance $\alpha/2$
Null hypothesis H_0: $\bar{x}_1 = \bar{x}_2$ Alternative H_1: $\bar{x}_1 < \bar{x}_2$	$t_c = \frac{\lvert\bar{x}_1-\bar{x}_2\rvert}{s\sqrt{\frac{1}{n_1}+\frac{1}{n_2}}}$ $df = n_1 + n_2 - 2$	$t_c = \frac{\lvert\bar{x}_1-\bar{x}_2\rvert}{\sqrt{\frac{s_1^2}{n_1}+\frac{s_2^2}{n_2}}}$ $df = \frac{(n_1-1)(n_2-1)}{(n_1-1)(1-c)^2+(n_2-1)c^2}$	One-sided test. H_1 can be true if $\bar{x}_1 \geq \bar{x}_2$. Accept H_0 if $t_c < t_t$ for df degrees of freedom and level of significance α
Null hypothesis H_0: $r = 0$ Alternative H_1: $r \neq 0$	$r = \frac{\sum_{i=1}^{n}(x_{1i}-\bar{x}_1)(x_{2i}-\bar{x}_2)}{(n-1)s_1s_2}$ $t_c = \frac{r\sqrt{n-2}}{\sqrt{1-r^2}}$; $df = n - 2$		Student's test for the significance of the correlation coefficient. Reject H_0 if $t_c > t_t$ for df degrees of freedom and level of significance α

Table 1.5 ANOVA for one-factor experiment

Source of variation	Sum of squares	Degrees of freedom	Mean square	*F*-value
Independent Variable	SS_1	$df_1 = I - 1$	$MS_1 = SS_1/df_1$	$F_c = MS_1/MS_E$ Find a tabulated value F_t for df_1 and df_E degrees of freedom and level of significance α. If $F_c > F_t$ then it can be concluded that the independent variable has a significant effect on the output
Error	SS_E	$df_E = I(n-1)$	$MS_E = SS_E/df_E$	
Total	SS_T	$df_T = In - 1$		

$$S_i = \sum_{j=1}^{n} y_{ij}\ ;\ \bar{S}_i = S_i/n;\ i=1, 2, \ldots, I$$

$$S_T = \sum_{i=1}^{I} S_i\ ;\ \bar{S}_T = S_T/I$$

$$SS_T = \sum_{i=1}^{I}\sum_{j=1}^{n} y_{ij}^2 - \frac{S_T^2}{nI}$$

$$SS_1 = \frac{1}{n}\sum_{i=1}^{I} S_i^2 - \frac{S_T^2}{nI}$$

$$SS_E = SS_T - SS_1$$

Note: I is the number of levels of independent variable.
n is the number of repeats.

component and the mean sum of squares due to errors. This ratio yields the *F*-value given in Table 1.5. If the calculated *F*-value related to a variable is greater than the tabulated *F*-value associated with corresponding degrees of freedom and probability level, this indicates that the effect of the variable is significant. The calculations are summarized in Tables 1.5 and 1.6. The application of ANOVA is an important stage in the analysis of experimental data (Chudoba *et al.*, 2007; Harpa, 2010a, b).

Regression analysis is one of the main tools in generating mathematical models by fitting a model equation to experimental data. In general, the regression analysis is based on the application of the least squares method for the estimation of unknown coefficients in the model equation. This method minimizes the sum of squares of the differences between the experimental values of the dependent variable, y_{ue}, and those estimated by the model, y_u. Polynomials of various degrees are often used to describe complex non-linear relationships between the dependent and independent variables because the model equation is linear with respect to the unknown coefficients, and therefore the procedure for the calculation of the coefficients reduces to the solution of a system of linear simultaneous equations.

For example, for a second degree polynomial model with k variables the estimated value of y for observation u is:

Table 1.6 ANOVA for 2^2 factorial design

Source of variation	Sum of squares	Degrees of freedom	Mean square	*F*-value
Variable 1	SS_1	$df_1 = l_1 - 1$	$MS_1 = SS_1/df_1$	$F_{1c} = MS_1/MS_E$
Variable 2	SS_2	$df_2 = l_2 - 1$	$MS_2 = SS_2/df_2$	$F_{2c} = MS_2/MS_E$
Interaction	SS_{12}	$df_{12} = (l_1 - 1)(l_2 - 1)$	$MS_{12} = SS_{12}/df_{12}$	$F_{12c} = MS_{12}/MS_E$
Error	SS_E	$df_E = l_1 l_2(n - 1)$	$MS_E = SS_E/df_E$	Find tabulated values F_{1t} for df_1 and df_E degrees of freedom, F_{2t} for df_2 and df_E degrees of freedom, F_{12t} for df_{12} and df_E degrees of freedom, and level of significance α. If any of F_{1c}, F_{2c}, and F_{12c} are greater than the respective tabulated value, then it can be concluded that the independent variable in question (or their interaction) has a significant effect on the output
Total	SS_T	$df_T = l_1 l_2 n - 1$		

$$S_{ij} = \sum_{k=1}^{n} y_{ijk} \; ; i = 1, 2, \ldots, l_1; \; j = 1, 2, \ldots, l_2$$

$$S_i = \sum_{j=1}^{l_2} S_{ij} \; ; S_j = \sum_{i=1}^{l_1} S_{ij} \; ; S_T = \sum_{i=1}^{l_1} \sum_{j=1}^{l_2} S_{ij}$$

$$SS_T = \sum_{i=1}^{l_1} \sum_{j=1}^{l_2} \sum_{k=1}^{n} y_{ijk}^2 - \frac{S_T^2}{n l_1 l_2}$$

$$SS_1 = \frac{1}{n l_1} \sum_{i=1}^{l_1} S_i^2 - \frac{S_T^2}{n l_1 l_2}$$

$$SS_2 = \frac{1}{n l_2} \sum_{j=1}^{l_2} S_j^2 - \frac{S_T^2}{n l_1 l_2}$$

$$SS_{12} = \frac{1}{n} \sum_{i=1}^{l_1} \sum_{j=1}^{l_2} S_{ij}^2 - \frac{S_T^2}{n l_1 l_2} - SS_1 - SS_2$$

$$SS_E = SS_T - SS_1 - SS_2 - SS_{12}$$

Note: l_1 is the number of levels of variable 1.
l_2 is the number of levels of variable 2.
n is the number of repeats.

$$y_u = a_0 + \sum_{i=1}^{k} a_i x_{iu} + \sum_{i=1}^{k} a_{ii} x_{iu}^2 + \sum_{i=1}^{k-1} \sum_{j=i+1}^{k} a_{ij} x_{iu} x_{ju} \qquad [1.8]$$

This can be transformed into linear equation by introducing new variables

$$v_0 = 1; \; v_1 = x_1; \; v_2 = x_2, \ldots, v_{11} = x_1^2; \; v_{22} = x_2^2, \ldots, v_{12} = x_1 x_2; \; v_{13} = x_1 x_3, \ldots, v_{k-1,k} = x_{k-1} x_k \qquad [1.9]$$

and then the least squares function is:

$$L = \sum_{u=1}^{n} \left[y_{ue} - \left(a_0 v_0 + \sum_{i=1}^{k} a_i v_{iu} + \sum_{i=1}^{k} a_{ii} v_{iiu} + \sum_{i=1}^{k-1} \sum_{j=i+1}^{k} a_{ij} v_{iju} \right) \right]^2 \qquad [1.10]$$

where n is the number of observations.

This function must satisfy the set of conditions that its partial derivatives with respect to unknown model coefficients are equal to zero:

$$\frac{\partial L}{\partial a_0}=0;\ \frac{\partial L}{\partial a_1}=0;\ \frac{\partial L}{\partial a_2}=0,\ldots,\frac{\partial L}{\partial a_{11}}=0;$$
$$\frac{\partial L}{\partial a_{22}}=0,\ldots,\frac{\partial L}{\partial a_{12}}=0;\ \frac{\partial L}{\partial \boldsymbol{a}_{13}}=0,\ldots,\frac{\partial L}{\partial a_{k-1,k}}=0 \qquad [1.11]$$

This leads to a system of linear equations as follows:

$$\boldsymbol{X}^T\boldsymbol{X}\boldsymbol{A}=\boldsymbol{X}^T\boldsymbol{Y}, \qquad [1.12]$$

where $\boldsymbol{X}=\begin{vmatrix}1 & v_{11} & \ldots & v_{k-1,k1}\\ 1 & v_{12} & \ldots & v_{k-1,k2}\\ \ldots & \ldots & \ldots & \ldots\\ 1 & v_{1n} & \ldots & v_{k-1,kn}\end{vmatrix}$ is the matrix of observed values of new independent factors according to Equation [1.9]; $\boldsymbol{A}=\begin{vmatrix}a_0\\ a_1\\ \ldots\\ a_{k-1,k}\end{vmatrix}$ is the matrix of unknown model coefficients; $\boldsymbol{Y}=\begin{vmatrix}y_1\\ y_2\\ \ldots\\ y_n\end{vmatrix}$ is the matrix of observed values of the dependent factor.

Note that because v_0=1 the first column in $\boldsymbol{X}$ consists entirely of 1s.

The solution of this system yields the model coefficients

$$\boldsymbol{A}=(\boldsymbol{X}^T\boldsymbol{X})^{-1}\boldsymbol{X}^T\boldsymbol{Y} \qquad [1.13]$$

The coefficients should be tested for significance using F-test.

It is often the case that a one-factor non-linear regression equation has to be fitted to the experimental data. However, a simple change of variable can transform it into a linear equation: see Table 1.7, where the method of selection of the best fit model is also presented.

One of the important factors affecting properties of textiles is the fibre composition. Mixtures of fibres of different origins and in various proportions are widely used in order to achieve specific combination of properties (El-Mogahzy, 2004; El-Mogahzy *et al.*, 2004, 2005; Baykal and Erol, 2006; Reis, 2009; Zhang *et al.*, 2011). Typical examples are cotton/polyester, wool/acrylic, wool/nylon/elastane, glass/polyester, *p*-phenylene terephthalamide/polycarboxylate viscose/PPTA/PCA, multi-coloured melange mixtures, and many others. The studies of the composition–property relationships form specific cases of experimental designs where constraints on the proportions of the mixture components should be taken into account:

Table 1.7 Non-linear models reducible to a linear model

Model No	Model in terms of original variables	Transformation function	Model in terms of transformed variables	Middle point x_m	y_m
1	$y = ax^b$	$y_t = \lg y$ $x_t = \lg x$ $a_t = \lg a$	$y_t = a_t + bx_t$	$\sqrt{x_1 x_n}$	$\sqrt{y_1 y_n}$
2	$y = ab^x$	$y_t = \lg y$ $a_t = \lg a$ $b_t = \lg b$	$y_t = a_t + b_t x$	$\frac{x_1 + x_n}{2}$	$\sqrt{y_1 y_n}$
3	$y = a + \frac{b}{x}$	$x_t = \frac{1}{x}$	$y = a + bx_t$	$\frac{2x_1 x_n}{x_1 + x_n}$	$\frac{y_1 + y_n}{2}$
4	$y = \frac{1}{a + bx}$	$y_t = \frac{1}{y}$	$y_t = a + bx$	$\frac{x_1 + x_n}{2}$	$\frac{2y_1 y_n}{y_1 + y_n}$
5	$y = \frac{1}{a + \frac{b}{x}}$	$x_t = \frac{1}{x}$ $y_t = \frac{1}{y}$	$y_t = a + bx_t$	$\frac{2x_1 x_n}{x_1 + x_n}$	$\frac{2y_1 y_n}{y_1 + y_n}$
6	$y = a + b \lg x$	$x_t = \lg x$	$y = a + bx_t$	$\sqrt{x_1 x_n}$	$\frac{y_1 + y_n}{2}$
7	$y = ae^{bx}$	$y_t = \ln y$ $a_t = \ln a$	$y_t = a_t + bx$	$\frac{x_1 + x_n}{2}$	$\sqrt{y_1 y_n}$

1. Let the experimental data be a set of n pairs of observations $\{x_1, y_1; x_2, y_2, \ldots, x_i, y_i; x_{i+1}, y_{i+1}, \ldots, x_n, y_n\}$.
2. Calculate the middle point (x_m, y_m) for each of the models using formulae in the last two columns.
3. In the experimental data set find points i and $i + 1$ such that either (a) $x_i = x_m < x_{i+1}$ or (b) $x_i < x_m < x_{i+1}$.
4. In the case (a) above assume $y_{mc} = y_i$ otherwise in case (b) calculate y_{mc} by linear interpolation $y_{mc} = y_i + (y_{i+1} - y_i)(x_m - x_i)/(x_{i+1} - x_i)$.
5. Calculate differences $\Delta = |y_{mc} - y_m|$ between the estimated middle point and that calculated in (2).
6. The model which gives the smallest value of Δ can be accepted for further analysis.

Source: Adapted from Sevostianov, 1980.

$$0 \leq x_i \leq 1, \ i = 1, 2, \ldots q \qquad [1.14]$$

$$\sum_{i-1}^{q} x_i = 1 \qquad [1.15]$$

where x_i is the proportion of a component i in the mixture, and q is the number of components.

In these designs a specific simplex coordinate system is used to identify levels of variables in these designs; Fig. 1.3 gives an example of such system for three and four variables. Examples for three- and four-component mixtures, together with the location of experimental points for fitting polynomials of degree $m = 2$ and $m = 3$, are shown in Fig. 1.4.

The constraints [1.14] and [1.15] enable the number of coefficients in the polynomial models to be reduced, resulting in canonical polynomials (Cornell, 1990), often referred to as Scheffe polynomials. For example, in the case of three variables, the second degree polynomial

$$y = a_0 + a_1x_1 + a_2x_2 + a_3x_3 + a_{12}x_1x_2 + a_{13}x_1x_3 + a_{23}x_2x_3 + a_{11}x_1^2 + a_2x_2^2 + a_3x_3^2 \quad [1.16]$$

can be reduced to

$$y = b_1x_1 + b_2x_2 + b_3x_3 + b_{12}x_1x_2 + b_{13}x_1x_3 + b_{23}x_2x_3 \quad [1.17]$$

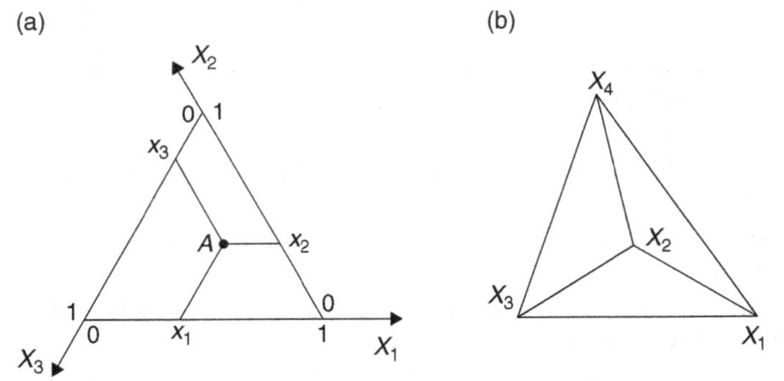

1.3 Simplex coordinate systems: (a) $q = 3$; (b) $q = 4$.

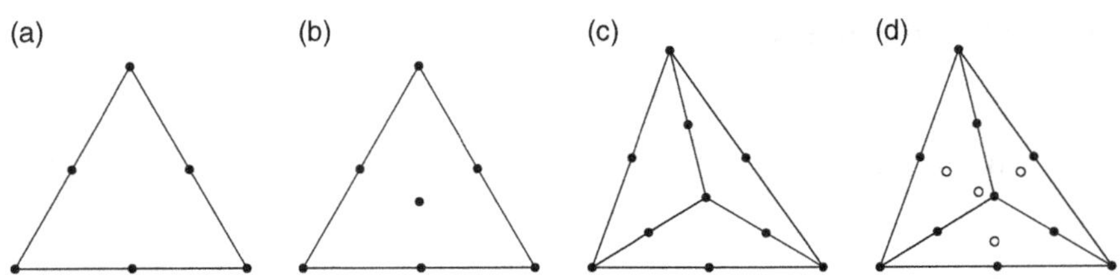

1.4 Experimental points in simplex coordinate systems: (a) $q = 3$, $m = 2$; (b) $q = 3$, $m = 3$; (c) $q = 4$, $m = 2$; (d) $q = 4$, $m = 3$.

where the new coefficients b_i and b_{ij} can be found through the experimental data as follows:

$$\left.\begin{aligned} &b_1 = y_1; b_2 = y_2; b_3 = y_3 \\ &b_{12} = 4y_{12} - 2y_1 - 2y_2 \\ &b_{13} = 4y_{13} - 2y_1 - 2y_3 \\ &b_{23} = 4y_{23} - 2y_2 - 2y_3 \end{aligned}\right\} \quad [1.18]$$

A detailed analysis of the approaches and methods used in experiments with mixtures can be found in Cornell (1990). The multivariate statistical methods, such as multivariate hypothesis testing, principal component analysis, cluster analysis, and discriminant analysis are discussed in Dillon and Goldstein (1984).

1.4 Modelling the geometry of fibre, yarn and fabric

Modelling the geometry and structure of textiles is important for the analysis of the mechanical and physical properties of textiles because it provides necessary information on the spatial shapes, orientation, and mutual position of fibres and threads. Models of textile mechanics require this information in order to define deformations and stresses in fibres and threads that develop under the applied forces. Models of air and water filtration, dyeing and finishing, thermal and acoustic insulation and conductivity, and clothing comfort, in addition to the above, require information on the dimensions and orientation of pores. In medical applications, such as tissue scaffolds, the distribution of pore dimensions is critical for the provision of nutrient and oxygen to maintain cell growth and tissue regeneration.

Geometrical models of woven, knitted, and braided fabrics are relatively simpler than those of semi-regular and random fibrous assemblies, due to the regularity of their structure. This enables only a small unit cell to be considered. However, the models attempting to describe natural shapes of loops and threads in the absence of external forces must take into consideration the mechanical properties of threads and their interaction. The threads' curvatures present in the knitted loops and at the contact regions of warp and weft threads cannot be considered as small, thus making the problem geometrically non-linear. The mutual compression of the threads at the contact regions changes the geometry of the unit cell; friction at contact regions introduces mechanical non-linearity, even without considering visco-elastic properties.

Thus, there are basically three options for the development of geometrical models. One is to assume a purely geometrical approach based on experimentally observed shape of the threads. This is defined by several fixed points and their piece-wise approximation by suitable functions explicitly describing the three-dimensional coordinates of the thread (Onal and Adanur, 2004). This method can only be considered at the initial stages of investigation, in order to obtain the first estimate of the real shape and can only be recommended for a fast geometrical simulation. The method is not sufficiently general because it does not provide any understanding of how the mechanical properties of the thread affect the geometry, e.g. the results obtained for a cotton knitted loop can be different from those for wool.

The second method is to derive loop and thread geometry from the analysis of equilibrium of forces and couples acting on the threads, and between the threads, at the contact regions. However, this method still requires assumptions to be made about the general geometry of the thread and the cross-sectional shape at the points of contacts. The force method requires different formulation of the problem if a new type of deformation is introduced, or if different assumptions about the fabric geometry are made.

The third method is based on the minimization of energy of elastic deformation. The advantage of the energy minimization method over the force method is that the energy is a scalar, whereas forces and couples are vectors. For this reason the formulations for energy components due to different types of deformation can be made and then modified independently to include new features if they appear to be important. The last two methods do not always produce explicit equations for the threads' geometry, because they often require application of numerical methods.

The complexity of geometrical models of semi-regular structures, e.g. yarns and random fibrous assemblies such as non-wovens, can be explained by a high level of randomness in their structure and a large number of elements, usually individual fibres that have to be considered. Furthermore, it is necessary to consider that fibres do not overlap one another, yet at the same time provide the required fibre orientation distribution and fibre packing density.

In modelling the geometry of fibres, yarns, and fabrics, it is important to understand how the model will be used. In this respect it is possible to indicate two different applications:

1. The aim of model development is to use it purely for visualization. In this case for the sake of model simplicity and speed of image generation, individual threads, and hence the fabric itself, can be assumed to be two-dimensional, i.e. they are flat and do not have any thickness. The fibrous structure of individual yarns can be simulated by applying textured patterns to the flat surface being simulated. The simulated yarn and fabric, however, may look very similar to reality due to the effect of shadows applied to appropriate parts of the image. With this method it is even possible to simulate the appearance of the material from different

points of view in three dimensions. This approach is known as the '2.5 dimensional' approach. The disadvantage of this is that it does not provide information which is necessary for the analysis of technical characteristics and performance.

2. The model is to be used in modelling structure-dependent properties, such as mechanical behaviour, filtration, air permeability, or heat insulation. This type of model is universal in the sense that it can be used for visualization but, depending on the degree of details, can be computationally intensive. The main factor here is the scale of the model where structural hierarchy plays a decisive role.

1.4.1 Geometrical models of knitted fabric

There are several known models of plain knitted loop geometry as follows:

Peirce's model (1947) is a purely geometrical model that approximates the loop shape on its projection to the fabric plane by two circular arcs of radius r and a straight line tangent to both arcs:

$$\left.\begin{aligned} &x^2+(y-c)^2=r^2\\ &(x-w/2)^2+y^2=r^2\\ &y=(x-w/4)\tan\theta+c/2\\ &r=w/4+d/2\\ &\theta=\frac{\pi}{2}+\arccos\left(2r/\sqrt{(w/2)^2+c^2}\right)-\arctan(2c/w) \end{aligned}\right\}\qquad[1.19]$$

The side projection is assumed to be an arc of radius $R = 4.172d$. In these equations c, w, and d are course spacing, wale spacing, and yarn diameter, respectively. These parameters define the location of the arcs and thus the straight line.

Leaf and Glaskin (1955) suggested parametric equations for the loop axis in three dimensions:

$$\left.\begin{aligned} &x=ad(1-cas\theta)\\ &y=ad\sin\theta\\ &z=hd(1-\cos\pi\theta/\varphi)/2 \end{aligned}\right\}\qquad[1.20]$$

where a is a coefficient, $\varphi = \pi-\arcsin(2/3)$, d is the yarn diameter, and hd is the maximum height of the axis above the fabric plane.

Leaf's 'elastica' model (1960) takes into account the thread mechanics in bending under the force, P, applied to form a loop:

$$\left.\begin{aligned}x &= b\left[2E(\varepsilon,\varphi)-F(\varepsilon,\varphi)\right]\\ y &= R\sin\left[2b\varepsilon\cos(\varphi/R)\right]\\ z &= R\cos\left[2b\varepsilon\cos(\varphi/R)\right]\end{aligned}\right\} \quad [1.21]$$

where $b^2 = B/P$, B is bending rigidity of the thread, $\varepsilon = \cos\alpha/2$, $\cos\theta/2 = \varepsilon\sin\varphi$, α and θ are the angles between the tangent to the loop and the direction of force P, $E(\varepsilon, \varphi)$ and $F(\varepsilon,\varphi)$ are elliptical integrals of the first and second kind, respectively, and R is the radius of a cylinder which defines the curvature of the loop in side projection.

The use of parametric equations for the loop shape has been made by Choi and Lo (2003):

$$\left.\begin{aligned}x(t) &= at^3 - 1.5at^2 + 0.5(a-w)t\\ y(t) &= 0.5(c+2e)(1-\cos\pi t)\\ z(t) &= 0.5(t_h - d)(1-\cos\pi t)\end{aligned}\right\} \quad [1.22]$$

where t is an independent parameter; a is a coefficient; d is yarn diameter; and c, w, and t_h are defined through course and wale spacing and the fabric thickness. These equations were then used in the energy minimization method.

There have been many publications on the application of force and energy minimization methods to plain, 1 × 1 rib, interlock, and warp-knitted fabrics (Postle and Munden, 1967a, b; Hepworth and Leaf, 1976; de Jong and Postle, 1977a, b, c; Hart *et al.*, 1985a, b) which produced predictive models with a good agreement with experimental data and gradually advanced understanding of factors governing this complex problem. In the recent years, however, there has been a return to geometrical methods used for the description of complex knitted structures such as 1 × 1 purl, balanced rib, Milano rib, single pique, double lacoste, and half and full cardigan (Kurbak and Alpyildiz, 2008, 2009a, b, c; Kurbak and Soydan, 2008). Geometrical modelling was applied to braided structures as well (Rawal *et al.*, 2005; Zhang and Li, 2007). Critical analysis of these models is required in particular to make sure that the loop shape equations provide continuity not only to the three-dimensional coordinates but also to the curvature and the torsion along the loop.

1.4.2 Models of yarn geometry in woven fabric

Yarn geometry in woven fabrics follows a repetitive pattern but it is usually difficult to work out a single equation for the yarn path that would fit to the whole length of the period. This is because in woven fabrics long straight sections of yarns (floats) are often present. The exception is probably the plain woven fabric where a sinusoidal shape of the yarn central axis can be assumed. This leads to equations adopted from Hu and Xin (2008, pp 55–56) and generalized in terms of the warp and weft yarn number as follows:

Warp yarn

$$\left.\begin{aligned} & z=(-1)^{j+1}\sin\left(y/2(d_t+t_t)\right) \\ & \left(x-\left[\frac{(2j-1)d_p}{2}+jt_p\right]\right)^2+\left(z-(-1)^{j+1}\sin\left[\frac{y}{2(d_t+t_t)}\right]\right)^2=\left(\frac{d_p}{2}\right)^2 \\ & y\in(0,4t_t+4d_t);\ x\in\left(jt_p+(j-1)d_p, j(t_p+d_p)\right);\ j\in[1,n_p] \end{aligned}\right\} \quad [1.23]$$

Weft yarn

$$\left.\begin{aligned} & z=(-1)^{j+1}\sin\left(y/2(d_p+t_p)\right) \\ & \left(y-\left[\frac{(2i-1)d_t}{2}+it_t\right]\right)^2+\left(z-(-1)^{i+1}\sin\left[\frac{x}{2(d_p+t_p)}\right]\right)^2=\left(\frac{d_t}{2}\right)^2 \\ & x\in(0,4t_p+4d_p);\ y\in\left(it_t+(i-1)d_t, i(t_t+d_t)\right);\ i\in[1,n_t] \end{aligned}\right\} \quad [1.24]$$

where j and i are the ordinal numbers of warp and weft yarns, respectively; n_p and n_t are the total number of warp and weft yarns to be modelled, respectively; d_p and d_t are the diameters of warp and weft yarns, respectively; and t_p and t_t are the distances between two neighbouring warp and weft yarns, respectively.

This model, however, assumes no lateral compression of warp and weft yarns at the contact regions so that the yarns' cross-sections are treated as circular. There are well-known models that proposed racetrack (Kemp, 1958) and lenticular (Shanahan and Hearle, 1978) cross-sectional shapes, where in the latter an energy analysis of fabric mechanics has been used. Despite recent advances in generation of woven fabric models (Sherburn *et al.*, 2012) the general problem of the yarn three-dimensional shape in woven fabric is far from being resolved. The main problem is that the description of the yarn geometry requires a number of fabric parameters which are affected by the weaving loom set-ups, the dynamics of the weaving process, and the yarn relaxation that follows.

Modelling the geometry of fibres in non-wovens is necessarily based on the application of various stochastic algorithms due to random disposition and shape of fibres in the structure. Numerical methods are used in these models for the generation of the values of random parameters according to the modelling distributions (Press *et al.*, 2002). These algorithms can produce a set of straight fibres randomly positioned in a given volume with the specified fibre packing fraction and fibre orientation distribution function (Coelho *et al.*, 1997; Williams and Philipse, 2003; Bezrukov and Stoyan, 2006; Lee and Jasiuk, 2013). There are algorithms that generate realistically looking curved fibres in three dimensions (Faessel *et al.*, 2005; Altendorf and Jeulin, 2011; Gaiselmann *et al.*, 2013). However, these algorithms do not satisfy all required conditions because they either do not produce a high enough fibre packing fraction, show poor performance in terms of achieving the necessary orientation distribution, do not work well enough for long fibres, or generate overlapping fibres.

1.4.3 Fibre migration in staple fibre yarn

Modelling the internal yarn structure and geometry is one of the most difficult problems, due to a large number of factors affecting the location of hundreds of fibres that have to be taken into account. At the same time, this problem is important for the analysis of yarn mechanics, unevenness, hairiness, and general appearance. Models of yarn geometry have been in continuous development since as early as the 1880s (Muller, 1880), when an equation for the strength of staple fibre yarn was obtained based on assumption of helical fibre path, the equation for which can be conveniently presented in a parametric form

$$\left.\begin{aligned} x(t) &= r\cos t \\ y(t) &= r\sin t \\ z(t) &= h/2\pi t \end{aligned}\right\} \qquad [1.25]$$

where r is the helix radius, h is the helix pitch, and t is an independent parameter.

This assumption has been used in early publications on yarn mechanics (Gegauff, 1907) until Peirce (1947) suggested that fibres actually have a much more complicated shape due to fibre migration. Fibre migration then received much attention first by attempts to characterize the phenomenon (Morton, 1956; Hamilton, 1958; Riding, 1964; Hearle *et al.*, 1965) and then to build a model of yarn geometry (Hearle and Bose, 1965; Hearle and Gupta, 1965; Treloar, 1965; Grishanov *et al.*, 1999, 2011; Morris *et al.*, 1999; Neckar *et al.*, 2006; Sreprateep and Bohez, 2006; Sriprateep and Bohez, 2009; Siewe *et al.*, 2009) and mechanics (Hearle *et al.*, 1969; El-Sheikh and Backer, 1972; van Luijk *et al.*, 1984a, b, 1985; Postle *et al.*, 1988; Tao, 1996).

In the process of yarn formation on conventional ring-spinning machines, individual fibres emerging from the nip of the front rollers change their radial position (migrate) until it is fixed by friction between the fibres generated through centripetal forces arising from the yarn twist and tension. This produces an entangled fibre assembly with a high degree of orientation along the yarn axis. In a discrete model the yarn structure is divided into a number of cross-sections along the yarn axis, where each cross-section is divided into K concentric ring zones of equal width, which in turn consist of a fixed number of virtual locations (VL); each VL can be occupied by at most one fibre at any time (Grishanov *et al.*, 1997). In this model, the radial and angular positions of an individual fibre in the yarn are defined by coordinates of a VL occupied by this fibre; the longitudinal position of the fibre is defined by the coordinate of the cross-section. The migration process can be represented by fibres moving from their current VL to a different VL in the next cross-section. There are many random factors that govern this complex process, such as the distribution of external forces acting on fibre, current spatial shape of the fibre, its mechanical properties, and interaction with the surrounding fibres and their shapes and mechanical properties. The fibre migration from a VL (i, r) in a cross-section L to a VL (j, s) in a cross-section $L + 1$ can be characterized by probability which is linked to the energy of fibre elastic deformation, $U^L_{ir;js}$, associated with this migration

$$p^L_{ir;js} = \frac{\sum_{k=0}^{K}\sum_{l=0}^{|2\pi l|} U^L_{ir;kl} - U^L_{ir;js}}{(N-1)\sum_{k=0}^{K}\sum_{l=0}^{|2\pi l|} U^L_{ir;kl}} \qquad [1.26]$$

where $$U^L_{ir;js} = \int_{\phi_L}^{\phi_{L+1}} \left[\frac{EI_x}{2}b^2(\phi) + \frac{EI_y}{2}q^2(\phi) + \frac{GI_0}{2}w^2(\phi)\right]\frac{ds}{d\phi}d\phi + \frac{P^2_{ir,js}l}{2EF} \qquad [1.27]$$

ϕ_L and ϕ_{L+1} are the angular positions of the fibre cross-section at cross-sections L and $L + 1$, respectively; E and G are Young's and shear module of fibre material, respectively; I_x, I_y, and I_0 are moments of inertia of the fibre cross-section with respect to the principal axes and the central axis, respectively; b and q are curvature components of the fibre axis in the principal directions of the fibre cross-section; w is the fibre torsion; s is the curvilinear coordinate along the fibre axis; $P_{ir;js}$ is the tensile force applied to the fibre; l is the fibre length; F is the area of the fibre cross-section; and N is the number of neighbouring VLs to the VL(i, r).

It is assumed that the distances between the leading ends of the fibres are distributed exponentially with parameter λ^0_{ir} and the fibre length follows a normal distribution. Fibre remains in the VL(i, r) for a migration distance, which is distributed exponentially with parameter μ_{ir}, and after that it either migrates to a neighbouring VL(j, s) with probability $p^L_{ir;js}$ or terminates with probability p^0_{ir}. The process of fibre migration thus can be regarded as a

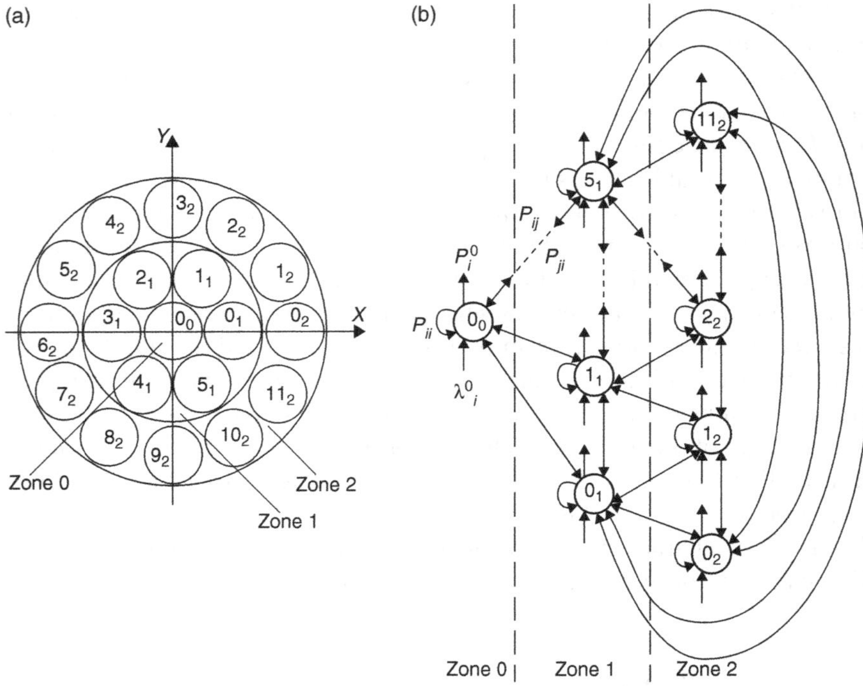

1.5 Fibre migration model: (a) VLs in the yarn cross-section; (b) an open network of queuing systems.

homogeneous Poisson process; this enables fibre migration to be modelled by a Jackson open network of *M*/*M*/1 queuing systems (Gross, 2008) as illustrated in Fig. 1.5 for a simple case of a yarn with three zones.

The number of fibres in the yarn cross-section and in the ring zones changes due to the fibre migration, termination, and new fibres coming. These changes happen in a way that provides a dynamic equilibrium to the yarn structure. The number of fibres in each ring zone can be expressed through the probabilities by a system of simultaneous equations as follows (Grishanov *et al.*, 2011):

$$n_0^{L+1} = n_0^L + p_{10} n_1^L \frac{z_0 - n_0^L}{z_0} + p_{0n}\left(z_0 - n_0^L\right) - p_{0t} n_0^L \frac{z_1 - n_1^L}{z_1} \qquad [1.28]$$

$$\begin{aligned} n_i^{L+1} &= n_i^L + p_{i-1,i} n_{i-1}^L \frac{z_i - n_i^L}{z_i} + p_{i+1,i} n_{i+1}^L \left(\frac{z_i - n_i^L}{z_i}\right) + p_{in}\left(z_i - n_i^L\right) \\ &\quad - p_{i,i+1} n_i^L \frac{z_{i+1} - n_{i+1}^L}{z_{i+1}} - p_{i,i-1} n_i^L \frac{z_{i-1} - n_{i-1}^L}{z_{i-1}} - p_{it} n_i^L \end{aligned} \qquad [1.29]$$

$$n_{K-1}^{L+1} = n_{K-1}^{L} + p_{K-2,K-1} n_{K-2}^{L} \frac{z_{K-1} - n_{K-1}^{L}}{z_{K-1}} + p_{K-1n}\left(z_{K-1} - n_{K-1}^{L}\right) \\ - p_{K-1,K-2} n_{K-1}^{L} \frac{z_{K-2} - n_{K-2}^{L}}{z_{K-2}} - p_{K-1i} n_{K-1}^{L} \qquad [1.30]$$

where n_i^L and n_i^{L+1} are the number of fibres in zone i at the cross-sections L and $L + 1$, respectively; p_{ij}, p_{it}, and p_{in} are the probabilities for the fibre to migrate from zone i to zone j, to terminate in zone i, and to come into zone i, respectively; z_i is the number of VL in zone i; and K is the number of ring zones.

The steady-state number of fibres in the ring zones $\left(n_0^*, n_1^*, \ldots, n_{K-1}^*\right)$ can be found by an iterative application of Equations [1.28]–[1.30]; note that these numbers depend on the values of probabilities p_{ij}, p_{it}, and p_{in}.

It is not possible to find a solution for this system in a conventional way because for a yarn with K zones there are K equations, but the number of unknown probabilities is $4K$–2. The probabilities p_{ij}, p_{it}, and p_{in} were found as a solution to a multi-dimensional minimization problem that provides minimum difference between the experimentally known average number of fibres in the ring zones $\left(n_0^e, n_1^e, \ldots, n_{K-1}^e\right)$ and the steady-state number of fibres $\left(n_0^*, n_1^*, \ldots, n_{K-1}^*\right)$ obtained from the iterations. Powell's method (Himmelblau, 1972) was used in order to find the minimum of the objective function:

$$F\left(p_{ij}, p_{it}, p_{in}\right) = \sum_{m=0}^{K-1} \left\{n_m^e - n_m^*\left(p_{ij}, p_{it}, p_{in}\right)\right\}^2 + G\left(n_0^*, n_1^*, \ldots, n_{K-1}^*\right) \\ + H\left(p_{ij}, p_{it}, p_{in}\right) \rightarrow \min \qquad [1.31]$$

where $G\left(n_0^*, n_1^*, \ldots, n_{K-1}^*\right)$ and $H\left(p_{ij}, p_{it}, p_{in}\right)$ are penalty functions.

The application of this model has shown very good agreement between the predicted and experimentally known average number of fibres in the ring zones. The implementation of this model enabled realistic three-dimensional images of staple fibre yarn to be generated.

This model has been recently extended by Li *et al.* (2013) to remove limitations related to the use of VLs of equal size which cannot accommodate fibres that are different in diameter, such as those in inhomogeneous fibre blends.

1.5 Modelling the structure of textiles

The topological approach has been used in many areas of physics and chemistry, such as the dynamics of DNA supercoiling and recombination properties (Ernst and Sumners, 1990; Flapan, 2000), fluid mixing (Thiffeault and Finn, 2006), discrimination between protein chain structures (Rogen and Bohr, 2003), molecules' behaviour in colloids (Evans and Haw, 2002; Evans and Starrs, 2002), structure of plasmas (Moffatt, 1990), and many others. Work on the application of knot theory to textiles has only started quite recently.

From a mathematical point of view, the structural properties of textiles can be considered as their topological properties. Knot theory is a branch of low-dimensional topology that studies properties of idealized objects, such as knots, braids, links, and tangles, which depend on mutual position of their constituting elements. These mathematical objects can be quite naturally associated with physical objects such as textile fibrous assemblies, because both types essentially are spatial thread-based objects. The differences between the two are in fine details of the properties of the threads which must be assumed to be infinitely thin, frictionless, and continuously deformable without breaking or passing through each other. Under these assumptions, standard mathematical definitions of knots, braids, links, and tangles can be used, and indeed have been used, in application to textile structures.

Basic definitions in a simplified form can be given and illustrated in Fig. 1.6 (Prasolov and Sosinskii, 1997; Cromwell, 2004; Chmutov *et al.*, 2010). A knot is a smooth closed curve embedded in three-dimensional space without self-intersections; a knot which is equivalent to a circle is called an 'unknot'. A link is a collection of knots chained together where each individual knot is a link's component. A braid is a set of ascending simple non-intersecting strings which connect points $A_1, A_2, \ldots, A_n$ on a line with points $B_1, B_2, \ldots, B_n$ on a parallel line. A tangle is a generalization of knots, links, and braids which can contain an arbitrary collection of closed and opened strings with their ends fixed in space.

Knots are represented by diagrams which are planar projections where the intersections of the threads are called crossings and the mutual position of the threads is identified as over-pass and under-pass. Two knots are considered equivalent if they can be transformed into each other by continuous deformation (isotopy) without self-intersection and breaking. These transformations should follow a set of rules (Reidemeister moves) which prevent the generation of geometrically ambiguous configurations shown in Fig. 1.7. Knots are classified in terms of crossing number, which is the minimum possible number of crossings in the knot diagram; in addition to this, links are classified by the number of their components.

Knot invariants are used in knot theory in order to characterize, distinguish, and classify topological properties of knots. A knot invariant is a function of a knot which takes the same value for all equivalent knots. There are numerical, matrix, polynomial, and finite-type invariants. In this section, the application of some numerical and polynomial invariants to textiles will be

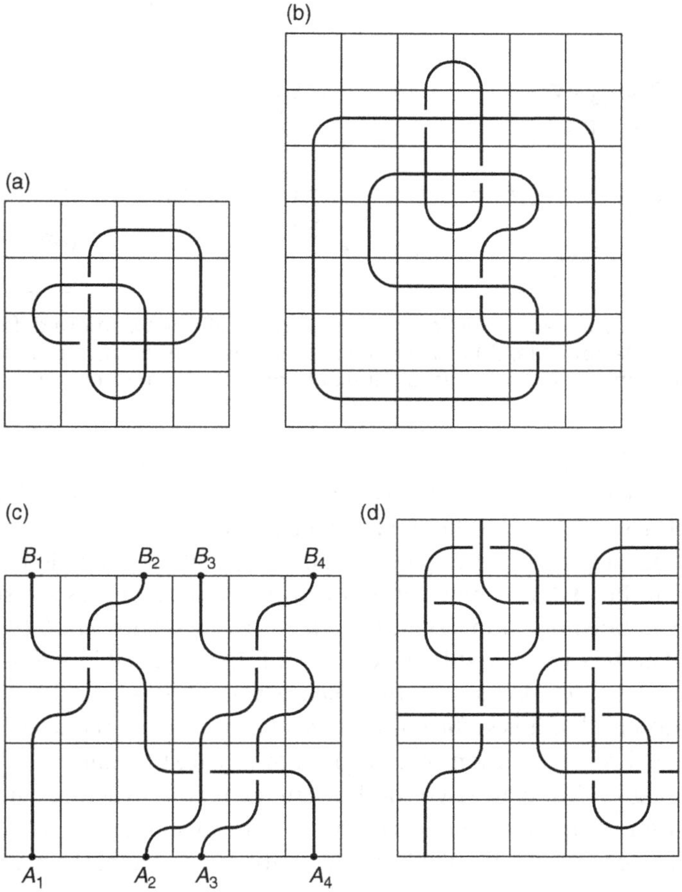

1.6 Examples of basic topological objects: (a) knot; (b) link; (c) braid; (d) tangle.

considered. The problem of classification of textile structures is important in the context of automatic generation of structures where it is necessary to have a general method that would be able to identify topologically different and physically coherent structures which do not contain closed loops.

The structure of regular textiles can be defined by a unit cell periodically repeating in two directions across the fabric. In conjunction with this, the term doubly-periodic structure, or 2-structure for short, has been introduced (Grishanov *et al.*, 2007). In order to characterize the structural properties of a 2-structure, it is sufficient to consider a planar diagram of a unit cell. It has been shown that it is always possible to transform a planar diagram of a unit cell into a unit square. The periodic nature of 2-structures can be modelled by a torus diagram; this can be obtained by joining the opposite sides of the unit

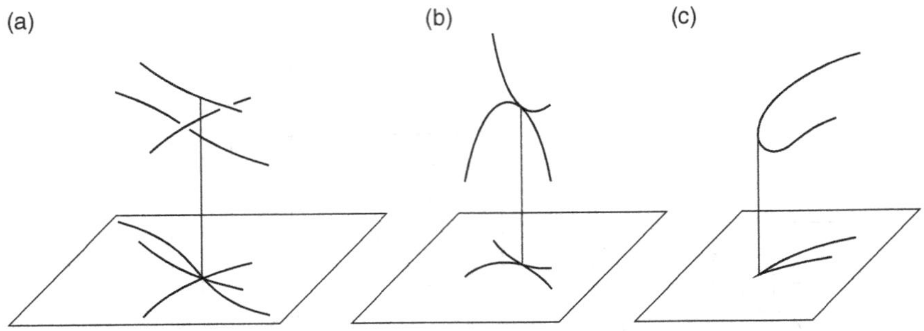

1.7 Prohibited projections: (a) projections of three or more different points coincide; (b) projections of tangents to different curves coincide; (c) tangent to the curve is parallel to the projection direction.

square. The diagram of a unit cell should contain the minimum possible number of crossings. This approach has led to the identification of unit cells for woven fabrics which were different from the conventional definition of repeating patterns based on the periodic manner of operation of textile machinery as shown in Fig. 1.8. Examples of unit cells for knitted fabrics are shown in Fig. 1.9. The polynomial invariant generated by Grishanov *et al.* (2007) has been applied to various textile structures and showed its ability to recognize geometrical characteristics of textiles such as yarn twist – see Table 1.8.

It was found that it is sufficient to have eleven prime structural elements for graphical representation of a diagram of any textile material; these are shown in Fig. 1.10, and are used for drawing the textile structures in this section. This set of prime elements can be used for a simple matrix coding of textile structures, and any 3D fibrous assemblies. For example, matrices M_1 and M_2 respectively for a plain weave and a single jersey are as follows:

$$M_1 = \begin{vmatrix} G & G \\ J & I \end{vmatrix} \quad [1.32]$$

$$M_2 = \begin{vmatrix} D & J & J & C \\ J & F & E & J \\ E & C & D & F \end{vmatrix} \quad [1.33]$$

A new Kauffman-type polynomial invariant has been generated by Grishanov *et al.* (2009a, b) based on Grishanov *et al.* (2007) for the classification of textile structures by considering torus diagrams of unit cells. The invariants for woven and knitted fabrics are given in Table 1.9, where a is the polynomial variable. Terms in brackets are referred to as a *winding* and denoted as (m, n), where m and n are the number of intersections of the windings with a torus meridian and longitude, respectively. It can be

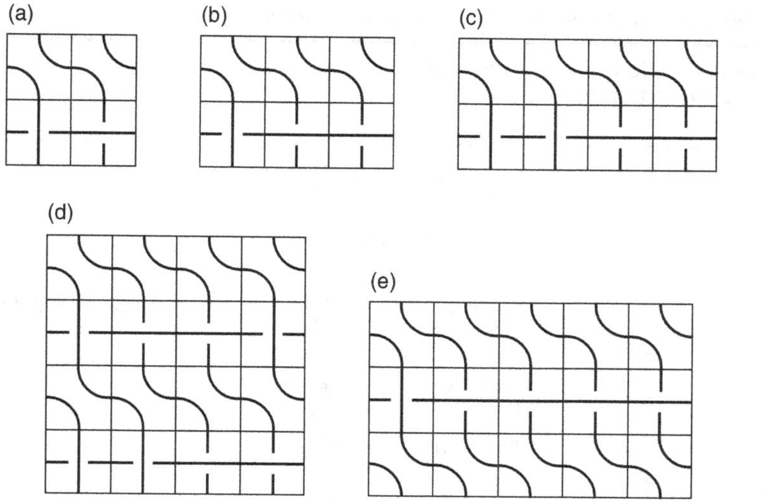

1.8 Minimal unit cells of woven fabrics: (a) plain weave; (b) 1/2 twill; (c) 2/2 twill; (d) 2/2 hopsack; (e) 5/3 sateen.

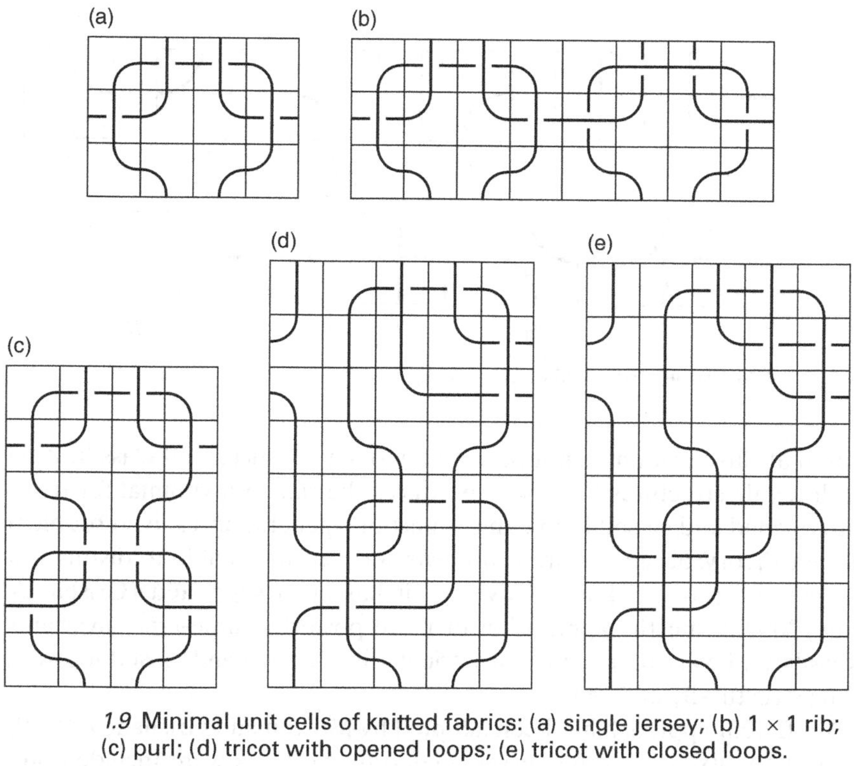

1.9 Minimal unit cells of knitted fabrics: (a) single jersey; (b) 1 × 1 rib; (c) purl; (d) tricot with opened loops; (e) tricot with closed loops.

Table 1.8 Polynomial invariants of twisted yarns

Twist direction Z/S	No of strands	Polynomial
Z	2	$\sum_{i=1}^{2k}(-1)^i a^{-6k+2+4(i-1)} + a^{2k}t^2$
Z	3	$2(a^{-6k} - a^{6k})t + a^{6k}t^3$
Z	4	$2\sum_{i=1}^{2k}(-1)^i a^{12k-2+4(i-1)} + \sum_{i=1}^{4k}(-1)^{i+1}a^{-4k+2+4(i-1)} + 3(a^{-4k} - a^{12k})t^2 + a^{12k}t^4$
S	2	$\sum_{i=1}^{2k}(-1)^{i+1}a^{-2k+2+4(i-1)} + a^{-2k}t^2$
S	3	$-2(a^{-6k} - a^{6k})t + a^{-6k}t^3$
S	4	$\sum_{i=1}^{4k}(-1)^i a^{-12k+2+4(i-1)} + 2\sum_{i=1}^{2k}(-1)^{i+1}a^{4k+2+4(i-1)} - 3(a^{-12k} - a^{4k})t^2 + a^{-12k}t^4$

Note: k is twist level in turns per unit cell.

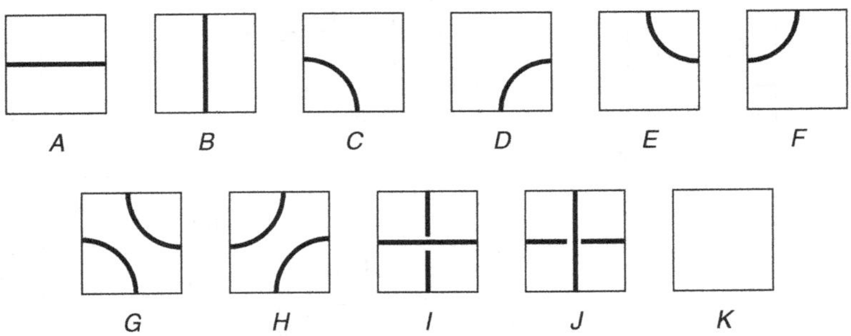

1.10 Prime structural elements.

seen that this invariant is not powerful enough for successful classification of all textile structures because it produces the same polynomial for tricot with opened and closed loops. This is part of a general unresolved problem in knot theory, because there is no invariant yet that is able to distinguish between all possible knots. However, it has been suggested (Grishanov *et al.*, 2009c) that the application of more powerful finite-type invariants (Vassiliev's invariants) can classify tricot with opened and closed loops as being structurally different.

A different approach has been adopted by Morton and Grishanov (2009) where two auxiliary closed curves related to the torus were included into torus diagrams for 2-structures, thus producing a multicomponent link. The

Table 1.9 Polynomial invariants of doubly-periodic textile structures

Textile structure	Polynomial invariant
Plain weave	$(a^{-2} + a^{2})(\mathbf{0, 0}) + (\mathbf{2, 0}) + (\mathbf{0, 2})$
1/2 twill	$a(\mathbf{3, 0}) - (a^{-3} + a + a^{5})(\mathbf{1, 0}) + a^{-1}(\mathbf{1, 2})$
2/2 twill	$(\mathbf{4, 0}) - (a^{-4} + 2 + a^{4})(\mathbf{2, 0}) + (\mathbf{0, 2})$
2/2 hopsack	$-(a^{-10} + a^{-6} - 2a^{-2} - 2a^{2} + a^{6} + a^{10})(\mathbf{0, 0}) + (\mathbf{4, 0}) - (a^{-8} + 2 + a^{8})(\mathbf{2, 0}) + 2(\mathbf{2, 2}) + (\mathbf{0, 4}) - (a^{-8} + 2 + a^{8})(\mathbf{0, 2}) + 2(\mathbf{-2, 2})$
5/3 sateen	$a^{-3}(\mathbf{3, 1}) - (a^{-5} - 2a^{-1} + a^{3})(\mathbf{1, 1}) + a^{3}(\mathbf{-1, 3}) - (a^{-3} - 2a + a^{5})(\mathbf{-1, 1})$
Single jersey	$-(a^{-4} - 3 + a^{4})(\mathbf{1, 0}) + (a^{2} - a^{6})(\mathbf{1, 2}) - (a^{-6} - a^{-2})(\mathbf{-1, 2})$
1 × 1 rib	$(a^{-12} - a^{-8} - 5a^{-4} + 11 - 5a^{4} - a^{8} + a^{12})(\mathbf{1, 0}) - (a^{-4} - 2 + a^{4})(\mathbf{1, 4}) + (a^{-10} - 4a^{-6} + 3a^{-2} + 3a^{2} - 4a^{6} + a^{10})(\mathbf{1, 2}) - (a^{-4} - 2 + a^{4})(\mathbf{-1, 4}) + (a^{-10} - 4a^{-6} + 3a^{-2} + 3a^{2} - 4a^{6} + a^{10})(\mathbf{-1, 2})$
Purl	$(2a^{-10} - 6a^{-6} + 4a^{-2} + 4a^{2} - 6a^{6} + 2a^{10})(\mathbf{0, 0}) + (a^{-12} - 2a^{-8} + a^{-4} + a^{4} - 2a^{8} + a^{12})(\mathbf{2, 0}) - (a^{-4} - 2 + a^{4})(\mathbf{2, 2}) + (a^{-8} - 4a^{-4} + 7 - 4a^{4} + a^{8})(\mathbf{2, 0}) - (a^{-4} - 2 + a^{4})(\mathbf{-2, 2})$
Tricot with opened loops	$-(a^{-4} - 2 + a^{4})(\mathbf{3, 0}) + (a^{-8} - 2a^{-4} + 3 - 2a^{4} + a^{8})(\mathbf{1, 0}) - (a^{-6} - a^{-2} - a^{2} + a^{6})(\mathbf{1, 2}) - (a^{-6} - a^{-2} - a^{2} + a^{6})(\mathbf{-1, 2})$
Tricot with closed loops	$-(a^{-4} - 2 + a^{4})(\mathbf{3, 0}) + (a^{-8} - 2a^{-4} + 3 - 2a^{4} + a^{8})(\mathbf{1, 0}) - (a^{-6} - a^{-2} - a^{2} + a^{6})(\mathbf{1, 2}) - (a^{-6} - a^{-2} - a^{2} + a^{6})(\mathbf{-1, 2})$

Table 1.10 Multi-variable Alexander polynomials for textile structures

Textile structure	Polynomial invariant
Plain weave	$-1 + 2px + 2py - p^{2}x^{2} - e^{2}y^{2} + ((1 - e)^{2}(1 - p)^{2} - 4ep)xy + 2p^{2}ex^{2}y + 2pe^{2}xy^{2} - p^{2}e^{2}x^{2}y^{2}$
Single jersey	$(1 - y)(1 - x - t)(x + t - tx)$
1 × 1 rib	$(tx - t - x)(t + x - 1)(t^{2}x - tx + 1)(t^{2}x - t + 1)(y - 1)$
$M \times N$ rib	$((tx - t - x)(t + x - 1))^{M}((t^{2}x - tx + 1)(t^{2}x - t + 1))^{N}(y - 1)$
Purl	$(1 + (1 - t_1)(1 - t_2)x)((1 - t_1)(1 - t_2) - x)(y - 1)^{2}$
Warp-knitted Pillar-stitch	$(x - 1)(xt + 1 - t)(xt - x + 1)$
Single jersey with a closed thread	$(tx - t - x)(t + x - 1)(t - 1)(y - 1)$
Four non-interlaced threads	$(py - 1)(p^{2}y - 1)(ex - 1)(e^{2}x - 1)$

Note: e is warp and p is weft.

link was then transformed into a braid, and a multicomponent Alexander polynomial, $\Delta_L(x, y, t_1, t_2, \ldots, t_n)$, has been used for the classification of textiles (Morton, 1998) where variables x and y represent two closed curves and variables t_i are for the fabric's threads, $i = 1, 2, \ldots, n$; n is the number of independent components (threads) in the fabric. Table 1.10 gives examples of the polynomial for textile structures. The polynomial can be used for the identification of closed loops in the fabric using the rule that if $t_i = 1$ for all i then

$$\Delta_L(x, y, 1, 1, \ldots, 1) = 0 \qquad [1.34]$$

This is illustrated in Table 1.10 for a single jersey with a closed thread added around the fabric's thread.

The polynomial is also sensitive to the presence of non-interlaced threads, disconnected strips, and layers. In this case, the polynomial factorizes into factors containing the variables related to the individual threads and layers. It can be seen from Table 1.10 that the polynomial for a plain weave cannot be factorized in this way, whereas the polynomial for four non-interlaced threads contains factors depending only on x and e or y and p, where e is warp and p is weft. A different and more powerful method for the identification of disjoined elements in textile structures has been developed in Grishanov and Vassiliev (2011).

One of the numerical invariants, the splitting number, has been used in an attempt to directly estimate the degree of entanglement of non-woven structures (Grishanov *et al.*, 2012a, b) in contrast to indirect methods (Ghassemieh *et al.*, 2002; Mao and Russell, 2006; Xiang *et al.*, 2007). The SEM images of a hydroentangled non-woven fabrics were considered as projections of tangles, and a recursive algorithm has been developed to estimate the splitting number and the degree of entanglement for the samples. The splitting number was defined as the minimum number of crossings that must be switched so that the fibre assembly becomes a disjoined collection of fibres lying in parallel planes while some of these fibres may be knotted to themselves. Similar to the unknotting number (Taniyama, 2009), the splitting number is always less than the crossing number divided by 2. This estimated degree of entanglement can be used as an indicator of how much the fibres in the structure are entangled in comparison to the theoretical maximum of 0.5. For example, the splitting number for a simple structure in Fig. 1.11 with 6 fibres and 22 crossings is 6, and the degree of entanglement is very low at $S = 6/22 = 0.273$. Assuming that the strength of the non-woven fabric is proportional to the degree of entanglement, subject to all other parameters being equal, this structural characteristic can be used for non-destructive estimation of the fabric's strength of real and simulated samples. The correlation between the degree of entanglement and the tensile strength of non-wovens can be established experimentally.

Interest in this new approach of modelling textile structures using knot theory is growing. Irvine (2012) reported an application of braids to bobbin lace design. This led to the generation of an algorithm for enumeration of possible lace patterns, which helped in the identification of four new lace patterns that have never been reported before.

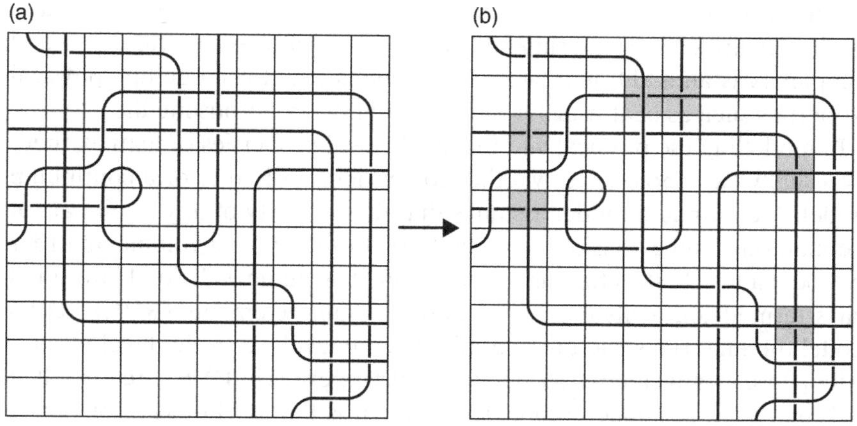

1.11 Example of splitting the fibrous structure: (a) original structure; (b) split structure. Grey unit cells indicate the switched crossings.

1.6 Modelling the mechanical properties of textiles

Mechanical and physical properties of textiles are of great importance in all stages of their production and all areas of application. Strength, extensibility, and frictional properties, together with fibre length and cross-sectional geometry, affect the design of working parts and production rates of spinning equipment. The same properties define fibre cohesion and ultimately the mechanical properties of yarns. Yarn geometry and mechanics are important in the manufacture of woven, knitted, and braided fabrics where again they have effects on the design and productivity of the machinery (Metzkes *et al.*, 2013). The geometry and mechanics of yarns in turn are decisive factors affecting the structural and the mechanical characteristics of the fabrics, which then should be taken into account at further stages of manufacture. At the clothing assembly line curvilinear patterns made from different materials are stitched together; the compatibility of their mechanical properties plays an important role in preventing seams puckering (Hu *et al.*, 2006). The conformity of the textile materials to the human body shape and to the shapes of technical components for automobile and airspace applications is governed by a combination of their structural and mechanical properties (Gibson, 2010). In biomedical applications structural, mechanical, chemical, and surface compatibility of composites with the host tissues are of paramount importance (Ramakrishna *et al.*, 2001).

These examples explain the continuous practical and theoretical interest in modelling all aspects of properties of textiles. In this section, approaches to modelling mechanical and physical properties of textiles are outlined.

It has been mentioned in Section 1.4 that models of textile geometry and mechanics are closely interlinked, where the level of structural detail and complexity is an additional factor; the same is true for modelling physical properties such as heat transfer or air/liquid flow in porous media.

In modelling the physical and the mechanical properties of textiles, there are usually two choices, i.e. whether to consider a discrete or a continuum model. A continuum model assumes that the property of any small part of the material can be considered equal to that of the whole volume. In order to model a textile structure as a continuum, its volume is divided into small parts termed unit cells or representative volume elements (RVE). RVEs model the material structure at a microscopic level, i.e. at the level of individual fibre arrangements. The mechanical properties of RVEs are modelled and then used at a macroscopic level, which is the level of yarn or fabric, under the assumption that the whole volume of the material can be re-constructed from a number of RVEs.

The decision in favour of a continuum model can be made only if there are a sufficiently large number of structural elements in the RVEs of the material or, in other words, the dimensions of the RVE are much greater than the dimensions of the structural elements. The effect of each individual element on the property in question would be small, so that a continuous or even a uniform distribution of the properties over the representative volume can be considered and the average value is used in analysis. The limit when it is no longer possible to use the continuum model is different for different types of materials and different types of properties. The question of how the unit cell can be defined, depending on statistical characteristics of geometrical arrangement of structural elements, has been considered by Zeman and Sejnoha (2007) where examples of statistically equivalent unit cells were given.

Textile materials can often be characterized by preferential orientation and symmetry in fibre arrangement, so that they can be considered not as general anisotropic, but orthotropic or even transversely isotropic materials (Fig. 1.12). This simplifies the models' development and experimental verification where there would be a smaller number of parameters to be measured. For example, the linear elastic behaviour of anisotropic material can be described in a matrix form as follows:

$$\sigma = S\varepsilon \qquad [1.35]$$

$$\varepsilon = C\sigma \qquad [1.36]$$

where $\varepsilon = (\varepsilon_x, \varepsilon_y, \varepsilon_z, \gamma_{yz}, \gamma_{xz}, \gamma_{xy})^T$ and $\sigma = (\sigma_x, \sigma_y, \sigma_z, \tau_{yz}, \tau_{xz}, \tau_{xy})^T$ are strain and stress vectors, respectively; ε and γ with indices are tensile and shear strains,

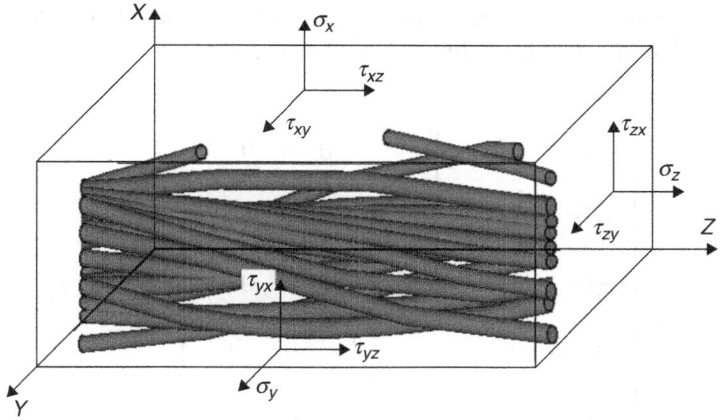

1.12 Fibre assembly in a unit cell.

respectively; σ and τ with indices are tensile and shear stresses, respectively; and $\boldsymbol{S}$ and $\boldsymbol{C}$ are 6×6 stiffness and compliance matrices, respectively.

Out of 36 elements in matrices $\boldsymbol{S}$ and $\boldsymbol{C}$ above, there are only 21 independent elements due to symmetry in stresses and strains. For orthotropic materials, which have different properties in orthogonal directions, there are nine independent constants in matrix $\boldsymbol{C}$

$$\boldsymbol{C} = \begin{vmatrix} \frac{1}{E_x} & -\frac{\nu_{xy}}{E_y} & -\frac{\nu_{xz}}{E_z} & 0 & 0 & 0 \\ -\frac{\nu_{xy}}{E_x} & \frac{1}{E_y} & -\frac{\nu_{yz}}{E_z} & 0 & 0 & 0 \\ -\frac{\nu_{xz}}{E_x} & -\frac{\nu_{yz}}{E_y} & \frac{1}{E_z} & 0 & 0 & 0 \\ 0 & 0 & 0 & \frac{1}{G_{yz}} & 0 & 0 \\ 0 & 0 & 0 & 0 & \frac{1}{G_{xz}} & 0 \\ 0 & 0 & 0 & 0 & 0 & \frac{1}{G_{xy}} \end{vmatrix} \quad [1.37]$$

Transversally orthotropic materials are characterized by isotropic behaviour in the XY plane but display different properties in the z direction. For example, it can be assumed that the fibre orientation and packing density are symmetrical with respect to the yarn's central axis leading to isotropy of the yarn cross-section. Due to this, the number of independent constants in matrices $\boldsymbol{S}$ and $\boldsymbol{C}$ can be reduced to five as follows:

$$\boldsymbol{C} = \begin{vmatrix} \frac{1}{E_t} & -\frac{\nu_{tt}}{E_t} & -\frac{\nu_{tz}}{E_z} & 0 & 0 & 0 \\ -\frac{\nu_{tz}}{E_t} & \frac{1}{E_t} & -\frac{\nu_{tz}}{E_z} & 0 & 0 & 0 \\ -\frac{\nu_{tz}}{E_t} & -\frac{\nu_{tz}}{E_t} & \frac{1}{E_z} & 0 & 0 & 0 \\ 0 & 0 & 0 & \frac{1}{G_{tz}} & 0 & 0 \\ 0 & 0 & 0 & 0 & \frac{1}{G_{tz}} & 0 \\ 0 & 0 & 0 & 0 & 0 & \frac{1}{G_{tt}} \end{vmatrix}, \qquad [1.38]$$

where index t stands for transverse direction, i.e. x or y.

It can be seen from a number of publications on textile mechanics that the modelling methods progressed from force and torque equilibrium analysis to energy methods and to the finite-element and RVE analysis. This trend can be traced through the entire range of textile products, from yarns to woven, knitted, braided and non-woven fabrics, and to composites.

Analysis of the tensile behaviour of continuous filament and staple fibre single yarns has been presented in a fundamental publication by Hearle *et al.* (1969); this analysis was based on a discrete model with a detailed consideration of fibre geometry and fibre migration in the yarn. The distribution of strains and stresses in the yarn have been analysed at small and large extensions where the effects of transverse forces and fibre slippage were taken into account. It was noted that in this method the stress–strain relations for fibres were assumed to be linear and it would be difficult to apply the same approach in the non-linear case; energy-based analysis is simpler, but it does not give the stress distribution over the yarn cross-section. Treloar and Riding (1963) used strain-energy analysis to predict the stress–strain behaviour of a continuous filament yarn over large range of strains. Van Luijk *et al.* (1984a, b) assumed yarn as continuum and developed a specific finite-element model that made it possible to consider fibre migration. Postle *et al.* (1988) considered yarn bending, torsion, and lateral compression using both discrete and continuous models of the yarn structure. The most general finite-element approach to yarn mechanics has been proposed by Munro *et al.* (1997a, b) where a cubic element having 48 different modes of deformation was introduced. The model enabled twisting, bending, extension, and compression deformations to be simulated on a universal basis.

Hearle and Shanahan (1978a, b) applied the energy method to the analysis of the tensile and bending behaviour of woven fabrics. The discrete models of a plain woven fabric have been examined by Leaf and Kandil (1980) and Leaf *et al.* (1993) using the energy method. A unit cell of the plain woven fabric with one warp and one weft thread was analysed, resulting in the formulation of energy components related to different types of deformation. The expression for total energy was then used to derive tensile modulus and bending rigidity in initial deformations. However, recently it has been shown (Li *et al.*, 2011) that through the analysis of symmetries it is possible to reduce the size of the unit cell to 1/16 of the unit cell usually considered in plain woven fabric mechanics. Cavallaro *et al.* (2007) analysed plain woven fabric performance under the combined actions of biaxial tensile, bending, and shear deformations in inflatable structures using finite-element analysis. Mishra (2013) applied the finite-element method to triaxial woven fabrics.

The mechanics of knitted fabrics received similar amount of attention where energy minimization methods (Hart *et al.*, 1985a, b; Hu *et al.*, 1998; Adanur and Onal, 2004; Kyosev *et al.*, 2005; Choi and Lo, 2006) and finite-element method (Loginov *et al.*, 2002a, b, c) addressed the problems of loop configuration and fabric tensile deformation.

There has been a rapidly growing interest in structural applications of textile reinforced composites over the past few decades. This interest can be explained by advantages that composites have over other materials used in structural engineering, such as the high stiffness and strength that can be achieved at low weight and manufacturing cost. The prediction of mechanical properties of composites is complicated by the necessity to take into account:

- differences in mechanical properties of fibres and matrix,
- complex geometry of fibres and yarns in the reinforcing layers,
- statistical distribution of material properties,
- initiation and propagation of failure both in matrix and in the fibrous assembly accompanied by energy dissipation (Miravete, 1999; Bunsell and Renard, 2005).

This stimulated advances in the development of the finite-element method in combination with a multi-scale homogenization procedure. In general, the multi-scale modelling of structure–property relationships of textile reinforced composites can be summarized in the diagram in Fig. 1.13 as follows:

1. Model the properties of constituting elements at structural level (i–1), e.g. fibres and matrix, or use experimentally measured/theoretically predicted values.
2. Model the geometry and structural arrangement of constituting elements. At this stage either periodic or random unit cell with equivalent

periodic boundary conditions can be considered depending on whether the textile material in question has a repetitive or a random structure. In the unit cell nodal points should be identified where the displacements will be calculated.

3. Establish modelling equations which link properties of constituting elements and their structural arrangement to the properties of the assembly of structural elements. Modelling equations must include: (a) constitutive equations which describe material properties in terms of the relations between the deformations and the stresses where additional damage variables influence the elastic module and energy dissipation; (b) kinematic equations which provide links between deformations and displacements at the nodal points; and (c) depending on the type of problem equations of static or dynamic equilibrium of all forces.
4. Find solution to modelling equations using appropriate numerical methods; this provides property characteristics of the unit cell at the current structural level. Estimate the effective stiffness matrix using one of the homogenization methods (Klusemann and Svendsen, 2010).
5. Increment the structural level and repeat the process from (1) now considering that the properties obtained at previous level are readily available.

A detailed discussion of various homogenization methods can be found in Bardella (2000), Kouznetsova (2002), and Klusemann and Svendsen (2010).

There are a number of published papers in the area of composites, probably exceeding those in the area of textiles themselves; examples of a few recent publications on the application of the homogenization method in modelling the mechanical properties of fibre-reinforced composites are Saidpour and Oscar (2006), Maimi *et al.* (2008), Massart *et al.* (2011), and Melro *et al.* (2012).

1.7 Modelling the physical properties of textiles

Modelling physical properties of textiles, e.g. fluid or gas flow, is important for the engineering design of filters in automotive, medical, and food industry applications, for the production of better quality resin-impregnated composites (Mao and Russell, 2000a, b; Sobera and Kleijn, 2006; Zhong and Pan, 2007; Brochocka *et al.*, 2012; Payen *et al.*, 2012; Xiao *et al.*, 2012). Modelling is also important for the prediction of the ability of textile structures to transfer moisture and heat (Li *et al.*, 2002; Li and Zhu, 2003; Cimilli *et al.*, 2008; Barauskas and Abraitiene, 2011); these are essential properties that affect comfort in everyday wear, of sport, and protective clothing in a variety of environmental conditions.

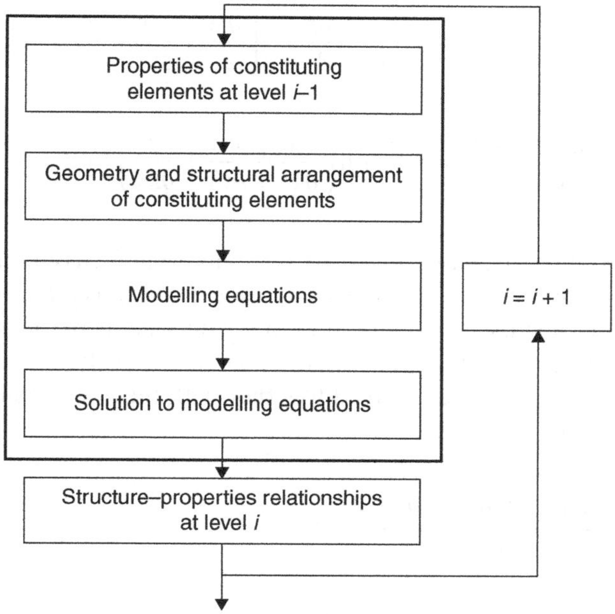

1.13 Structural hierarchy in modelling structure–property relationships.

Modelling physical properties has many common points with that of the textile mechanics. First of all, the structural arrangements at micro- (fibre), meso- (yarn), and macro-levels (fabric) need to be modelled. Similar to Section 1.6, the structure can be considered at different levels of detail and a choice should be made between discrete and continuous models. In contrast to modelling the textile mechanics where the structure modelling is concentrated on fibres and yarns, the distribution of dimensions and orientation of voids (pores) between the fibres and yarns is important for models of fluid flow. Closely related to this are models of filtration where in addition to the distribution of dimensions and shapes of particles, their interactions with the fibrous structure should be considered (Chernyakov *et al.*, 2011).

Second, constitutive equations for the fluid flow should be formulated. These are Navier-Stokes equations, which basically are the Newton's equations of motion written for fluid movement under the action of forces of external pressure and gravity and taking into account shear forces inside the fluid:

$$\rho\left(\frac{\partial u}{\partial t}+u\frac{\partial u}{\partial x}+v\frac{\partial u}{\partial y}+w\frac{\partial u}{\partial z}\right)=-\frac{\partial p}{\partial x}+\mu\left(\frac{\partial^2 u}{\partial x^2}+\frac{\partial^2 u}{\partial y^2}+\frac{\partial^2 u}{\partial z^2}\right)+\rho g_x \qquad [1.39]$$

$$\rho\left(\frac{\partial v}{\partial t}+u\frac{\partial v}{\partial x}+v\frac{\partial v}{\partial y}+w\frac{\partial v}{\partial z}\right)=-\frac{\partial p}{\partial y}+\mu\left(\frac{\partial^2 v}{\partial x^2}+\frac{\partial^2 v}{\partial y^2}+\frac{\partial^2 v}{\partial z^2}\right)+\rho g_y \quad [1.40]$$

$$\rho\left(\frac{\partial w}{\partial t}+u\frac{\partial w}{\partial x}+v\frac{\partial w}{\partial y}+w\frac{\partial v}{\partial z}\right)=-\frac{\partial p}{\partial z}+\mu\left(\frac{\partial^2 w}{\partial x^2}+\frac{\partial^2 w}{\partial y^2}+\frac{\partial^2 w}{\partial z^2}\right)+\rho g_y \quad [1.41]$$

where ρ is the volume density of the fluid; u, v, and w are the components of flow velocity vector in x, y, and z directions, respectively; p is the pressure; μ is the viscosity of the fluid; and g_x, g_y, and g_z are components of the acceleration vector due to gravity in x, y, and z direction, respectively.

These equations should be resolved together with the equation expressing the continuity of the flow:

$$\frac{\partial \rho}{\partial t}+\frac{\partial(\rho u)}{\partial x}+\frac{\partial(\rho v)}{\partial y}+\frac{\partial(\rho w)}{\partial z}=0 \quad [1.42]$$

A particular case of unidirectional laminar fluid flow (at low Reynolds numbers) through the porous media is described by Darcy's law:

$$-\frac{\partial p}{\partial x}=\frac{\mu}{k}V \quad [1.43]$$

where μ is the fluid viscosity; k is specific permeability of the porous structure; and V is the flow velocity.

Equation [1.43] for the case of high Reynolds numbers has a correction factor known as Forchheimer coefficient, F:

$$-\frac{\partial p}{\partial x}=\frac{\mu}{k}V+\frac{\rho F}{\sqrt{k}}V^2 \quad [1.44]$$

This coefficient is known to be difficult for evaluation, but one recent publication (Tamayol *et al.*, 2012) has suggested a way for its estimation.

The solution to the model of the flow interaction with a fibrous assembly requires application of numerical methods collectively known as computational fluid dynamics (CFD) implemented in computer software packages which use finite-element methods.

A relatively new approach to the resolution of the fluid dynamics problem is the lattice Boltzmann method (LBM). LBM models the fluid flow as a movement of imaginary particles interacting with the obstacles (fibres) presented as a discrete lattice mesh. This method has an advantage over the finite-element method in handling irregular boundaries present in porous materials, turbulent flows, and multi-phase cases; the method can be more effectively implemented using parallel computing (Succi, 2001). Examples of application of LBM to modelling permeability of yarns and fabrics can be found in Koponen *et al.* (1998), Nabovati *et al.* (2009), Turan *et al.* (2012), and Turan and Okur (2013).

1.8 Modelling textile processes: continuous models

Modelling the processes of textile technology is perhaps the most difficult and complex area of textile research. Models of this type involve an additional layer of complexity associated with the difficulties in describing a direct link between the properties of the input product, the parameters of the process, and the properties of the output product.

Continuous models of the processes of textile technology can be considered if it is necessary to understand the dynamic behaviour of the process as a continuous function of time, or if a continuous transformation of one product into another by a specific process is of interest. These models can be applied to modelling fibre flow through opening, carding, doubling, and spinning machines, to controlling devices such as warp tension and letting-off mechanism on weaving and warp-knitting machines; also to the movement of yarns and threads in weaving, knitting and breading, to the movement of machine parts, and many other similar cases.

One of the methods applied for modelling a continuous transformation of an input product to the output product by a processing unit, where both are characterized by a variation in their properties, is based on the application of control theory and signal processing theory. The processing unit, e.g. carding machine, can be modelled as a linear time-invariant system where the transformation process is described by a dynamic characteristic called a transfer function:

$$W(s) = \frac{L\{q_{\text{out}}(t)\}}{L\{q_{\text{in}}(t)\}} = \frac{Q_{\text{out}}(s)}{Q_{\text{in}}(s)} \qquad [1.45]$$

where $L\{\}$ is the Laplace transform; $Q_{\text{in}}(s)$ and $Q_{\text{out}}(s)$ are Laplace transforms (images) of the functions $q_{\text{in}}(t)$ and $q_{\text{out}}(t)$ that describe the properties of the input and output products, respectively; t is time and $s = \rho + \omega i$ is a complex variable; and ρ and ω are real numbers.

The one-to-one correspondence between many typical functions and their images can be found in mathematical handbooks (Korn and Korn, 1968).

Through the knowledge of the transfer function it is possible to model the process's response to the periodic, random, and step variations in the properties of the input product and to predict the resultant variations in the output product. The modelling can be done in three steps as follows:

1. Find the Laplace transform of the input $Q_{in}(s)=L\{q_{in}(t)\}$;
2. Find the Laplace transform of the output $Q_{out}(s)=Q_{in}(s)W(s)$;
3. Apply the inverse Laplace transform $q_{out}(t)=L^{-1}\{Q_{out}(s)\}$ to find the response as a function of time.

Additional information on the process behaviour can be obtained from the analysis of amplitude response function as follows:

$$A(\omega)=\frac{Q_{out}(\omega)}{Q_{in}(\omega)}=\sqrt{R^2(\omega)+I^2(\omega)} \qquad [1.46]$$

where R and I are real and imaginary parts of transfer function $W(i\omega)=R(\omega)+iI(\omega)$.

It is also possible to obtain the spectral density, $S_{out}(\omega)$, the variance, D_{out}, and the variation coefficient, CV_{out}, of the linear density of the output product using the relationships as follows:

$$\left.\begin{aligned} S_{out}(\omega) &= S_{in}(\omega)\left|W(i\omega)\right|^2 \\ D_{out} &= \int_0^{\infty} S_{out}(\omega)\,d\omega \\ CV_{out} &= \frac{\sqrt{D_{out}}}{\bar{q}_{out}}100\% \end{aligned}\right\} \qquad [1.47]$$

where $|W(i\omega)|$ is the module of the transfer function.

In order to find the transfer function for the process performed by a specific machine or device, it is necessary first to consider the sequence of sub-processes (stages) performed by the machine and generate a flow diagram which reflects the interaction between the individual stages. The original diagram should be re-arranged and simplified in order to produce a final diagram where the transfer function for the whole process will be expressed through the functions for the individual stages. The transfer function for each stage can be found by applying the Laplace transform to the (differential) equations describing the transformations taking place at that stage.

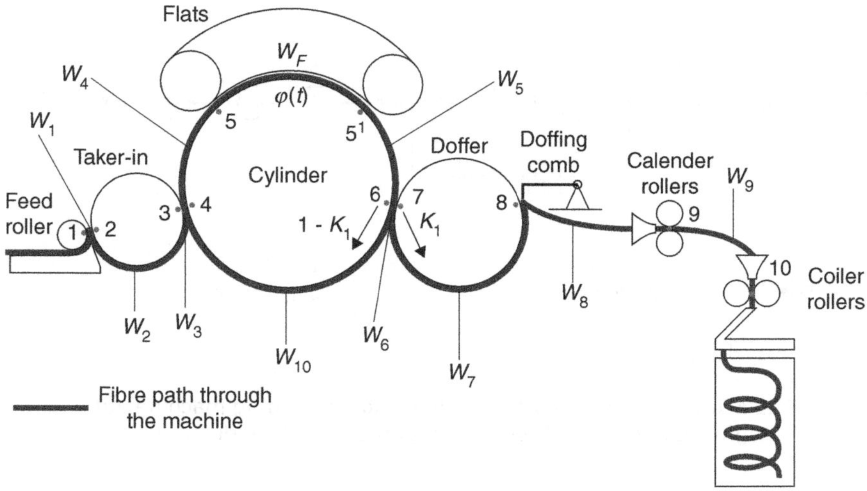

1.14 Basic arrangement of a revolving flat carding machine.

In the model that follows the analysis in Sevostianov and Sevostianov (1984) and Ashnin (1985), a revolving flat cotton carding machine is considered (see Fig. 1.14). A detailed description of the functionality of the carding machine can be found in Lawrence (2003, pp 103–112). It is assumed that the carding machine is fed with a fibrous web which is considered as a continuous flow of fibres where no distinction between individual fibres is made. In this way, the properties of individual fibres such as length, linear density, and fineness are ignored. This is a reasonable simplification, because it can be estimated that, if the linear density of the web is $T_w = 400$Ktex and the linear density of a cotton fibre is $T_f = 0.2$tex, then on average there can be as many as $N_f = T_w/T_f = 400\,000/0.2 = 2\,000\,000$ fibres in any cross-section of the incoming web. In this condition, the influence of an individual fibre on the whole process is negligibly small.

Let $q_{\text{in}}(t)$ be the mass of fibres coming into the machine per unit of time. The resultant product is a sliver which in a similar way can be considered as a continuous flow of fibres so that $q_{\text{out}}(t)$ is the mass of fibres going out of the machine per unit of time. In this simple model, it is assumed that there is no fibre loss during the process, i.e. all material which enters the machine leaves it as an output product. The distances between the working parts of the machine are small, in the order of 0.25 mm; this together with high velocities of taker-in and cylinder makes it possible to assume that the fibre transfer between the working parts happens instantly. The only time delay that has to be accounted for is the fibre travel time, τ, from point 1 to point 10. The fibre flow splits into two parts at point 6, after which a portion of the total flow is transferred to point 7 on the doffer; this portion is defined by

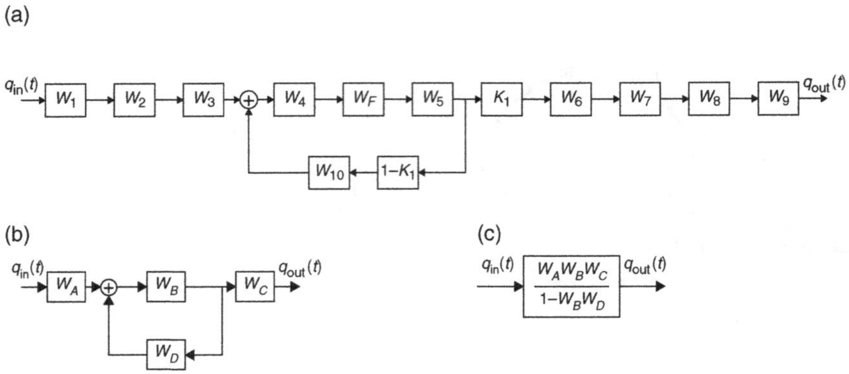

1.15 Transfer functions of the revolving flat carding machine: (a) original diagram; (b) first re-arrangement; (c) final diagram.

a transfer coefficient, K_1. A $(1-K_1)$ portion returns to point 2 where it joins the incoming flow of fibres. It is assumed that the wire clothing of the cylinder and that of the flats in zone $5\text{–}5^1$ has a retaining ability so that they can accumulate certain amount of fibres $\varphi(t)$ which varies with time. The fibre flow $q_5^1(t)$ at point 5^1 can be assumed to be proportional to the amount of fibres accumulated by the cylinder and flats with a coefficient c_c:

$$q_5^1(t) = c_c\varphi(t) \qquad [1.48]$$

The fibre flow through the machine can be represented by a diagram in Fig. 1.15a where transfer functions W_1 to W_{10} and W_F characterize the transformation of the fibre flow at points 1–10 and by cylinder-flats interaction, respectively. The first re-arrangement produces a simpler diagram, Fig. 1.15b, where,

$$W_A = W_1W_2W_3;\ W_B = W_4W_FW_5;\ W_C = K_1W_6W_7W_8W_9;\ W_D = (1-K_1)W_{10} \qquad [1.49]$$

The final diagram (Fig. 1.15c) gives the transfer function for the carding machine as follows:

$$W = \frac{W_AW_BW_C}{1-W_BW_D} \qquad [1.50]$$

Subject to the above-mentioned assumptions introduced to the model, the transfer function of the carding machine can be found in the form:

$$W(s) = \frac{1}{1+Ts}\frac{e^{-\tau s}}{E} \quad [1.51]$$

where τ is the total time for a fibre to travel from point 1 to point 10; $T = \dfrac{1}{c_c K_1}$ is the time constant; and E is the total draft which is the ratio of the velocity of coiler roller to the velocity of feed roller.

The knowledge of transfer function enables the performance of the carding machine to be investigated using Equation [1.47].

The linear density of the web which is fed into the carding machine randomly varies with time. This variation is commonly known as unevenness, which is a universal phenomenon across all textile products. Unevenness can be classified as random and periodical. The production processes should be optimized in order to reduce the level of unevenness and in particular the amplitude of its periodic components. The usual approach to the analysis of the unevenness is to consider the product's linear density as a random function. The application of correlation and spectral analysis enables the periodic components of the unevenness and their contribution to the total level of unevenness to be identified.

The spectral density function can be obtained by integration as follows:

$$S(\omega) = \frac{2}{\pi}\int_0^\infty r(\delta)\cos\omega\delta d\delta \quad [1.52]$$

where $r(\delta)$ is correlation function for the product's linear density; δ is the distance between the product cross-sections where its linear density was measured; and ω is the frequency.

Figure 1.16 shows examples of correlation function and spectral density for various typical cases of unevenness.

Let the input web have only random unevenness characterized by the correlation function as follows:

$$r_{in}(\tau) = D_{in}e^{-\alpha\delta} \quad [1.53]$$

where D_{in} is the variance of the linear density and α is a coefficient.

The spectral density then is:

$$S_{\text{in}}(\omega) = \frac{2}{\pi}\int_0^\infty D_{\text{in}}e^{-\alpha\delta}\cos\omega\delta; \quad d\delta = \frac{2D_{\text{in}}\alpha}{\pi(\alpha^2+\omega^2)} \quad [1.54]$$

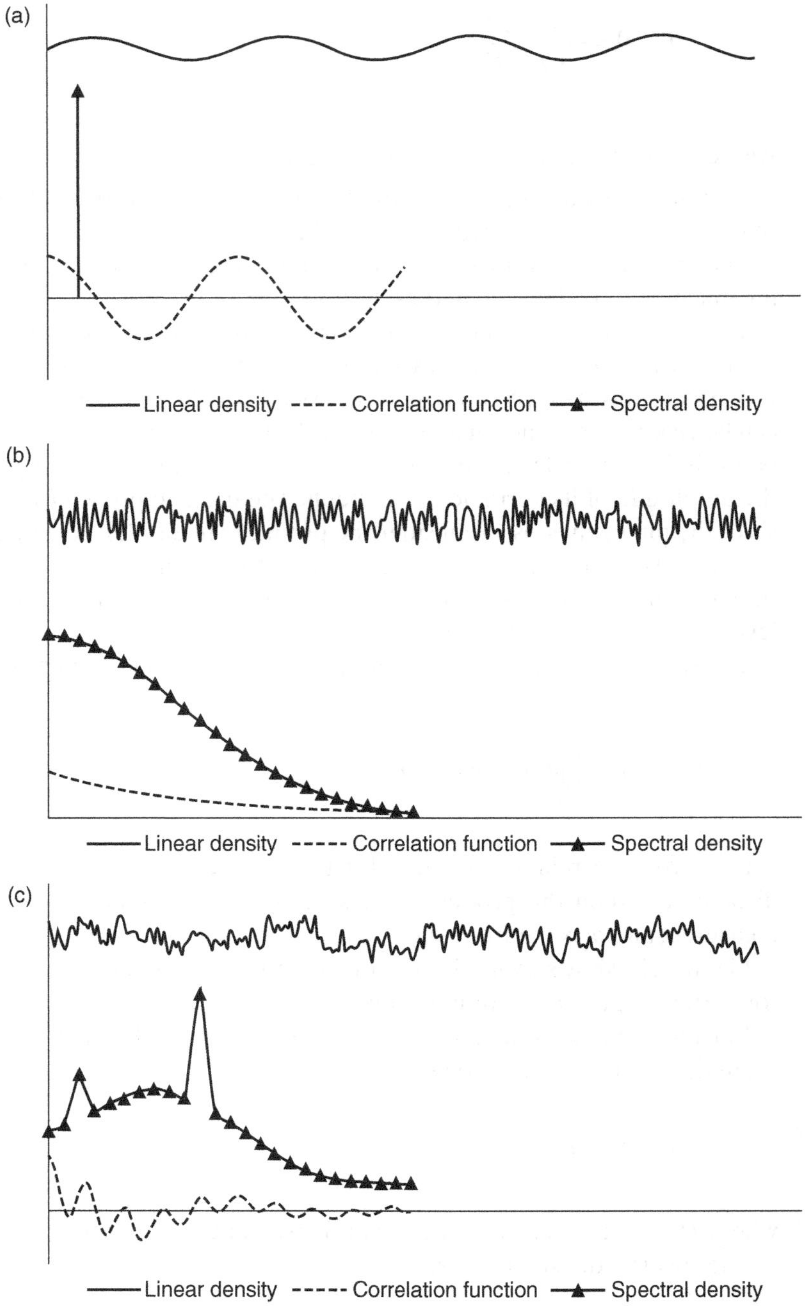

1.16 Examples of linear density, correlation function $r(\delta)$, and spectral density function $S(\omega)$: (a) periodic unevenness; (b) random unevenness; (c) combination of two-harmonic periodical and random unevenness.

The squared module of the transfer function of the carding machine can be found as follows:

$$|W(s)|^2 = \frac{1}{E^2(1+\omega^2 T^2)} \qquad [1.55]$$

The spectral density for the output sliver can be obtained by substituting Equations [1.54] and [1.55] into Equation [1.47]:

$$S_{\text{out}}(\omega) = \frac{2D_{\text{in}}\alpha}{\pi(\alpha^2+\omega^2)} \frac{1}{E^2(1+\omega^2 T^2)} \qquad [1.56]$$

Now the variance of the sliver linear density can be obtained as:

$$D_{\text{out}} = \int_0^{\infty} S_{\text{out}} d\omega = \frac{D_{\text{in}}}{E^2(1+\alpha T)} \qquad [1.57]$$

It can be seen that the carding machine has a levelling-out effect which reduces the variation coefficient of the linear density of input web. The variation coefficients for the web and the sliver can be worked out as follows:

$$C_{\text{in}} = \sqrt{\frac{D_{\text{in}}}{\bar{g}_{\text{in}}^2}}$$

$$\bar{g}_{\text{out}} = \frac{\bar{g}_{\text{in}}}{E}$$

$$C_{\text{out}} = \sqrt{\frac{D_{\text{out}}}{\bar{g}_{\text{out}}^2}} = \sqrt{\frac{D_{\text{in}}E^2}{\bar{g}_{\text{in}}^2 E^2(1+\alpha T)}} = \frac{C_{\text{in}}}{\sqrt{1+\alpha T}} \qquad [1.58]$$

where $\bar{g}_{\text{in}}$ and $\bar{g}_{\text{out}}$ are average linear densities of input and output product, respectively.

The reduction coefficient,

$$c = \frac{C_{\text{in}}}{C_{\text{out}}} = \sqrt{1+\alpha T} \qquad [1.59]$$

does not depend on the frequency of the input. This effect can be increased by:

- Increasing α. This means that for the input web the short waves of unevenness should dominate the unevenness spectrum and that changes in the linear density should have a high degree of randomness. This can be achieved by properly adjusting the opening and cleaning units at the preparation stage of fibre processing.
- Increasing T by increasing the amount of fibres $\varphi(t)$ accumulated on the surface of the cylinder. This can be achieved by increasing the diameter of the cylinder and by extending the area of interaction between the flats and the cylinder. As an alternative, a tandem carding system can be used which employs two carding cylinders one after another. This system has the additional advantage of causing less damage to the fibres but tandem machines are expensive to run (Lawrence, 2003, pp 103–112).

The simplifications in this model can be relaxed by taking into account the wastes generated by taker-in, flats, and cylinder as a proportion of the total amount of the fibre mass being processed by the machine (Sevostianov and Sevostianov, 1984). These proportions can be characterized by experimentally measured coefficients c_1, c_2, and c_3, respectively.

The transfer function in this case will be:

$$W(s) = \frac{(1-c_1)(1-c_c)K_1}{1+Ts-(1-c_2)(1-c_3)(1-K_1)\mathrm{e}^{-\tau_1 s}}\frac{\mathrm{e}^{-\tau_1 s}}{E} = \frac{C_1}{1+Ts-C_2\mathrm{e}^{-\tau_1 s}}\frac{\mathrm{e}^{-\tau_1 s}}{E} \quad [1.60]$$

The squared module of the transfer function of the carding machine can be found as follows:

$$|W(s)|^2 = \frac{C_1^2}{E^2\left[1+\omega^2T^2+C_2^2+2C_2\sqrt{1+\omega^2T^2}\sin(\omega\tau_1-\vartheta)\right]} \quad [1.61]$$

It can be seen that in comparison to Equation [1.56] the spectral density for the output sliver will have a periodic component and a phase shift:

$$S_{\mathrm{out}}(\omega) = \frac{2D_{\mathrm{in}}\alpha}{\pi(\alpha^2+\omega^2)} = \frac{C_1^2}{E^2\left[1+\omega^2T^2+C_2^2+2C_2\sqrt{1+\omega^2T^2}\sin(\omega\tau_1-\vartheta)\right]}$$

[1.62]

It can also be seen that the reduction in the amplitude of unevenness depends on the frequency, i.e. amplitudes of short waves of unevenness will be reduced to a greater extent than those of long waves.

This approach has been extended to two-dimensional analysis of the carding process (Cherkassky, 1994) which, in addition to the fibre movement through the machine, included lateral diffusion of fibres across the carding machine; this is important for the prediction of properties of non-woven materials made from carded webs. This approach has been used to model the joint actions of 'swift-worker-stripper' (Rust and Gutierrez, 1994) and to the whole roller card (Sevostianov and Sevostianov, 1984; Ashnin, 1985). The results of more recent publication (Kuo *et al.*, 2013a) have been used in engineering design of a roller card control system (Kuo *et al.*, 2013b).

Experimental verification of this modelling method has shown good agreement between the theoretically predicted and experimentally obtained results. However, the application of this method relies on the knowledge of several constants such as coefficients c_c and K_1 in Equations [1.48] and [1.51] that are difficult to measure experimentally because of their dependence on fibre properties, roller settings, geometry of wire clothing, and the speed of working parts. A comprehensive investigation of fibre behaviour in carding has been reported by Lee (2001), who probably for the first time established theoretical framework and generated analytical models for the analysis of forces acting on fibres and the effects of the above-mentioned parameters. This work should be continued in order to advance understanding of this important process.

1.9 Modelling textile processes: discrete models

In contrast to the continuous models, the discrete models consider the processes at the level of individual structural elements, e.g. individual fibres, threads or loops, or individual stages of the process. In these models the processes are modelled as a series of states where the transition from one state to another happens with a probability. The underpinning theories for these models are theory of Markov processes (Kemeny and Snell, 1960), queuing theory (Gross *et al.*, 2008), and finite automata theory (Anderson, 2006; Hopcroft *et al.*, 2007).

Modern textile industry employs electronically controlled machinery that has the advantage of being easily programmable for the production of a wide range of products according to the current market demands. Discrete models are often used when designing digital electronic devices. This section gives an example of the application of a discrete model to the knitting process, thus bridging the gap between the continuous and the discrete processes.

The knitted structure can be made from three basic structural elements, which are a knitted loop, a tuck loop, and a float (Spencer, 2001). The formation of a knitted structure from these elements on a knitting machine is a result of a sequence of operations with the threads and loops performed

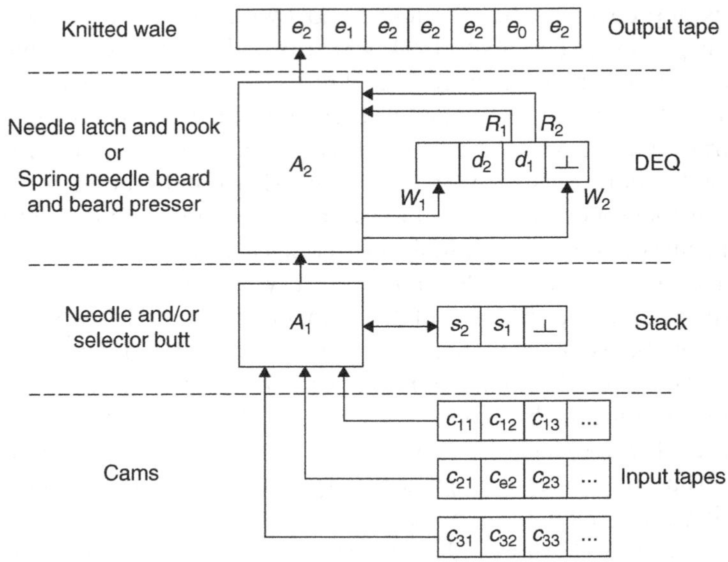

1.17 A discrete model of the knitting machine.

by the needles due to their reciprocal movement. The needles are moved in the tricks by the cams acting on the butts situated on the needle or a selector. The sequence of movements is defined by the sequence of the cams, their configuration, and the length of the butt. It can be said that the needle is a program-controlled device that recognizes and performs an algorithm determined by the sequence of cams (Grishanov *et al.*, 1997). The functionality of the needle can be modelled by a finite automaton – a device that recognizes and functions according to an algorithm (Anderson, 2006; Hopcroft *et al.*, 2007).

In this model (Fig. 1.17), each cam track can be considered as an input tape containing input symbols of an algorithm written on the tape. The needle butt is the reading head of the controlling automaton A_1 with a stack memory which reads the input symbols and produces an output symbol. The output symbol, which is the direction and the distance of the needle movement, determines the operation to be performed by the needle latch represented by the operational automaton A_2. The space under the hook and on the needle stem provides storage of the knitted structural elements and their transfer to the wale or to another needle. The needle can have a hook and a latch at the opposite ends; these type of needles is used for the manufacture of purl structures. The structural elements therefore can be moved to the space under the hook from both sides of the needle. According to this, automaton A_2 has a memory for the storage of symbols, i.e. the structural elements, in the form of a double-ended queue (DEQ) (Knuth, 1997). The automaton A_2 performs the operation defined by the A_1, transforms the contents of its memory, and produces an output symbol which is the corresponding structural element. In this way, the modelling of the knitting process is reduced to the analysis of the sequence of input symbols and their transformation to the output symbol.

The finite automaton A_2 can be defined by the ordered set as follows (Bratchikov, 1975):

$$A_2 = (Q, V_i, V_s, V_o, \mu, \delta, z, q, F) \qquad [1.63]$$

where Q is the finite set of automaton states; V_i, V_s, and V_o are alphabets of input, stack, and output symbols, respectively; μ and δ are the transition and output functions that determine a new automaton state and the symbols which are written into automaton memory or to an output tape; z is the initial symbol in the memory; q is the initial state; and F is the finite set of final states.

The basic needle positions relative to its resting position constitute the set of states Q given in Table 1.11 where superscript $i = 1, 2$ denotes the needle bed number. The alphabet of the output symbols can be defined as $V_o = (e_0^i, e_1^i, e_2^i)$ where e_0^i is a float, e_1^i is a tuck loop, and e_2^i is a knitted loop. The alphabet of stack symbols is $V_s = (d_0^i, d_1^i, d_2^i, \varepsilon, \perp)$ where d_0^i is a yarn being fed prior to loop formation, d_1^i is the tuck loop held by the

Table 1.11 Automaton states

Automaton state	Needle position
q_0^i	Rest position. This is both the initial state q and the final state F
q_1^i	Intermeshing, knocking over, and sinking have been finished
q_2^i	The latch has been closed and the landing has been performed
q_3^i	Tuck height has been reached and the new yarn was fed onto the needle
q_4^i	Full clearing height has been reached and the new yarn was fed onto the needle
q_5^i	Needle position when receiving the loop in a loop transfer process
q_6^i	Needle position when transferring the loop in a loop transfer process
q_7^i	The engagement of both hooks of a double-headed purl needle with the sliders of opposite needle beds (used only for purl machines)

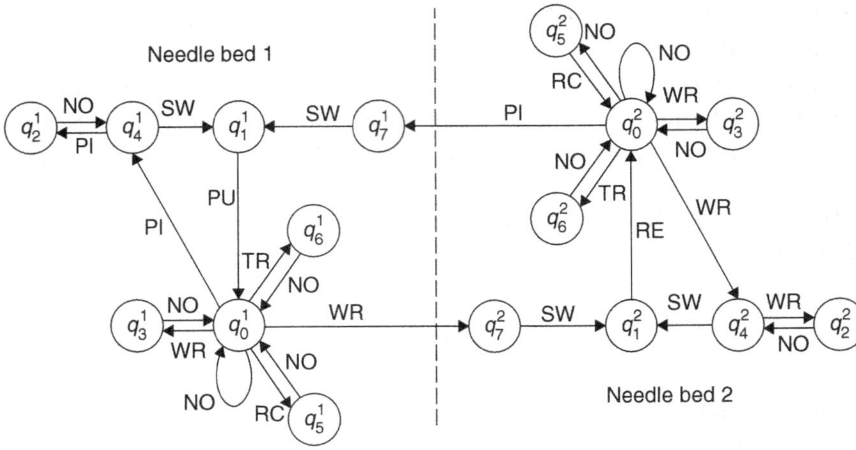

1.18 The graph of states and automaton operations. PI – push in; PU – pop up; WR – write; RE – read; TR – transfer; RC – receive; SW – swap; NO – no operation.

needle, d_2^i is the old loop that has not been knocked over, ε is an empty symbol which denotes the absence of yarn, and $\perp$ is the stack bottom symbol.

The relationship between the cam actions, the change in the needle position, and the fabric structural element produced corresponds to the relationship between the input symbol, the automaton state, and the output symbol. The alphabet V_i and the transition function μ are presented in Table 1.12 and as a graph of states in Fig. 1.18. Note that for single-bed machines the needle functions correspond to either left or right part of the graph. For double-bed flat and circular knitting machines the needle functions correspond to both parts of the graph but without the links related to the states q_7^1 and q_7^2. The functionality of a purl machine is described by the whole graph.

Table 1.13 illustrates the formation of three basic structural elements, i.e. loop, tuck, and float; it defines the automaton output function δ and operations with the memory. A more detailed description of this model is presented in Grishanov *et al.* (1997) where the automaton operations with the memory and other functions are explained together with the coding systems used. This model describes the functionality of a knitting machine under a general assumption of external actions on knitting working parts of the machine that result in the formation of a fabric. The physical arrangement of the machine can be different and so can be the means for controlling the machine's functions. This model therefore can be applied to different types of knitting machines. In a similar way, the formation of a woven fabric on a weaving loom or a braided structure can be modelled.

Table 1.12 Alphabet of input symbols V_i and transition function μ

Input symbol	Transition
h_0^i	$q_0^i \to q_0^i$
h_1^i	$q_0^i \to q_3^i$
h_2^i	$q_0^i \to q_4^i$
h_3^i	$q_0^i \to q_6^i$
h_4^i	$q_0^i \to q_7^{3-i}$
h_5^i	$q_0^i \to q_5^i$
h_6^i	$q_3^i \to q_0^i$
h_7^i	$q_4^i \to q_2^i$
h_8^i	$q_2^i \to q_4^i$
h_9^i	$q_4^i \to q_1^i$
h_A^i	$q_1^i \to q_0^i$
h_B^i	$q_5^i \to q_0^i$
h_C^i	$q_6^i \to q_0^i$
h_D^i	$q_7^i \to q_1^i$

1.10 Optimization of performance characteristics: linear and non-linear methods

Optimization of performance characteristics of textile products and processes is often the final stage in the engineering design where the results of experimental and theoretical modelling are combined in order to achieve the best possible combination of properties. In many cases this also includes consideration of economic and logistic aspects. It can be seen from the sections above that the process–structure–property system is governed by complex multi-factor relationships. In these conditions, the optimization problem can be formulated as a problem of finding a combination of independent factors $X^* \{x_1^*, x_2^*, \ldots, x_m^*\}$ (optimal point) that provide an extremum f^* (a maximum or a minimum) of a function $f(x_1, x_2, \ldots, x_m)$ (called the objective function) of many variables.

Optimization problems can be classified as unconstrained, where no limitations are imposed on the range of possible values of independent factors, and constrained, where additional conditions (constraints) define the range of admissible values of the factors. The objective function and the

Table 1.13 Formation of basic structural elements

Initial state	Input symbol	Memory content	Operation	New memory content	New state	Output symbol
Plain loop formation						
q_0^i	h_2^i	$d_2^i \perp$	PI	$d_0^i\ d_2^i \perp$	q_4^i	-
q_4^i	h_9^i	$d_0^i d_2^i \perp$	SW	$d_2^i\ d_0^i \perp$	q_1^i	-
q_1^i	h_A^i	$d_2^i d_0^i \perp$	PU	$d_2^i \perp$	q_0^i	e_2^i
Tuck loop formation						
q_0^i	h_1^i	$d_2^i \perp$	PI	$d_0^i\ d_2^i \perp$	q_3^i	-
q_3^i	h_6^i	$d_0^i\ d_2^i \perp$	NO	$d_1^i\ d_2^i \perp$	q_0^i	-
q_0^i	h_2^i	$d_1^i d_2^i \perp$	PI	$d_0^i d_1^i d_2^i \perp$	q_4^i	-
q_4^i	h_9^i	$d_0^i d_1^i d_2^i \perp$	SW	$d_2^i d_1^i d_0^i \perp$	q_1^i	-
q_1^i	h_A^i	$d_2^i d_1^i\ d_0^i \perp$	PU	$d_2^i \perp$	q_0^i	$e_2^i\ e_1^i$
Float formation						
q_0^i	h_0^i	$d_2^i \perp$	NO	$d_2^i \perp$	q_0^i	e_0^i

constraints can be linear or non-linear. According to this, the particular case when both the objective function and the constraints are linearly dependent on the factors is known as a linear programming problem (LPP). The LPP can be formulated in its canonical form as follows:

Minimize the linear objective function $y = f(\boldsymbol{X}) = \boldsymbol{c}^{\mathrm{T}}\boldsymbol{X}$ of variables $\boldsymbol{X}(x_1, x_2,\ldots, x_m)$ subject to k linear constraints $a^T\boldsymbol{X} \geq b$ and m constraints $\boldsymbol{X} \geq 0$, where $\boldsymbol{a}$ is $m \times k$ matrix of coefficients, $\boldsymbol{b}$ is $k \times 1$, and $\boldsymbol{c}$ is $m \times 1$ column vectors of coefficients (Fig. 1.19a).

The general non-linear programming (NLP) problem can be formulated in a similar way as follows:

Minimize the objective function $y = f(\boldsymbol{X})$ of variables $\boldsymbol{X}(x_1, x_2,\ldots, x_m)$ subject to linear and/or non-linear equality constraints $h_i(\boldsymbol{X}) = 0$, $i = 1, 2,\ldots, k$, and linear and/or non-linear inequality constraints $g_j(\boldsymbol{X}) \geq 0$, $j = 1, 2, \ldots, l$ (Fig. 1.19b). This definition excludes specific cases when variables take only

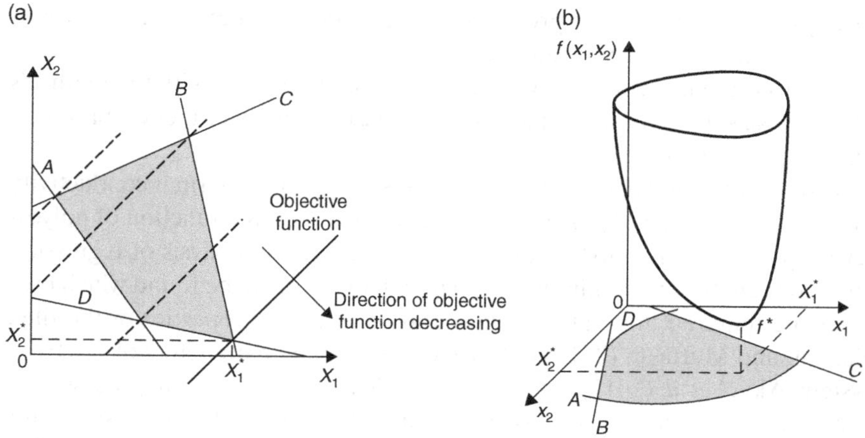

1.19 Linear (a) and non-linear (b) optimization. *A*, *B*, *C*, and *D* are constraints; x_1^* and x_2^* are coordinates of the optimum; grey areas show admissible range of factors.

integer values and the constraints are expressed in the form of differential equations.

The solution of the optimization problem sometimes can be found in analytical form by resolving the system of simultaneous equations

$$\frac{\partial f}{\partial x_1} = 0; \ \frac{\partial f}{\partial x_2} = 0, \ldots, \frac{\partial f}{\partial x_m} = 0 \qquad [1.64]$$

In the majority of cases the search for the optimal solution requires the application of numerical methods. There is a general method for resolving the LPP, known as Dantzig's simplex algorithm (or simplex method) – please refer to Sultan (1993). It has been shown that the solution of the LPP can be found at one of the corners of a polygonal area that defines the admissible values of factors (Fig. 1.19a).

However, there is no general method for non-linear optimization. There are a number of sophisticated algorithms which use information on the behaviour of the objective function in the factorial space in order to reach the optimum point; these are known as direct methods. In addition to this gradient methods make use of the objective function's derivatives. A detailed description of numerical methods and their implementations can be found in Himmelblau (1972), Luenberger (1984), Press *et al.* (2002).

There are numerous examples where non-linear optimization methods have been applied to the design of textile materials and technological processes due to non-linear nature of the objective function; there are only few examples where the linear programming method has been used, e.g. in cotton yarn engineering (Stjepanovic, 1997).

The non-linear methods have been used in combination with the methods of response surface design which generated the objective function based on experimental data.

Yasinskaya *et al.* (2011) studied flexural strength and thermal conductivity of polyester/acrylic and polyester/wool insulating tiles as a function of polyester proportion in the mixture and volume density. The analysis of regression equations enabled the optimum combination of factors to be found which provides the required thermal insulation and mechanical properties of the tiles. Karthik and Murugan (2013) optimized the spinning parameters of DREF-3 system. Akyol *et al.* (2013) considered an industrial textile drying process.

Non-linear optimization has been applied in woven fabric design with the specified mechanical properties at minimum cost (Chen, 1991). The production cost was expressed through technical parameters of the fabric, such as yarn linear density, twist, warp and weft spacing, and included material prices, labour costs, and overheads. The constraint functions served to limit the fabric weight and cover factor, tensile, bending, and shear module. Gradient methods have been used to find the optimum. Ghiasi *et al.* (2008) used a direct search method of Nelder and Mead (see Himmelblau, 1972) for simultaneous non-linear optimization of structural and manufacturing parameters of a composite material. Dong and Leung (2009) suggested an optimization method of a replenishment strategy for the textile manufacturers. Choi *et al.* (2011) suggested a non-linear inventory model for the optimization of the profits in fashion retail sector. Non-linear formulation has been used in the development of a control system for the roller carding machine (Kuo *et al.*, 2013b).

Recent years have seen an increasing use of artificial neuron networks (ANN) and genetic algorithms (GA) for the optimization problems; a good description of these methods can be found in Patterson (1996) Rojas (1996). These methods do not always require, as linear and NLP methods do, a mathematical formulation of the objective and constraints functions. In this case it is, however, necessary to have a large experimental database for these methods to produce reliable results. Sette *et al.* (1997) applied a combination of ANN and GA to the optimization of the spinning process based on a massive experimental investigation where the influence of five machine setting parameters and 14 fibre parameters on yarn strength and elongation was studied resulting in 1382 data points. Subramanian *et al.* (2007) used this approach for the optimization of yarn properties. Further development of this approach (Sette and van Langenhove, 2002), where a combination of ANN, linear, and non-linear constrained optimization has been employed, has enabled yarn price to be included in the model as well. Furferi *et al.* (2013) investigated fabric coating with polyurethane compound where six process parameters and four fabric parameters have been included into an ANN model.

Hansel *et al.* (2002) optimized the weight of a laminate structure by optimizing its topology using a GA. Wisniewski (2004) considered a homogeneous orthotropic model of fibre-reinforced multilayered composite material and applied a generic algorithm for the maximization of the material's stiffness, which resulted in the identification of optimal orientation of fibres in each layer. Kim *et al.* (2005) developed a semi-geodesic path algorithm for the design of filament wound pressure vessels. The finite-element method was then used to model the mechanical behaviour of the vessels up to the point of failure. The GA was used to optimize the technical parameters of the design. In a similar fashion, Awad *et al.* (2011) used a combination of finite-element method for modelling the mechanics of fibre composite sandwich panels and GA for the cost-optimal design.

Non-linear methods are used not only for the optimization as illustrated above but also in regression analysis when fitting functions which are non-linear with respect to their coefficients. For example, an application of the least squares method for the estimation of coefficients a and b of the function $y = a(1-exp(-bx))$ leads to NLP. In Section 1.4.3 NLP has been used for the estimation of unknown parameters of a fibre migration model.

1.11 Sources of further information and advice

There is an increasing demand for new materials with multifunctional properties, meeting high technical specifications, produced at low cost and in a reduced time. These demands intensified the development of engineering design methods of textile materials with an extensive use of mathematical modelling and computer simulation of geometry, structure, and properties. This trend will continue towards the generation of new more effective models implemented in intelligent computer-aided design/computer-aided manufacturing (CAD/CAM) systems and should stimulate the development of new technologies.

Textile modelling is a multidisciplinary area of research based on a combination of fundamental and applied science. Information on modelling textile structures can be found in textile-orientated journals as follows:

Textile Research Journal
The Journal of the Textile Institute
Research Journal of Textile and Apparel
Fibres and Textiles in Eastern Europe
Journal of Technical Textiles
Association of Universities for Textiles (AUTEX) Research Journal

A number of journals are dedicated to specific areas of research with more attention to fundamental theories where textiles are one of the topics of interest:

Composite Structures
Composites Science and Technology

Journal of Composite Materials
Computational Materials Science
Journal of Materials Science
Journal of Polymer Science
Physical Review E
Journal of Applied Physics

For information on fundamental aspects of mathematical and computational methods in modelling the following titles can be recommended:

Anderson J A (2006), *Automata Theory with Modern Applications*, Cambridge, Cambridge University Press.

Cromwell P R (2004), *Knots and Links*, Cambridge, Cambridge University Press.

Gross D, Shortle J F, Thomson J M and Harris C M (2008), *Fundamentals of Queueing Theory*. 4th edn., Hoboken, NJ, USA, John Wiley and Sons.

Himmelblau D M (1972), *Applied Nonlinear Programming*, London, McGraw-Hill.

Kauffman L H (1993), *Knots and Physics*, 2nd edn., London, World Scientific.

Knuth D E (1997), *The Art of Computer Programming. Vol I: Fundamental Algorithms*, 3rd Edn., London, Addison-Wesley.

Patterson D W (1996), *Artificial Neural Networks: Theory and Applications*, London, Prentice-Hall.

Press W H, Teukolsky S A, Vetterling W T and Flannery B P (2002), *Numerical Recipes in C++. The Art of Scientific Computing*, Cambridge, Cambridge University Press.

Succi S (2001), *The Lattice Boltzmann Equation for Fluid Dynamics and Beyond*, New York, Oxford University Press.

Timoshenko S P and Goodier J N (1970), *Theory of Elasticity*, Auckland, McGraw-Hill.

Zienkiewicz O C and Taylor R L (2000), *The Finite-Element Method: The Basis*, Vol. **1**, 5th edn., Oxford, Butterworth-Heinemann.

Zienkiewicz O C and Taylor R L (2005), *The Finite Element Method for Solid and Structural Mechanics*, Vol. **2**, 6th edn., Amsterdam, Elsevier.

Zienkiewicz O C, Taylor R L and Nithiarasu P (2005), *The Finite Element Method for Fluid Dynamics*, Vol. **3**, 6th edn., Amsterdam, Elsevier.

1.12 References

Adanur S and Onal L (2004), 'Analysis of a novel 3d hybrid woven/knitted fabric structure: Part II: Mechanical model to predict modulus and extension', *Text Res J*, **74**(10), 865–871.

Akyol U, Kahveci K and Cihan A (2013), 'Determination of optimum operating conditions and simulation of drying in a textile drying process', *J Text Inst*, **104**(2), 170–177.

Altendorf H and Jeulin D (2011), 'Random-walk-based stochastic modelling of three-dimensional fibre systems', *Phys Rev E*, **83**, 041804.

Anderson J A (2006), *Automata Theory with Modern Applications*. Cambridge, Cambridge University Press, 255p.

ANSYSa http://www.ansys.com/Business+Initiatives, accessed 13/06/2013.

ANSYSb http://www.ansys.com/staticassets/ANSYS/staticassets/resourcelibrary/whitepaper/WP-Impact-of-Strategic-Simulation-on-Product-Profitability.pdf, accessed 01/10/2013.

Ashnin N M (1985), *Fibrous Materials Carding*, Moscow, Legprombitizdat, 144p.

Awad Z K, Aravinthan T and Zhuge Y (2011), 'Cost Optimum Design of Structural Fibre Composite Sandwich Panel for Flooring Applications' *Advances in FRP Composites in Civil Engineering* – Proceedings of the 5th International Conference on FRP Composites in Civil Engineering (CICE 2010), Ye L, Feng P and Yue Q, Editors, 27–29 September 2010, Beijing, China, 478–481.

Barauskas R and Abraitiene A (2011), 'A model for numerical simulation of heat and water vapor exchange in multilayer textile packages with three-dimensional spacer fabric ventilation layer', *Text Res J*, **81**(12), 1195–1215.

Bardella L (2000), Mechanical behaviour of glass-filled epoxy resins: Experiments, homogenization methods for syntactic foams, and applications, PhD, University of Brescia, Italy.

Baykal P D and Erol O B R (2006), 'Prediction of strength and elongation properties of cotton/polyester-blended OE rotor yarns', *Fibres Text East Eur*, **14**, No. 1(55), 18–21.

Bezrukov A and Stoyan D (2006), 'Simulation and statistical analysis of random packings of ellipsoids', *Part Part Syst Char*, **23**, 388–398.

Bratchikov I L (1975), *The Syntax of the Programming Languages*, Moscow, Nauka, 230p.

Brochocka A, Makowski K and Majchrzycka K (2012), 'Penetration of different nanoparticles through melt-blown filter media used for respiratory protective devices', *Text Res J*, **82**(18), 1906–1919.

Bunsell A R and Renard J (2005), *Fundamentals of Fibre Reinforced Composite Materials*, Bristol, Institute of Physics Publishing, 398p.

Cavallaro P V, Sadegh A M and Quigley C J (2007), 'Decrimping behavior of uncoated plain-woven fabrics subjected to combined biaxial tension and shear stresses', *Text Res J*, **77**(6), 403–416.

Chen X (1991), The design of woven fabrics to meet mechanical specifications at minimum cost, PhD thesis, University of Leeds.

Cherkassky A (1994), 'A two-dimensional mathematical model of the carding process', *Text Res J*, **64**(3), 169–175.

Chernyakov A L, Kirsch A A and Kirsch V A (2011), 'Elastic vibrations of a fiber due to impact of an aerosol particle and their influence on the efficiency of fibrous filters', *Phys Rev E*, **83**, 056303.

Chmutov S, Duzhin S and Mostovoy J (2010)), *Introduction to Vassiliev Knot Invariants: 1st Draft*, http://www.math.osu.edu/~chmutov/preprints/.

Choi K and Lo T (2003), 'An energy model of plain knitted fabric', *Text Res J*, **73**, 739–748.
Choi K F and Lo T Y (2006), 'The shape and dimensions of plain knitted fabric: A fabric mechanical model', *Text Res J*, **76**(10), 777–786.
Choi T M, Chiu C H and To K M C (2011), 'A fast fashion safety-first inventory model', *Text Res J*, **81**(8), 819–826.
Chudoba R, Vorechovsky M, Eckers V and Gries T (2007), 'Effect of twist, fineness, loading rate and length on tensile behavior of multifilament yarns (a multivariate study)', *Text Res J*, **77**(11), 880–891.
Cimilli S, Nergis F B U and Candan C (2008), 'Modeling of heat transfer measurement unit for cotton plain knitted fabric using a finite element method', *Text Res J*, **78**(1), 53–59.
Cochran W C and Cox G M (1957), *Experimental Designs*. New York, John Wiley and Sons, Inc.
Coelho D, Thovert J F and Adler P (1997), 'Geometrical and transport properties of random packings of spheres and aspherical particles', *Phys Rev E*, **55**, 1959–1978.
Cornell J A (1990), *Experiments with Mixtures*, 2nd edn., New York, John Wiley and Sons, 632p.
Cromwell P R (2004), *Knots and Links*, Cambridge, Cambridge University Press, 328p.
de Jong S and Postle R (1977a), 'Energy analysis of knitted fabric mechanics by means of optimal control theory. Part I: The nature of loop interlocking in the plain knitted structure', *J Text Inst*, **68** (10), 307–315.
de Jong S and Postle R (1977b), 'Energy analysis of knitted fabric mechanics by means of optimal control theory. Part II: Relaxed fabric dimensions and tensile properties of the plain knitted structure', *J Text Inst*, **68** (10), 316–323.
de Jong S and Postle R (1977c), 'Energy analysis of knitted fabric mechanics by means of optimal control theory. Part III: The 1x1 rib-knitted structure', *J Text Inst*, **68** (10), 324–329.
Dillon W R and Goldstein M (1984), *Multivariate Analysis. Methods and Applications*. Toronto, John Wiley and Sons, Inc., 608p.
Dong A H and Leung S Y S (2009), 'A Simulation-based replenishment model for the textile industry', *Text Res J*, **79**(13), 1188–1201.
El-Mogahzy Y (2004), 'An integrated approach to analyzing the nature of multicomponent fiber blending: Part I: Analytical aspects', *Text Res J*, **74**(8), 701–712.
El-Mogahzy Y, Farag R, Abdelhady F and Mohamed A (2004), 'An integrated approach to analyzing the nature of multicomponent fiber blending: Part II: Experimental analysis of structural and attributive blending', *Text Res J*, **74**(9), 767–775.
El-Mogahzy Y, Farag R, Abdelhady F and Mohamed A (2005), 'An integrated approach to the analysis of multi-component fiber blending. Part III: Analysis of interactive fiber blending', *Text Res J*, **75**(12), 833–845.
El-Sheikh A and Backer S (1972), 'The mechanics of fibre migration. Part I: Theoretical analysis', *Text Res J*, **42**(3), 137–146.
Evans R M L and Haw M D (2002), 'Correlation length by measuring empty space in simulated aggregates', *Europhys Lett*, **60**, 404–410.
Evans R M L and Starrs L (2002), 'Emergence of a stress transmission length scale in transient gels', *J Phys: Condens Matter*, **14**, 2507–2529.

Ernst C J and Sumners D W (1990), 'A calculus for rational tangles: Applications to DNA recombination', *Math Proc Camb Phi Soc*, **108**, 489–515.

Faessel M, Delisee C, Bos F and Castera P (2005), '3D modelling of random cellulosic fibrous networks based on X-ray tomography and image analysis', *Comp Sci Tech*, **65**(13), 1931–1940.

Feng J, Xu B G and Tao X M (2013), 'Systematic investigation and optimization of fine cotton yarns produced in a modified ring spinning system using statistical methods', *Text Res J*, **83**(3), 238–248.

Flapan E (2000), *When Topology Meets Chemistry: A Topological Look at Molecular Chirality*, Cambridge, Cambridge University Press, 241p.

Furferi R, Governi L and Volpe Y (2013), 'Modelling and simulation of an innovative fabric coating process using ANN', *Text Res J*, **83**(12), 1282–1294.

Gaiselmann G, Froning D, Totzke C, Quick C, Manke I, Lehnert W and Schmidt V (2013), 'Stochastic 3D modeling of non-woven materials with wet-proofing agent', *Int J Hydrogen Energ*, **38**, 8448–8460.

Ghassemieh E, Acar M and Versteeg H (2002), 'Microstructural analysis of non-woven fabrics using scanning electron microscopy and image processing. Part 2: Application to hydroentangled fabrics', *Proc Inst Mech Eng, Part L: J Mater Des Appl*, **216**(4), 211–218.

Gegauff C (1907), 'Force et elasticite des files en cotton', *Bulletin de la Societe Industrielle de Mulhouse*, **77**, 153–176.

Ghiasi H, Pasini D and Lessard L (2008), 'Constrained globalized Nelder-Mead method for simultaneous structural and manufacturing optimization of a composite bracket', *J Comp Mater*, **42**(7), 717–736.

Gibson R F (2010), 'A review of recent research on mechanics of multifunctional composite materials and structures', *Comp Struct*, **92**, 2793–2810.

Grishanov S A, Cassidy T and Spencer D J (1997a), 'A model of the loop formation process on the knitting machines using finite automata theory', *Appl Math Model*, **21**(7), 455–465.

Grishanov S A, Lomov S V, Cassidy T and Harwood R J (1997b), 'The simulation of the geometry of two component yarns. Part II: Fibre distribution in the yarn cross section', *J Text Inst*, **88**, 352–372.

Grishanov S, Tausif M and Russell S J (2012a), 'Characterization of fibre entanglement in nonwoven fabrics based on knot theory', *Comp Sci Tech*, **72**(12), 1331–1337.

Grishanov S, Tausif M and Russell S J (2012b), 'Application of dynamic splitting to estimate fibre entanglement in simulated nonwoven fibrous assemblies', *Comp Sci Tech*, **72**(12), 1318–1323.

Grishanov S A and Vassiliev V A (2011), 'Invariants of links in 3-manifolds and splitting problem of textile structures', *J Knot Theor Ramif*, **20**(3), 345–370.

Grishanov S, Meshkov V and Omelchenko A (2009a), 'A topological study of textile structures. Part I: An introduction to topological methods', *Text Res J*, **79**(8), 702–713.

Grishanov S, Meshkov V and Omelchenko A (2009b), 'A topological study of textile structures. Part II: Topological invariants in application to textile structures', *Text Res J*, **79**(9), 822–836.

Grishanov S, Meshkov V and Vassiliev V (2009c), 'Recognizing textile structures by finite type knot invariants', *J Knot Theor Ramif*, **18**(2), 209–235.

Grishanov S, Meshkov V and Omelchenko A (2007), 'Kauffman-type polynomial invariants of doubly-periodic structures', *J Knot Theor Ramif*, **16**(6), 779–788.

Grishanov S, Siewe F and Cassidy T (2011), 'An application of queuing theory to modelling of melange yarns. Part II: A method of estimating the fibre migration probabilities and a yarn structure simulation algorithm', *Text Res J*, **81**(8), 798–818.

Grishanov S A, Harwood R J and Bradshaw M S (1999), 'A model of fibre migration in staple fibre yarn', *J Text Inst*, 1999, **90**, 298–321.

Gross D, Shortle J F, Thomson J M and Harris C M (2008), *Fundamentals of Queueing Theory*. 4th edn., Hoboken, NJ, USA, John Wiley and Sons, 488p.

Hamilton J B (1958), 'The radial distribution of fibres in blended yarns. Part I: Characterisation by a migration index', *J Text Inst*, **49**(9), T411–T423.

Hansel W, Treptow A, Becker W and Freisleben B (2002), 'A heuristic and a genetic topology optimization algorithm for weight-minimal laminate structures', *Comp Struct*, **58**, 287–294.

Harpa R (2010a), 'Optimization of the drafting system for the ring spinning. Part I', *International Scientific Conference* 19–20 November 2010, Gabrovo, Bulgaria, 364–367.

Harpa R (2010b), 'Optimization of the drafting system for the ring spinning. Part II', International Scientific Conference 19–20 November 2010, Gabrovo, Bulgaria, 368–371.

Hart K, de Jong S and Postle R (1985a), 'Analysis of the single bar warp knitted structure using an energy minimization technique. Part I: Theoretical development', *Text Res J*, **55**(8), 489–498.

Hart K, de Jong S and Postle R (1985b), 'Analysis of the single bar warp knitted structure using an energy minimization technique. Part II: Results and comparison with woven and weft knitted analysis', *Text Res J*, **55** (9), 530–539.

Hearle J W S and Bose O N (1965), 'Migration of fibres in yarns. Part II: A geometrical explanation of migration', *Text Res J*, **35**, 693–699.

Hearle J W S and Gupta B S (1965), 'Migration of fibres in yarns. Part III: A study of migration in staple fibre rayon yarn', *Text Res J*, **35**, 788–795.

Hearle J W S, Gupta B S and Merchant V B (1965),'Migration of fibres in yarns. Part I: Characterisation and idealization of migration behaviour', *Text Res J*, **35**, 329–334.

Hearle J W S, Grossberg P and Backer S (1969), *Structural Mechanics of Fibres, Yarns and Fabrics*, New York, John Wiley & Sons, 469p.

Hearle J W S and Shanahan W J (1978a), 'An energy method for calculations in fabric mechanics. Part I: Principles and methods', *J Text Inst*, **69**(4), 81–91.

Hearle J W S and Shanahan W J (1978b), 'An energy method for calculations in fabric mechanics. Part II: Examples of applications of the method to woven fabrics', *J Text Inst*, **69**(4), 92–100.

Hepworth R B and Leaf G A V (1976), 'The Mechanics of an idealised weft-knitted structure', *J Text Inst*, **67**(8), 241–248.

Himmelblau D M (1972), *Applied Nonlinear Programming*, London, McGraw-Hill, 498p.

Hopcroft J E, Motwani R and Ullman J D (2007), *Introduction to Automata Theory, Languages, and Computation*, 3rd edn., Boston, Pearson/Addison Wesley, 535p.

Hu J, Jiang Y and Ko F (1998), 'Modeling uniaxial tensile properties of multiaxial warp knitted fabrics', *Text Res J*, **68**(11), 828–834.

Hu J L, Ma L, George B, Wong W S-K and Zhang W (2006), 'Modelling multi-layer seam puckering', *Text Res J*, **76**(9), 665–673.

Hu J and Xin B (2008), 'Structure and mechanics of woven fabrics' Schwartz P (2008), *Structure and Mechanics of Textile Fibre Assemblies*, Cambridge, Woodhead Publishing Limited, 248p.

Irvine V (2012), 'Broadening the palette for bobbin lace: A combinatorial approach', *Proceedings of Bridges 2012: Mathematics, Music, Art, Architecture, Culture*, R Bosch, D McKenna and R Sarhangi, Ed., Tessellations Publishing, USA, 191–198.

Karthik T and Murugan R (2013), 'Influence of spinning parameters on milkweed/cotton DREF3 yarn properties', *J Text Inst*, **104**(9), 938–949.

Kauffman L H (1993), *Knots and Physics*, 2nd edn., London, World Scientific.

Kemeny J G and Snell J L (1960), *Finite Markov Chains*, Princeton, NJ, USA, Van Nostrand, 210p.

Kemp A (1958), 'An extension of Pierce cloth geometry to the treatment of noncircular threads', *J Text Inst*, **49**, T44–T48.

Kim C-U, Kang J-H, Hong C-S and Kim C-G (2005), 'Optimal design of filament wound structures under internal pressure based on the semi-geodesic path algorithm', *Compos Struct*, **67**, 443–452.

Klusemann B and Svendsen B (2010), 'Homogenization methods for multi-phase elastic composites: Comparisons and benchmarks', *Technische Mechanik*, **30**(4), 374–386.

Knuth D E (1997), *The Art of Computer Programming. Vol I: Fundamental Algorithms*, 3rd edn., London, Addison-Wesley, 672p.

Koponen A, Kandhai D, Hellen E, Alava M, Hoekstra A, Kataja M, Niskanen K, Sloot P and Timonen J (1998), 'Permeability of three-dimensional random fiber webs', *Phys Rev Lett*, **80**, 716–719.

Korn G A and Korn T M (1968), *Mathematical Handbook for Scientists and Engineers*, New-York, McGraw-Hill Book Company.

Kouznetsova V G (2002), Computational homogenization for the multi-scale analysis of multi-phase materials, PhD, Eindhoven, Technische Universiteit, The Netherlands.

Kuo C-F J, Vu H Q, Chou Y-C and Li Y-C (2013a), 'Quality and uniformity control of fiber web by roller card system. Part I: Dynamic modeling formulation and experimental identification', *Text Res J*, **83**(7), 761–770.

Kuo C-F J, Vu H Q and Li Y-C (2013b), 'Quality and uniformity control of fiber web by roller card system. Part II: Design and validation of a roller card control system', *Text Res J*, **83**(9), 947–960.

Kurbak A and Alpyildiz T (2008), 'A geometrical model for the double lacoste knits', *Text Res J*, **78**(3), 232–247.

Kurbak A and Alpyildiz T (2009a), 'Geometrical models for balanced rib knitted fabrics. Part II: Applications of 1 × 1 rib model to presser-foot knitted 1 × 1 rib, interlock and half milano rib', *Text Res J*, **79**(6), 495–505.

Kurbak A and Alpyildiz T (2009b), 'Geometrical models for cardigan structures. Part I: Full cardigan', *Text Res J*, **79**(14), 1281–1300.

Kurbak A and Alpyildiz T (2009c), 'Geometrical models for cardigan structures. Part II: Half cardigan', *Text Res J*, **79**(18), 1635–1648.

Kurbak A and Soydan A S (2008), 'Basic studies for modeling complex weft knitted fabric structures. Part III: A geometrical model for 1 × 1 purl fabrics', *Text Res J*, **78**(5), 377–381.

Kyosev Y, Angelova Y and Kovar R (2005), '3D modeling of plain weft knitted structures of compressible yarn', *Res J Text App*, Hong Kong, **9**(1), 88–97.

Lawrence C A (2003), *Fundamentals of Spun Yarn Technology*, Woodhead Publishing Limited, 524p.

Leaf G A V and Glaskin A (1955), 'The geometry of plain knitted loop', *J Text Inst*, **46**, T587–T605.

Leaf G A V (1960), 'Models of the plain knitted loop', *J Text Inst*, **51**, T49.

Leaf G A V, Chen Y and Chen X (1993), 'The initial bending behaviour of plain-woven fabrics', *J Text Inst*, **84**(3), 419–428.

Leaf G A V and Kandil K H (1980), 'The initial load–extension behaviour of plain-woven fabrics', *J Text Inst*, **71**(1), 1–7.

Lee M E-M (2001), Mathematical models of the carding process, PhD thesis, St. John's College, University of Oxford, UK.

Lee Y and Jasiuk Y (2013), 'Apparent elastic properties of random fiber networks', *Comp Mat Sci*, **79**, 715–723.

Li P, Malengier B, Goessens T, Van Langenhove L and Ciocci M C (2013), 'An extended virtual location method for yarn cross-section construction', *J Text Inst*, **104**(5), 461–472.

Li S, Zhou C, Yu H and Li L (2011), 'Formulation of a unit cell of a reduced size for plain weave textile composites', *Comp Mat Sci*, **50**(5), 1770–1780.

Li Y, Zhu Q and Yeung K W (2002), 'Influence of thickness and porosity on coupled heat and liquid moisture transfer in porous textiles', *Text Res J*, **72**(5), 435–446.

Li Y and Zhu Q (2003), 'Simultaneous heat and moisture transfer with moisture sorption, condensation, and capillary liquid diffusion in porous textiles', *Text Res J*, **73**(6), 515–524.

Loginov A U, Grishanov S A and Harwood R J (2002a), 'Modelling load-extension behaviour of plain knitted fabric. Part I: A unit cell approach towards knitted fabric mechanics', *J Text Inst*, **93**, Part 1, No 3, 218–238.

Loginov A U, Grishanov S A and Harwood R J (2002b), 'Modelling load-extension behaviour of plain knitted fabric. Part II: Energy relationships in the unit cell', *J Text Inst*, **93**, Part 1, No 3, 239–250.

Loginov A U, Grishanov S A and Harwood R J (2002c), 'Modelling load-extension behaviour of plain knitted fabric. Part III: Model implementation and experimental verification', *J Text Inst*, **93**, Part 1, No 3, 251–275.

Luenberger D G (1984), *Linear and Nonlinear Programming*, 2nd edn., Reading, Massachusets, Addison-Wesley, 491p.

Maimi P, Mayugo J A and Camanho P P (2008) 'A three-dimensional damage model for transversely isotropic composite laminates', *J Comp Mater*, **42**(25), 2717–2745.

Mao N and Russell S (2000a), 'Directional permeability in homogeneous nonwoven structures Part II: Permeability in idealised structures', *J Text Inst*, **91**(2), 244–258.

Mao N and Russell S (2000b), 'Directional permeability in homogeneous nonwoven structures. Part I: The relationship between directional permeability and fibre orientation', *J Text Inst*, **91**(2), 235–243.

Mao N and Russell S J (2006), 'A framework for determining the bonding intensity in hydroentangled nonwoven fabrics', *Comp Sci Tech*, **66**(1), 80–91.

Massart T J, Mercatoris B C N, Piezel B, Berke P, Laiarinandrasanal L and Thionnet A (2011), 'Multi-scale modelling of heterogeneous shell structures', *Comp Assist Mech Eng Sci*, **18**, 53–71.

Melro A R, Camanho P P, Andrade Pires F M and Pinho S T (2012), 'Numerical simulation of the non-linear deformation of 5-harness satin weaves', *Comp Mater Sci*, **61**, 116–126.

Metzkes K, Schmidt R, Martin J, Hoffmann G and Cherif C (2013), 'Simulation of the yarn transportation dynamics in a warp knitting machine', *Text Res J*, **83**(12), 1251–1262.

Miravete A (1999), *3-D Textile Reinforcements in Composite Materials*, Cambridge, Woodhead Publishing Limited, 308p.

Mishra R (2013), 'Meso-scale finite element modeling of triaxial woven fabrics for composite in-plane reinforcement properties', *Text Res J*, **83**(17), 1836–1845.

Moffatt H (1990), 'The energy spectrum of knots and links', *Nature*, **347**, 367–369.

Montgomery D C (1991), *Design and Analysis of Experiments*, 3rd edn., Singapore, John Wiley and Sons, Inc., 649p.

Morris P J, Merkin J H and Rennell R W (1999), 'Modelling of yarn properties from fibre properties', *J Text Inst*, **59**, 536–547.

Morton W E (1956), 'The arrangement of fibres in single yarns', *Text Res J*, **26**, 325–331.

Morton H R and Grishanov S (2009), 'Doubly periodic textile structures', *J Knot Theor Ramif*, **18**(12), 1597–1622.

Muller E (1880), 'Über die festigkeitseigenschaften fadenförmiger fasergebilde in ihrer abhängigkeit vom drähte derselben', *Civilingenieur*.

Nabovati A, Llewellin E W and Sousa A C M (2009), 'A general model for the permeability of fibrous porous media based on fluid flow simulations using the lattice Boltzmann method', *Comp Part A: Appl Sci Manuf*, **40**(6–7), 860–869.

Neckar B, Soni M K and Das D (2006), 'Modelling of radial fiber migration in yarns', *Text Res J*, **76**(6), 486–491.

Onal L and Adanur S (2004), 'Analysis of a novel 3D hybrid woven/knitted fabric structure: Part I: Geometric model and verification', *Text Res J*, **74**, 827–832.

Patterson D W (1996), *Artificial Neural Networks: Theory and Applications*, London, Prentice-Hall, 477p.

Payen J, Vroman P, Lewandowski M, Perwuelz A, Calle-Chazelet S and Thomas D (2012), 'Influence of fiber diameter, fiber combinations and solid volume fraction on air filtration properties in nonwovens', *Text Res J*, **82**(19), 1948–1959.

Peirce F T (1947), 'Geometrical principles applicable to the design of functional fabrics', *Text Res J*, **17**(3), 123–147.

Postle R and Munden D L (1967a), 'Analysis of the dry relaxed knitted loop configuration. Part I: Two-dimensional analysis', *J Text Inst*, **58**, T329–T351.

Postle R and Munden D L (1967b), 'Analysis of the dry relaxed knitted loop configuration. Part II: Three-dimensional analysis', *J Text Inst*, **58**, T352–T365.

Postle R, Carnaby G A and de Jong S (1988), *The Mechanics of Wool Structures*, Chichester, Ellis Horwood, 462p.

Prasolov VV and Sosinskii AB (1997), *Knots, Links, Braids, and 3-Manifolds: An Introduction to the New Invariants in Low-Dimensional Topology*, Providence, R.I., American Mathematical Society. 1997.

Press W H, Teukolsky S A, Vetterling W T and Flannery B P (2002), *Numerical Recipes in C++. The art of Scientific Computing*, Cambridge, Cambridge University Press.

Ramakrishna S, Mayer J, Wintermantel E and Leong K W (2001), 'Biomedical applications of polymer-composite materials: A review', *Comp Sci Tech*, **61**, 1189–1224.

Rawal A, Potluri P and Steeleand C (2005), 'Geometrical modelling of the yarn paths in three-dimensional braided structures', *J Tech Text*, **35**(2), 115–135.

Reis J M L (2009), 'Effect of textile waste on the mechanical properties of polymer concrete', *Mat Res*, **12**(1), 63–67.

Riding G (1964), 'Filament migration in single yarns', *J Text Inst*, **55**, T9–T17.

Rogen P and Bohr H (2003), 'A new family of global protein shape descriptors', *Math Biosci*, **182**(2), 167–181.

Rojas R (1996), *Neural Networks*, Berlin, Springer-Verlag, 453p.

Rust J P and Gutierrez H Z (1994), 'Mathematical modelling and simulation for carding', *Text Res J*, **64**(10), 573–578.

Saidpour H and Oscar Q (2006), 'Stiffness properties of intraply woven hybrid composites by numerical homogenisation', *Proc Advances in Computing and Technology*, The School of Computing and Technology, 1st Annual Conference, University of East London, 184–195.

Sette S, Boullart L, van Langenhove L and Kiekens P (1997), 'Optimising the fibre-to-yarn process with a combined Neural Networks/Genetic Algorithm approach', *Text Res J*, **67**(2), 84–93.

Sette S and van Langenhove L (2002), 'Optimising the fibre-to-yarn production process: Finding a blend of fibre qualities to create an optimal price/quality yarn', *AUTEX Res J*, **2**(2), 57–63.

Sevostianov A G (1980), *Research Methodology in Textile Industry*, Moscow, Legkaya Industria, 392p.

Sevostianov A G and Sevostianov P A (1984), *Modelling of Technological Processes*, Moscow, Legkaya and Pischevaya Promishlennost, 344p.

Shanahan W J and Hearle J W S (1978), 'An energy method for calculations in fabric mechanics. Part II: Examples of application of the method to woven fabrics', *J Text Inst*, **69**(4), 92–100.

Sherburn M, Long A, Jones A, Crookston J and Brown L (2012), 'Prediction of textile geometry using an energy minimization approach', *J Ind Text*, **41**(4), 345–369.

Siewe F, Grishanov S, Cassidy T and Banyard G (2009), 'An application of queuing theory to modelling of melange yarns. Part I: A queuing model of melange yarn structure', *Text Res J*, **79**(16), 1467–1485.

Sobera M P and Kleijn C R (2006), 'Hydraulic permeability of ordered and disordered single-layer arrays of cylinders', *Phys Rev E*, **74**, 036301.

Spencer D J (2001), *Knitting Technology: A Comprehensive Handbook and Practical Guide*, Cambridge, Woodhead Publishing Limited, 386p.

Sreprateep K and Bohez E. L. J (2006), 'Computer aided modelling of fiber assemblies' *Computer-Aided Design and Applications*, **3**(1–4), 367–376.

Sriprateep K and Bohez E L J (2009), 'A new computer geometric modelling approach of yarn structures for the conventional ring spinning process', *J Text Inst*, **100**(3), 223–236.

Stjepanovic Z (1997), 'Development of an integrated system environment for cotton yarn engineering', *Proc. 3rd International Conference on Design to Manufacture in Modern Industry*, Portoroz, Slovenia, 1997, University in Maribor, Faculty of Mechanical Engineering, 1997, 479–486.

Succi S (2001), *The Lattice Boltzmann Equation for Fluid Dynamics and Beyond*, New York, Oxford University Press, 288p.

Sultan, A (1993), *Linear Programming: An Introduction with Applications*, London, Academic Press, 568p.

Tao X (1996), 'Mechanical properties of a migrating fiber', *Text Res. J.*, **66**(12), 754–762.

Tamayol A, Wong K W and Bahrami M (2012), 'Effects of microstructure on flow properties of fibrous porous media at moderate Reynolds number', *Phys Rev E*, **85**, 026318.

Taniyama K (2009), 'Unknotting numbers of diagrams of a given nontrivial knot are unbounded', *J Knot Theor Ramif*, **18**(8), 1049–1063.

Thiffeault J-L and Finn M D (2006), 'Topology, braids, and mixing in fluids', *Philosophical Transactions of the Royal Society A*, **364**, 3251–3266.

Treloar L R G (1965), 'A migrating-filament theory of yarn properties', *J Text Inst*, **56**, T359–T380.

Treloar *L R G* and Riding G (1963), 'A theory of the stress-strain properties of continuous filament yarn', *J Text Inst*, **54**, T156–T170.

Turan R B, Okur A, Deveci R and Acikel M (2012), 'Predicting the intra-yarn porosity by image analysis method', *Text Res J*, **82**(16), 1720–1728.

Turan R B and Okur A (2013), 'Prediction of the in-plane and through-plane fluid flow behavior of woven fabrics', *Text Res J*, **83**(7), 700–717.

van Luijk C J, Carr A J and Carnaby G A (1984a), 'Finite-element analysis of yarns. Part I: Yarn model and energy formulation', *J Text Inst*, **75**, 342–353.

van Luijk C J, Carr A J and Carnaby G A (1984b), 'Finite-element analysis of yarns. Part II: Stress analysis', *J Text Inst*, **75**, 354–362.

van Luijk C J, Carr A J and Carnaby G A (1985), 'The mechanics of staple fibre yarn. Part I: Modelling assumptions', *J Text Inst*, **76**, 11–18.

Williams S R and Philipse A P (2003), 'Random packings of spheres and spherocylinders simulated by mechanical contraction', *Phys Rev E*, **67**, 051301.

Wisniewski J (2004), 'Optimal design of reinforcing fibres in multilayer composites using Genetic Algorithms', *Fibres Text East Eur*, **12**, No 3(47), 58–63.

Xiang P, Kuznetsov A V and Seyam A-F M (2007), 'Simulation of fiber entanglement by modeling vorticity in water flow field', *Text Res J*, **77**(5), 312–329.

Xiao X, Zeng X, Long A, Lin H, Clifford M and Saldaeva E (2012), 'An analytical model for through-thickness permeability of woven fabric', *Text Res J*, **82**(5), 492–501.

Yasinskaya N N, Karpenya A M and Chukasova-Il'yushkina E V (2011), 'Experimental study and optimization of the composition of heat-insulating tiles made of textile production wastes', *Fibre Chem*, **42**(6), 388–390.

Zeisberg P (1990), 'Stitch-bonding technology for the production of innovative composites', *Melliand Textilber*, **71**(11), 872–874.

Zeman J and Sejnoha M (2007), 'From random microstructures to representative volume elements', *Modell Simul Mater Sci Eng*, **15**(4), S325–S335.

Zhang M-Z and Li H-J (2007), 'Automatically generated geometric description of 3D braided rectangle preform', *Comp Mater Sci*, **39**(4), 836–841.

Zhang Y F, Sun B K, Yang Y L, Han C F, Chen W and Li M M (2011), 'Multi-objective optimization of blending ratio of outlast fiber/cotton blending yarn', *Adva Mater Res*, **332–334**, 638–642.

Zhong W and Pan N (2007), 'Aerosol filtration by fibrous filters: A statistical mechanics approach', *Text Res J*, **77**(5), 284–289.

2
Novel surface treatments for high performance textiles

H. M. EL-DESSOUKY, University of Leeds, UK and Mansoura University, Egypt

DOI: 10.1533/9780857099075.70

Abstract: The use of two environmentally friendly technologies, the 'plasma' and 'laser' approaches, in treating the surface of textiles is discussed in this chapter. The surface of fibrous materials can be functionalized and structured on the nano- and micro-scales, and the results of the treatment of natural and synthetic fibres such as wool, cotton, polypropylene, polyester and polyamide which have been treated by different plasmas and lasers are detailed and are found to be similar to morphological modification of a textile surface using the 'ripple/roll-structure' technique.

Key words: plasma, laser, fibre, fabric, textile, surface, treatment, structure.

2.1 Introduction

The chemical and physical characteristics of plasmas and lasers make them an environmentally attractive option for surface treatment in the textile industry. Since the 1970s, lasers have been used for such textile applications as cutting and printing and, since the mid-1980s, plasmas have started to be used in the industry to oxidize fibre surfaces for improved dyeing.

This chapter will be presented in two sections: plasma technologies for textile applications (Section 2.2) and laser technologies for textiles (Section 2.4).

2.2 Plasma technologies for textiles

Explanations of the plasma technology and its application in textiles have been carried out by means of fundamental characteristics, types, generation and textile-surface treatment.

2.2.1 Plasmas

The term 'plasma' was first used by Irving Langmuir in 1926 to describe an ionized gas which represents a distinct state of matter different from that

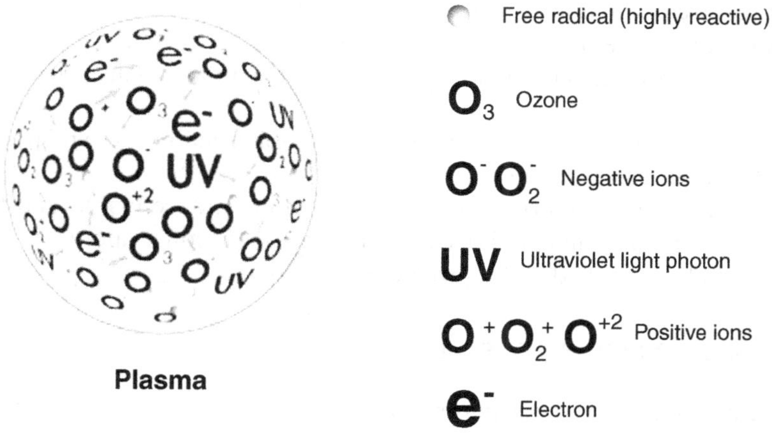

2.1 Plasma medium.

of gas, liquid or solid. A plasma is formed when matter is heated until the atoms lose all of their electrons, forming a mixture of ions, free electrons, neutrons, photons, free radicals, metastable excited species, and molecular and polymeric fragments (Fig. 2.1). Plasmas are highly reactive and can be used: to modify a substrate (plasma activation); to deposit chemical materials (plasma polymerization or plasma grafting); and to remove substances previously deposited (plasma cleaning or plasma etching) (Pane *et al.*, 2001).

2.2.2 Types of plasma

Plasma can be generated at three frequency ranges:

- LF – Low Frequency (50–450 kHz)
- RF – Radio Frequency (13.56 or 27.12 MHz)
- MW – Microwave (915 MHz or 2.45 GHz)

Depending on the size of plasma reactor and the required treatment, there are two further main types of plasma which are used for textile applications (Shishoo, 2007):

- low pressure plasma
- atmospheric pressure/cold plasma.

Low pressure plasmas, which have commonly been used in the microelectronics industry, have been developed for textile processing to achieve batch functionalization of fibrous products and flexible web-like materials. Low pressure plasmas are formed when a vacuum chamber is pumped down to a pressure in the range of 10^{-2} to 10^{-3} mbar.

The advantages of the low pressure plasma approach are:

- Fissure-penetration: even the most complex geometries and porous substrates can be treated.
- There are no thermal or mechanical strains induced in the substrates.
- Good environmental performance and operational safety.
- Well-controlled and reproducible results.

2.2.3 Plasma generation

Atmospheric pressure plasmas may be generated via a number of techniques including:

- Glow discharge (GD).
- Corona discharge (CD).
- Dielectric-barrier discharge (DBD).

GD is performed at reduced pressure and is characterized by high uniformity, stability and flexibility compared to alternative approaches. The

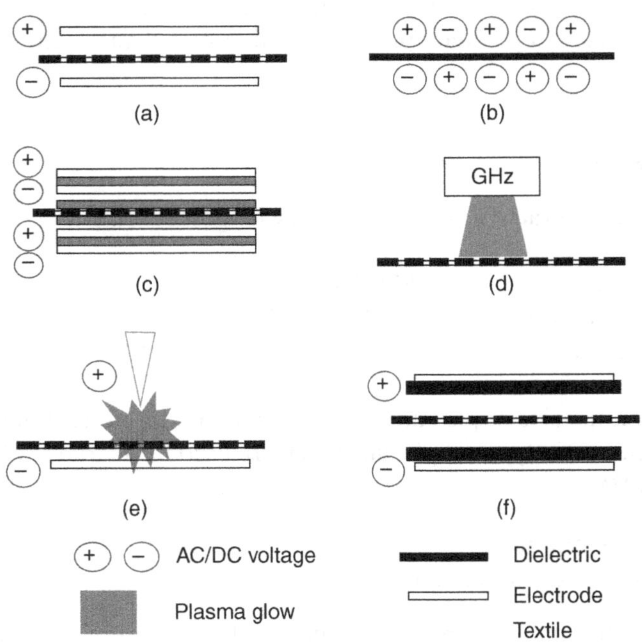

2.2 Schematic principle of different plasma processes, see text for explanation. (*Source*: Adapted from Verschuren and Kiekens, 2001.)

plasma is generated by application of a direct current at MW, LF (50 Hz) or RF (40 kHz, 13.56 MHz) across two parallel plates or, alternatively, a series of electrodes (Fig. 2.2a–2.2d).

CD is performed at atmospheric pressure by application of an LF or pulsed high voltage across a pair of electrodes. Typically, the electrodes have a large difference in size (Fig. 2.2e). The corona consists of small lightning-type discharges, although the high degree of inhomogeneity associated with the approach presents problems for textile applications. Corona-based approaches can only be applied with loose fibres, and cannot penetrate deeply into yarn or woven fabrics, limiting their use in the textile industry.

DBD is formed by application of a pulsed voltage at frequencies up to 100 kHz to an electrode pair at least one of which is coated with a dielectric material (Fig. 2.2f). The DBD generates a non-thermal plasma with a multitude of arcs formed between the electrodes. The DBD approach is a superior technique in terms of uniform textile treatment when compared to the CD approach.

2.3 Plasma treatments for high performance textiles

The textile industry requires innovative new finishing techniques in order to improve both product quality and environmental performance. Plasma surface treatments which generate characteristics such as waterproofing, oil repellence and antistatic behaviour have distinct advantages in this respect, as only small quantities of chemicals are required and the devices used can also be environmentally friendly. With textile surfaces, three main effects can be observed: the cleaning effect, an increase of micro-roughness (e.g. anti-pilling finishing of wool) and production of free radicals to obtain hydrophilic or hydrophobic properties.

2.3.1 Plasma/textile surface interactions

Plasma treatment of textiles generate desirable surface modifications including but not limited to surface activation, surface etching, cross linking, chain scission, de-crystallization and oxidation. The process for treatment is dependent on the selection of working gas (air, oxygen, argon, fluorine, helium, carbon dioxide or their mixtures) and plasma density and energy. Different types of fibres treated by the same gas will display different responses (Pastore and Kiekens, 2001; Poll *et al.*, 2001; McCord *et al.*, 2002; McCord *et al.*, 2003).

Plasmas can be used to generate functionalized polymeric surfaces by application of an electric field to the working gas. The free electrons in this discharge are accelerated by the electric field and collide with the neutral

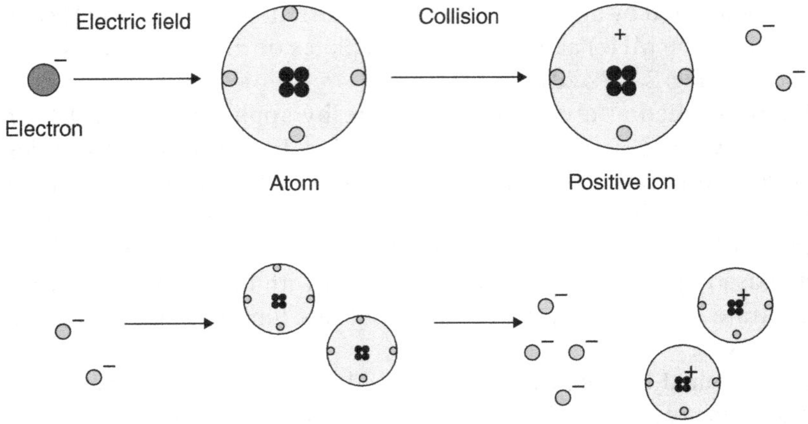

2.3 A schematic mechanism of plasma creation.

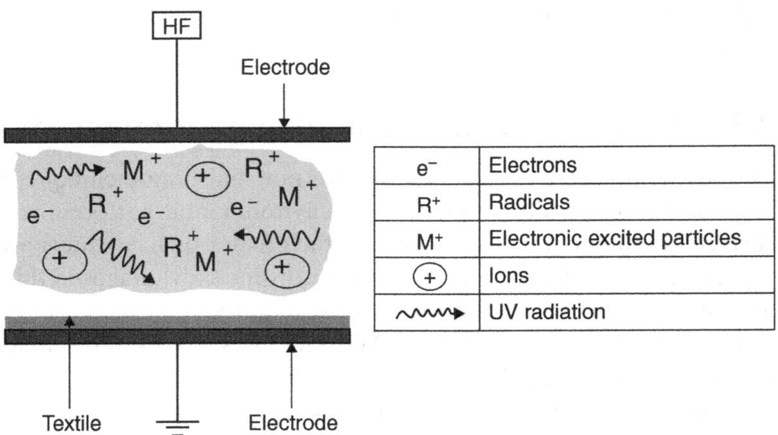

2.4 A schematic view of plasma device and different reactive species. (*Source*: Adapted from Yousefi *et al.*, 2003.)

atoms or molecules comprising the gas. Metastables, positive ions, electrons and free radicals are produced by consecutive collisions (Fig. 2.3).

Figure 2.4 shows a schematic diagram of a plasma device and different reactive species (Yousefi *et al.*, 2003). Plasma conversion of the feed gas produces chemically active particles that are able to modify the textile surface through chemical reactions at the surface. Various surface modifications can be obtained depending on process parameters such as:

- Noble gases (e.g., He, Ar) → Etching.
- Oxidizing gases (e.g., O_2, CO_2, H_2O) → Etching, cleaning and functionalization.

- Hydrocarbon containing gas (e.g., N or O-containing HC) → Plasma polymerization.

2.3.2 Characterization and properties of plasma-treated surfaces

Surface hydrophilicity of textiles can be increased by using various gases such as Ar and O_2, while the hydrophobicity can be increased by using CF_4, NF_3 or SiF_4. In addition to gas type, process parameters also impact on the final results; these parameters include plasma gas/gas mixtures, gas flow rate, operating temperature (T_g), operating pressure (low vacuum), operating power, contact time and cleanness of the operating chamber (vacuum).

Hwang *et al.* (2005) treated polypropylene (PP) non-woven fabrics using an atmospheric pressure GD plasma (Fig. 2.5). A stable and uniform plasma was obtained at a frequency of 5.0 kHz and the voltage across the plates reached 7.5 kVrms. Fabric treatments were conducted with He gas under atmospheric pressure, and the gas flow rate was 13 L/m with an exposure time of between 0 and 2 min at 30 s intervals. The resulting wettability of the plasma-treated PP fabrics was estimated by contact-angle measurements and was found to decrease with increased exposure time (Fig. 2.6), eventually levelling off. Longer plasma treatment results in a higher wettability than the shorter treatment. Note that longer plasma exposure times could induce more hydrophilic functional groups on the fabric surface due to the longer duration of the chemical interaction of plasma and PP.

Using scanning electron microscopy (SEM), the effect of the plasma treatment on the morphology of the textile surface was investigated and results for untreated and treated samples are shown in Fig. 2.7. The untreated PP

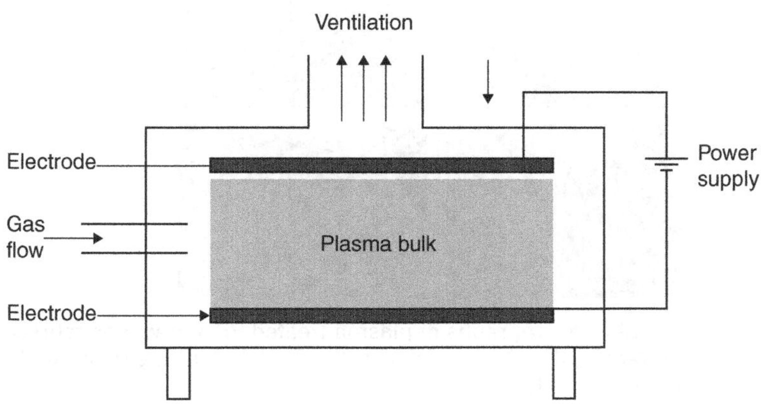

2.5 A schematic diagram of atmospheric pressure plasma system. (*Source*: Adapted from Hwang *et al.*, 2005.)

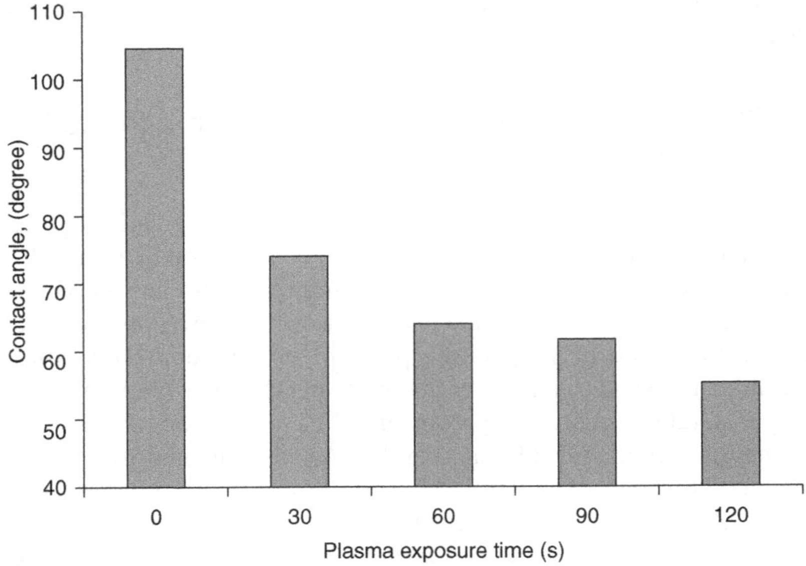

2.6 Wettability of plasma-treated PP non-woven fabric surface with different exposure times. (*Source*: Adapted from Hwang *et al.*, 2005.)

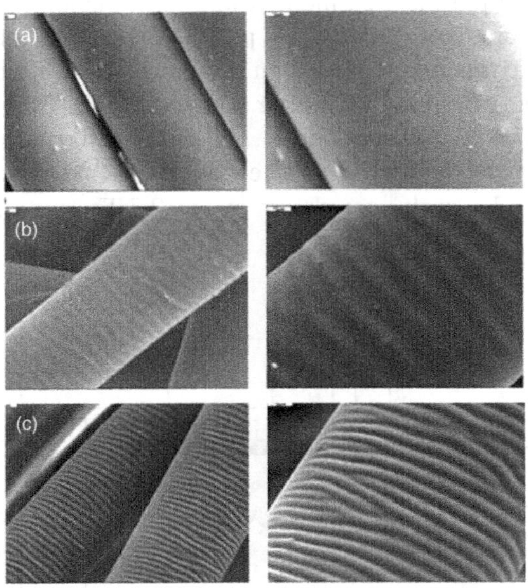

2.7 SEM micrographs of plasma-treated PP non-woven fabrics: (a) untreated, (b) 30 s and (c) 90 s; the RHS images are magnified. (*Source*: Courtesy of Hwang *et al.*, 2005.)

filament has a smooth surface, while the plasma-treated samples which were exposed for more than 30 s exhibited increased surface roughness resulting from the etching effect of the plasma.

Karahan and Özdogan (2008) used air plasma and an argon atmospheric plasma for enhancing the hydrophobicity of cotton. A DBD system and a sample of plain-weave cotton fabric were used. The samples were plasma-treated at a range of powers (50, 100 and 130 W) and for different exposure timescales (20, 40 and 60 s). The wettability of the samples was investigated via absorbency values, contact angles, wicking tests and X-ray photoelectron spectroscopy (XPS).

Table 2.1 gives the hydrophilicity as a function of exposure duration, power and the type of plasma treatment. It can be seen that the argon plasma was more effective than the air plasma, which can be attributed to the significant etching effect of the former. As seen in Table 2.2, the argon plasma was more effective than the air plasma, consistent with the hydrophilicity

Table 2.1 Hydrophilicity outcomes (s)

Treatment	Air plasma	Argon plasma
Untreated	Hydrophobic	Hydrophobic
50 W, 20 s	Hydrophobic	5
50 W, 40 s	Hydrophobic	1.2
50 W, 60 s	Hydrophobic	1
100 W, 20 s	8.6	1
100 W, 40 s	4.7	1
100 W, 60 s	3.5	1
130 W, 20 s	7.7	1
130 W, 40 s	1.4	1
130 W, 60 s	1	1

Source: Courtesy of Karahan *et al.* (2008).

Table 2.2 Contacts angles (°)

Treatment	Air plasma	Argon plasma
Untreated	107.83	107.83
50 W, 20 s	105.33	0
50 W, 40 s	103.41	0
50 W, 60 s	98.32	0
100 W, 20 s	71.72	0
100 W, 40 s	46.12	0
100 W, 60 s	32.82	0
130 W, 20 s	59.30	0
130 W, 40 s	0	0
130 W, 60 s	0	0

Source: Adapted from Karahan *et al.* (2008).

2.8 Contact angle images of (a) untreated fabric, (b) fabrics treated with air plasma under 130 W for 60 s and (c) treated with argon plasma under 130 W for 60 s. (*Source*: Courtesy of Karahan *et al.*, 2008.)

Table 2.3 Height (cm)

Treatment	Air plasma	Argon plasma
Untreated	0	0
50 W, 20 s	0.1	0.8
50 W, 40 s	0.2	4.6
50 W, 60 s	0.3	5.5
100 W, 20 s	2.2	7.1
100 W, 40 s	4.3	7.8
100 W, 60 s	4.8	9.0
130 W, 20 s	3.9	7.2
130 W, 40 s	7.8	8.4
130 W, 60 s	8.0	9.2

Source: Adapted from Karahan *et al.* (2008).

Table 2.4 Chemical composition of the untreated and plasma-treated cotton fabrics using air and argon plasmas at 130 W and 60 s

Treatment	C	O	O/C
Untreated	81.9	17.1	0.208
130 W, 60 s, air plasma	75.6	24.4	0.322
130 W, 60 s, argon plasma	73.3	26.7	0.364

Source: Adapted from Karahan *et al.* (2008).

outcomes. Figure 2.8 depicts these issues clearly. Wicking results gave comparatively better information about the surface modification resulting from the treatment than the hydrophilicity and contact angle measurements, as shown in Table 2.3.

Table 2.4 shows the results of XPS for the cotton fabric treated by air and argon plasmas. Due to the chemical effects of the plasma species, new functional groups were generated, which affected the surface composition.

A GD generator was used by Kan (2001) to study the oxygen plasma treatment of wool fibres of differing diameter. The power, pressure and time of treatment were 80 W, 10 Pa and 5 min, respectively. After the treatment, it

(a) (b)

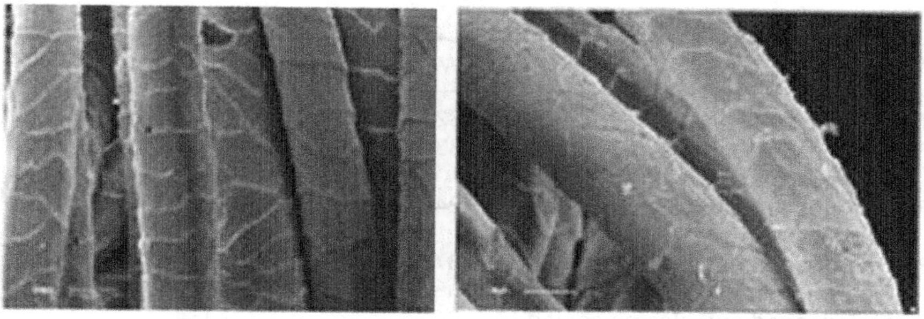

2.9 Plasma treatment of wool fibre: (a) sample 1 untreated; and (b) sample 2 argon plasma treated for 7 min. (*Source*: Courtesy of Masukuni and Norihiro, 2006.)

was observed that micro-pores and cleft lines parallel to the fibre axis were formed on the fibre surface. It was also found that the treated wool fabrics dyed faster than the untreated ones, and were deeper in colour than the fibres that were treated only with enzymes. The characteristics of the plasma-treated fabrics were further enhanced by addition of an enzyme treatment. Masukuni and Norihiro (2006) determined that the anti-felting behaviour of wool fabric was improved by use of an argon plasma treatment, as shown in Fig. 2.9.

The influence of plasma etching/cleaning and plasma coating processes, compared to more traditional methods of finishing (washing and coating), was investigated by Pane *et al.* (2001) with specific reference to waterproofing of acrylic fabrics for outdoor applications. Pre-treatment with 2000 W oxygen was used for the hydrophobic coating, and mixtures of argon, fluorocarbon and hydrocarbon gases were applied at time intervals of 3, 5 and 10 min. Some of the characteristics of the treated fabrics are given in Table 2.5. The abrasion (Martindale) and weathering (Xenotest) tests indicate that plasma coatings have good wear and UV resistance.

The creation of active centres within the cellulose chains used in initiating the copolymerization reactions with vinyl monomers was studied by Abidi and Hequet (2004). The aim was to impart hydrophobicity into lightweight cotton fabrics using N_2, O_2 and Ar MW plasmas (2.45 GHz under vacuum). Plasma treatment for 4 min with 500 W was sufficient to create active carbonyl groups, and the Ar plasma was observed to generate the most active groups. Before a second plasma treatment, the fabrics were impregnated with vinyl laurite, with optimal grafting efficiency seen at vinyl monomer concentrations of below 0.664 mol/L. Testing of treated fabric revealed that excellent water repellence was obtained.

Chen (1996) studied free radical formation on cotton and wool fibres treated with low temperature plasmas (LTP) of O_2, N_2, Ar, H_2, CO and CF_4 at the RF generator, with a power of 300 W and pressures between 0.3

Table 2.5 Effect of plasma coatings on the waterproof characteristics of fabrics with respect to traditional coating

Acrylic fabrics	Water column (cm)	Contact angle (°)
Traditionally finished	32	133
Plasma cleaned and plasma coated (Fluorocarbon 20%/Ar 80%)	33	133
Plasma cleaned and plasma coated (Fluorocarbon 90%/methane 10%)	24	123
Plasma cleaned and plasma coated (Fluorocarbon 93%/methane 7%)	37	140
Plasma cleaned and plasma coated (Fluorocarbon 90%/Ar 10%)	34	134

Source: Adapted from Pane *et al.* (2001).

Table 2.6 Relative free radical intensities detected by electron spin resonance (ESR) after the plasma treatments

Plasma gas	Cotton	Wool
O_2	0.5	0.4
N_2	0.6	0.5
Ar	1.6	0.6
H_2	1.8	0.6
CO	2.9	0.7
CF_4	3.1	0.9

Source: Adapted from Chen (1996).

and 1.5 Torr. Table 2.6 shows that free radical intensities were different for the various gases with forming occurring in the order $O_2 < N_2 < Ar < H_2 < CO < CF_4$. It was also noted that free radical formation was increased with increasing time.

2.4 Laser technologies for textiles

Explanations of the laser technology and its application in textiles have been carried out by means of fundamental characteristics and textile-surface treatment.

2.4.1 General characteristics of lasers

The word LASER is an acronym for light amplification by stimulated emission of radiation. The laser makes use of processes that increase/amplify light signals after those signals have been generated by other

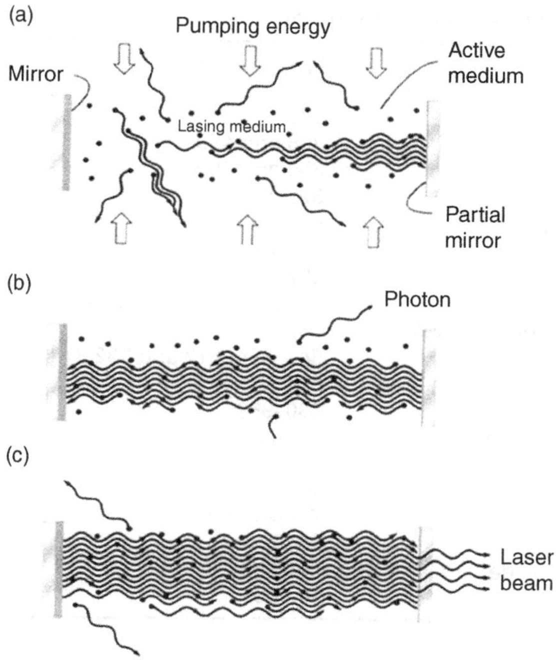

2.10 (a–c) The principle of lasing process, see text for explanation.

means. These processes include: *stimulated emission*, by which an atomic (excited) electron interacts with an electromagnetic wave of a certain frequency, transferring its energy (photon) to that field; and *optical feedback,* usually provided by mirrors. The laser consists of a gain or amplifying (lasing) medium where the stimulated emission takes place, and a set of mirrors (partially and fully reflecting) to feed the light back into the amplifier for continued growth of the developing laser beam, as shown in Fig. 2.10. Lasers have several special qualities when compared to ordinary light: (i) it comes in one colour; (ii) it comes in a thin beam; (iii) it can be made very intense or not intense at all; and (iv) it can be focused to a tiny spot.

A laser can be classified as operating in either continuous wave (CW) or pulsed (modulated) mode, depending on whether the power output is essentially continuous over time or whether its output takes the form of pulses of light. Some lasers can only be operated in pulsed mode, such as copper vapour, excimer, nitrogen, ruby and CO_2 lasers. Such lasers cover the wavelength range from the far infrared (F-IR) ($\lambda = 1000$ μm) to the soft X-ray region ($\lambda = 3$ nm) of the spectrum. Tables 2.7 and 2.8 give details of a range of common laser types.

Table 2.7 Examples of lasers and wavelengths

Laser type	Wavelength (nm)
Argon fluoride	193
Xenon chloride	308 and 459
Xenon fluoride	353 and 459
Helium cadmium	325–442
Rhodamine 6G	450–578
Copper vapour	511 and 578
Argon	457–528 514.5 and 488 most used
Frequency doubled Nd:YAG	532
Helium Neon	543, 594, 612 and 632.8
Ruby	694.3
Laser diodes	630–950
Ti:Sapphire or Ti:Al_2O_3	690–960
Alexandrite	720–780
Nd:YAG	1064
Hydrogen fluoride	2600–3000
Erbium:glass	1540
Carbon monoxide	5000–6000
Carbon dioxide	10 600

Table 2.8 Band wavelengths regions

Band	Range
Far infrared (F-IR)	10–1000 μm
Middle infrared (M-IR)	1–10 μm
Near infrared (N-IR)	0.7–1 μm
Visible (Vis)	0.4–0.7 μm or 400–700 nm
Ultraviolet (UV)	0.2–0.4 μm or 200–400 nm
Vacuum ultraviolet (V-UV)	0.1–0.2 μm or 100–200 nm
Extreme ultraviolet (E-UV)	10–100 nm
Soft X-rays	1 nm – approximately 20–30 nm Some overlap with E-UV

2.4.2 Laser treatment and functionalization of textiles surfaces

Surface modification/treatment of textile materials by chemical finishing processes is not environmentally friendly, as a result of which an alternative technology, 'laser irradiation', has been widely adopted. By using laser irradiation, the textile materials can be structurally modified and their surface can be functionalized and textured. Here, two different examples of textile materials are presented and discussed in terms of laser treatment; cotton, which is an example of natural material, and polyester (PET), which is a synthetic one.

2.4.3 Laser-treated cotton

Surface morphology and properties of laser-treated cotton fabrics were studied by Chow *et al.* (2010, 2011). Qualitatively, the sponge-like surface structure of cotton fibres were characterized after treatment with 10.6 μm CO_2 laser irradiation. Hundred percent plain-weave grey cotton fabric was treated using a pulsed CO_2 laser beam with a range of pulse energy (5–230 mJ), followed by desizing, scouring and bleaching treatments. SEM was used to investigate the surface morphology of the treated cotton specimens. Figure 2.11 shows the SEM images of: (a) untreated cotton; (b) laser-treated cotton; (c) laser-desized treated cotton; (d) laser-desized-scoured treated cotton; and (e) laser-desized-scoured-bleached treated cotton. It is clear that the untreated sample exhibits a smoother surface (see Fig. 2.11a), but that textured surfaces of various pores sizes were observed on the treated cotton samples. Thus it can be concluded that the CO_2 laser caused physical modification at the surface of cotton fabric.

Chow *et al.* (2011) determined numerous properties of the laser-treated cotton fabrics such as: fabric weight loss; yellowness index; breaking strength; tearing strength; dyeing; and colour measurement and fastness. Figure 2.12 shows a selection of the properties determined.

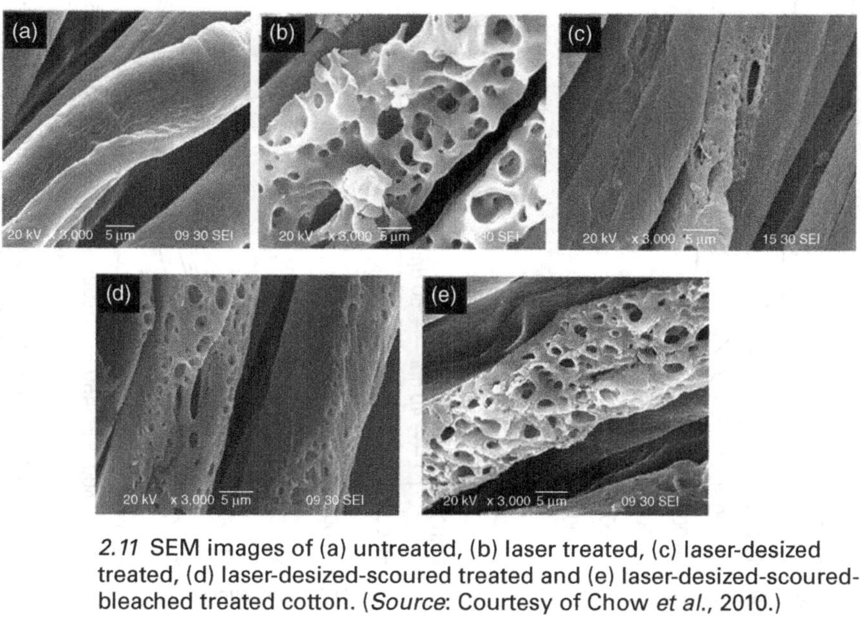

2.11 SEM images of (a) untreated, (b) laser treated, (c) laser-desized treated, (d) laser-desized-scoured treated and (e) laser-desized-scoured-bleached treated cotton. (*Source*: Courtesy of Chow *et al.*, 2010.)

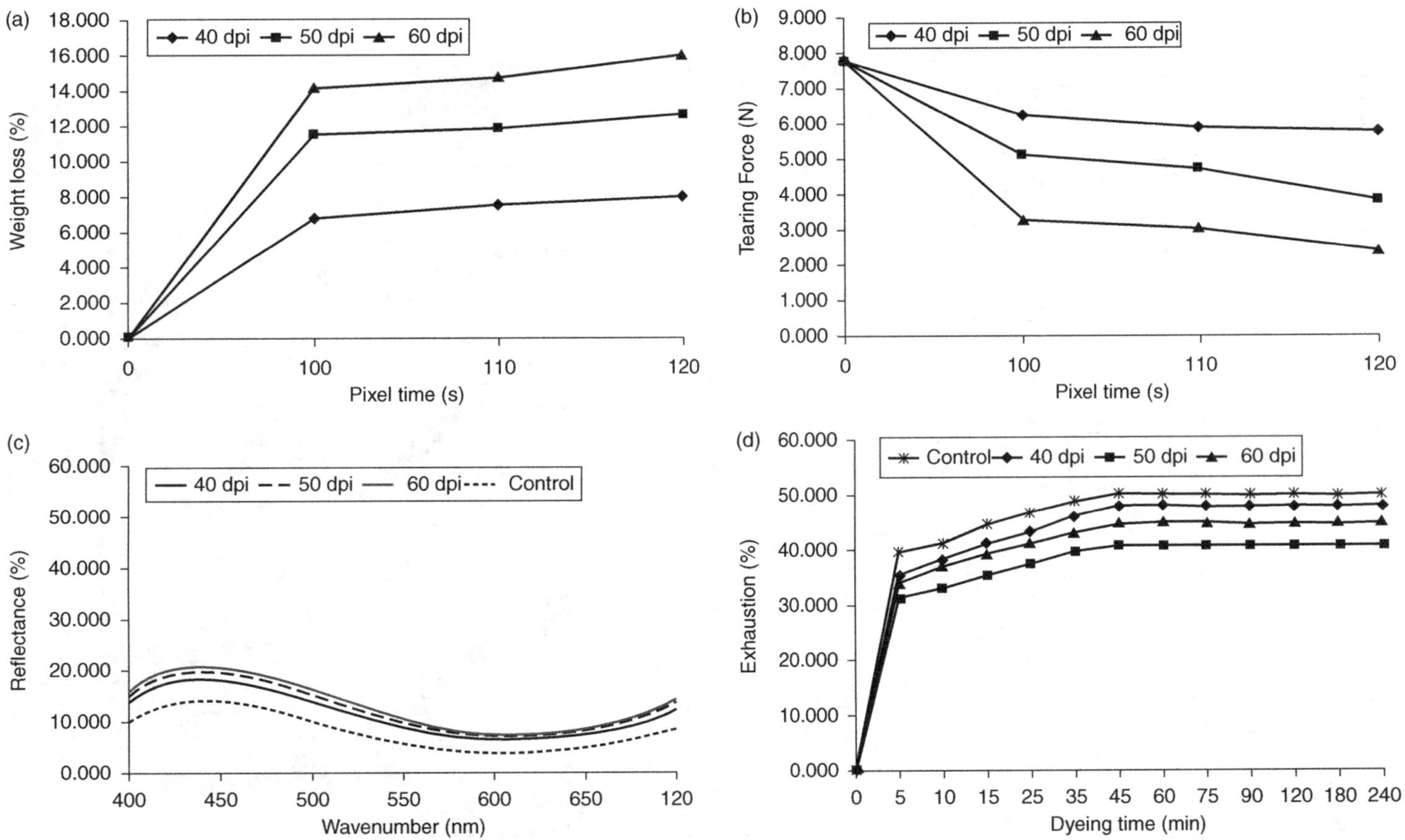

2.12 Properties of laser cotton treated (a) % weight loss, (b) tearing force, (c) % reflectance and (d) % dye bath exhaustion. (*Source*: Adapted from Chow *et al.*, 2011.)

2.4.4 Laser-treated polyester (PET)

Laser irradiation of polymeric fibres can be used to generate a morphological modification at the surface of the irradiated regions. The normally smooth surface of synthetic fibres can be modified by this technique to form a regular, ripple-like, structure, which has a significant effect on the physical properties of the fibre, e.g. adhesion of particles and coatings, wetting properties and optical appearance (Knittel and Schollmeyer, 1998).

The dyeability of polyester fabric is poor compared to that of other synthetic fibres, due to the lack of polar molecules and pendent groups. Lan *et al.* (1997) used an excimer laser at 248 nm, 1 Hz and fluence range (20–150 mJ/cm²) to treat the surface of polyester fabric in order to improve both the hydrophobicity and dyeability. Figure 2.13 shows that the contact angle for a water droplet on an untreated surface (Fig. 2.13a) is less than 90° while, after

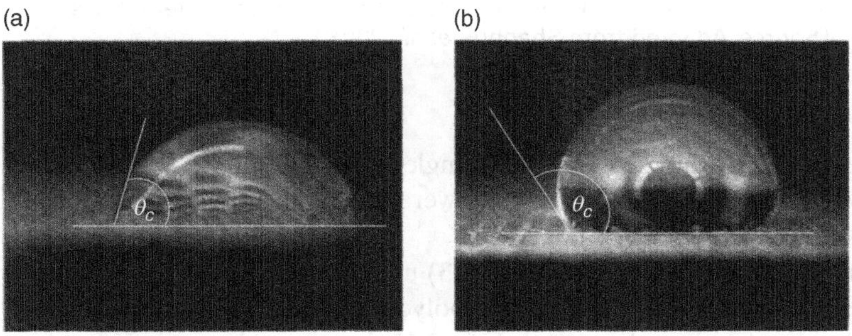

2.13 Contact angle between a drop of water and (a) untreated PET fabric and (b) laser treated PET (10 pulses of 60 mJ/cm²). (*Source*: Courtesy of Lan *et al.*, 1997.)

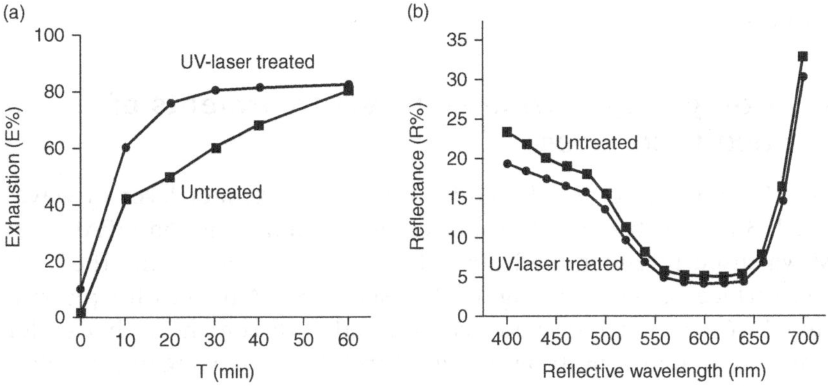

2.14 Dyeing properties: (a) the exhaustion rate and (b) the colour reflectance of untreated and laser treated polyester fabrics. (*Source*: Adapted from Lan *et al.*, 1997.)

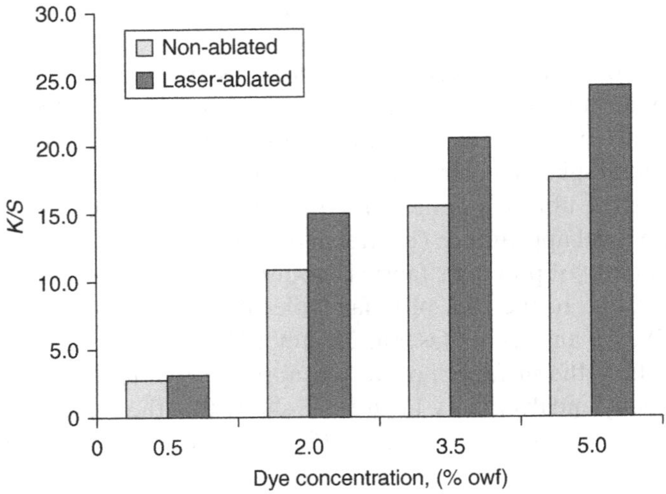

2.15 Influence of UV-laser on the K/S of red dyed polyester fabrics. (*Source*: Adapted from Shaohua *et al.*, 2003.)

excimer laser treatment, the contact angle is greater than 90° (Fig. 2.13b). The dyeing properties of PET fabrics were also improved by laser irradiation, as shown in Figs 2.14a and 2.14b.

In another study, Shaohua *et al.* (2003) used a UV-laser (10 pulses of 150 mJ/cm^2) to improve the dyeability of polyester fabric. They used three disperse dyes (red, blue and navy), and Fig. 2.15 shows the change in K/S values versus the red-dye concentration (% owf) of the untreated and laser-treated PET fabrics. It can be seen that the laser (ablated)-treated PET has a higher K/S than the untreated (non-ablated) fabric, indicative of greater uptake of the disperse dye.

2.5 Comparing plasma and laser treatments of textile surfaces

Yip *et al.* (2002) used an LTP and a UV (ArF) excimer laser of wavelength 193 nm to structure the surface of polyamide (nylon 6/PA6). An SEM was used to understand the influence of the plasma and the laser on the surface of the PA6. Figure 2.16 shows SEM images for plasma-treated PA6 (on the left) and laser-treated PA6 (on the right) under different treatment conditions. In Figs 2.16a–2.16c, showing oxygen plasma-treated nylon 6 fibres, it can be observed that change in the surface morphology correlates well with the power and treatment time, such that increasing the power and treatment time leads to an increase of

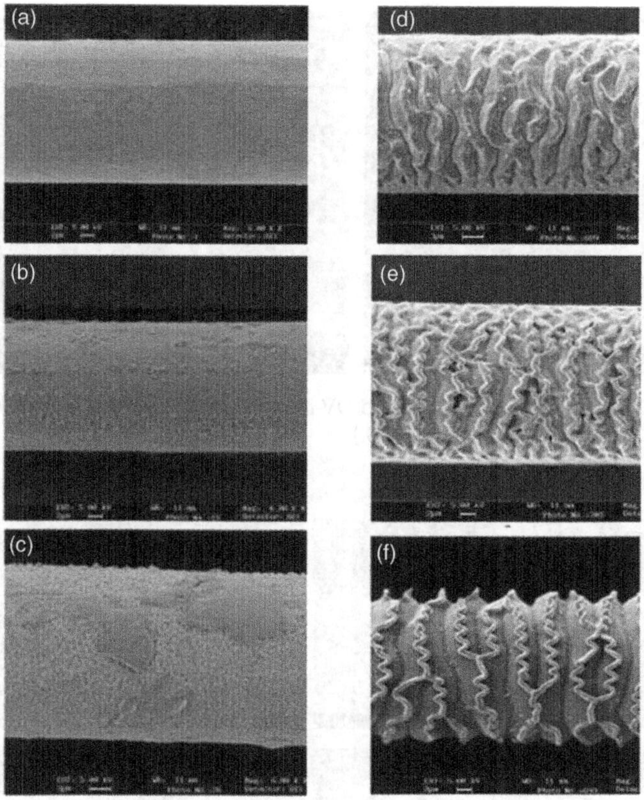

2.16 PA6 plasma- and laser-treated: (a) untreated PA6, (b) 100 W, 10 min, (c) 200 W, 30 min, (d) 50 mJ/cm², 20pulses, (e) 170 mJ/cm², 10 pulses and (f) 300 mJ/cm², 20 pulses. (*Source*: Courtesy of Yip *et al.*, 2002.)

the ripple-like effect (at the sub-micron level) perpendicular to the fibre axis. With the UV-laser-treated fibres, it was found that the ripple/roll structures (micron size) are also almost perpendicular to the fibre axis and the mean distance between rolls increases with increasing numbers of laser pulses and fluence (mJ/cm^2) applied.

Wong *et al.* (2000) used a pulsed UV (KrF) excimer laser and an oxygen gas discharge LTP to modify the surface structure of PET fibres. Although both plasma and laser-treated fibres exhibit similar ripple/roll-like structure surfaces, the morphological modifications are not identical, as shown in Fig. 2.17. The PET fibres were firstly treated by LTP (100 W, 5 Pa, 10 min) and then by laser irradiation under different fluence and pulses. It was found that the threshold for the formation of a ripple-like structure under laser irradiation was lower for PET with LTP pre-treatment (see Fig. 2.17). In addition,

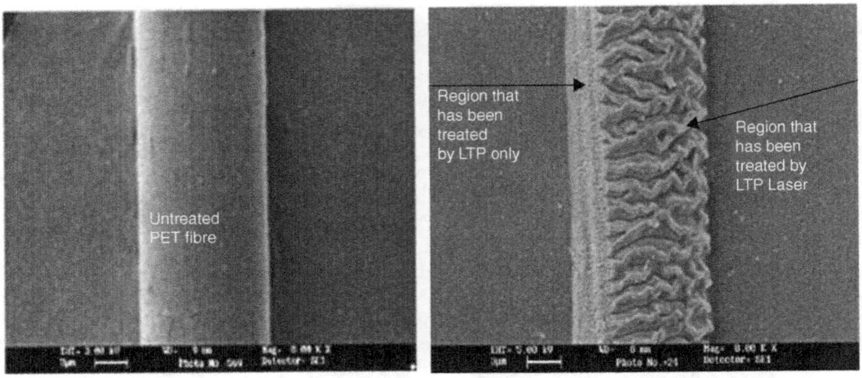

2.17 Effect of LTP plasma and UV-laser on polyester fibre. (*Source*: Courtesy of Wong *et al.*, 2000.)

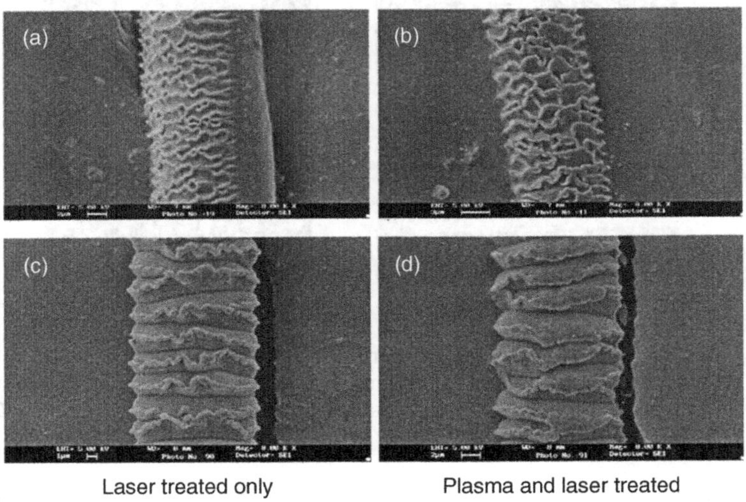

2.18 Effect of LTP plasma and UV-laser on polyester fibre: (a, b) 50 mJ, 3 pulses and (c, d) 150 mJ, 40 pulses. (*Source*: Courtesy of Wong *et al.*, 2000.)

the ripple-like structure was more prominent for LTP-pre-treated PET than for the non-LTP-treated counterpart under the same laser treatment. This is because there is more surface area developed on the LTP-pre-treated PET, which helps to increase the absorption of the laser energy (Fig. 2.18).

2.6 Conclusions

Improvement of the surface characteristics of textiles is required to meet economic, market and environmental needs. Textile products that require less energy, water and chemical treatments are preferred, and both plasma and laser technologies are promising in this respect.

2.7 References

Abidi N and Hequet E (2004), 'Cotton fabric copolymerization using microwave plasma: universal attenuated total reflectance-FITR study, *Journal of Applied Polymer Science*, **93**, 145–154.

Chen J R (1996), 'Study on free radicals of cotton and wool fibers treated with low-temperature plasma', *Journal of' Applied Polymer Science*, **62**, 1325–1329.

Chow Y L, Chan C K and Kan C W (2010), 'SEM analysis of CO_2 laser treated cotton grey fabric', *Microscopy: Science, Technology, Applications and Education, FORMATEX*, 1860–1867.

Chow Y L, Chan A and Kan C W (2011), 'Effect of CO_2 laser irradiation on the properties of cotton fabric', *Textile Research Journal*, **82**, 1220–1234..

Hwang Y J, Mccord M G, An J S, Kang B C and Park S W (2005), 'Effects of helium atmospheric pressure plasma treatment on low-stress mechanical properties of polypropylene nonwoven fabrics', *Textile Research Journal*, **75**, 771–778.

Kan C W (2001), 'The Effect of Low Temperature Plasma Treatment on Different Types of Wool Fibre' *The 6th Asian Textile Conference, Innovation & Globalization, Proceedings*, August 22–24, Hong Kong.

Karahan H A and Özdogan E (2008), 'Improvements of surface functionality of cotton fibers by atmospheric plasma treatment', *Fibers and Polymers*, **9**, 21–26.

Knittel D and Schollmeyer E (1998), 'Surface structuring of synthetic polymers by uv-laser irradiation, Part IV: applications of excimer laser induced surface modification of textile materials', *Polymer International*, **45**, 110–117.

Lan K S, Chan P W, Wong K H, Yeung K W, Chan K and Gong W Z (1997), 'Surface properties of polyester fabrics induced by excimer laser processing', *Journal of Materials Processing Technology*, **63**, 524–528.

Masukuni M and Norihiro I (2006), 'Relationship between anti-felting properties and physicochemical properties of wool fibers treated with Ar-plasma', *Textile Research Journal*, **89**, 687–694.

McCord M G, Hwang Y J and Hauser P J (2002), 'Modifying nylon and polypropylene fabrics with atmospheric pressure plasmas', *Textile Research Journal*, **72**, 491–498.

McCord M G, Hwang Y J, Qiu Y, Hughes L K and Bourham M A (2003), 'Surface analysis of cotton fabrics fluorinated in radio-frequency plasma', *Journal of Applied Polymer Science*, **88**, 2038–2047.

Pane S, Tedesco R and Greger R (2001), 'Acrylic fabrics treated with plasma for outdoor applications', *Journal of Industrial Textiles*, **31**, 135–145.

Pastore C M and Kiekens P (Eds.) (2001), '*Surface Characteristics of Fibers and Textiles*'. Marcel Dekker Inc., New York.

Poll H U, Schladitz U and Schreiter (2001), 'Penetration of plasma effects into textile structures', *Surface & Coatings Technology*, **142–144**, 489–493.

Shaohua L, Xiang Z, Tianmin T, Zhiqi X and Qihong L (2003), 'UV-laser treatment of PET fabric: changes in fibre structure and dyeing behaviour', *Color Technology*, **119**, 19–23.

Shishoo R (2007), '*Plasma Technologies for Textiles*', Woodhead Publishing Limited and CRC Press LLC, Cambridge.

Verschuren J and Kiekens P (2001), 'Plasma Technology for Textiles: Where Are We?' IX. *International Izmir Textile & Apparel Symposium*, October 2001, Izmir, Turkey.

Wong W, Chana K, Yeunga K W, Tsanga Y M and Lau K S (2000), 'Surface structuring of poly(ethylene terephthalate) fibres with a UV excimer laser and low temperature plasma', *Journal of Materials Processing Technology*, **103**, 225–229.

Yip J, Chan K, Sin K M and Lau K S (2002), 'Study of plasma-etched and laser-irradiated polyamide materials', *Materials Research Innovation*, **6**, 44–50.

Yousefi H R, Nakada Y, Ito H and Masugata K (2003), 'Investigation of the neutron production mechanism in a 20 kJ plasma focus device', *Journal of Fusion Energy*, **25**, 245–248.

3
High performance textiles for protective clothing

N. MAO, University of Leeds, UK

DOI: 10.1533/9780857099075.91

Abstract: In this chapter, the characteristic physical properties of high performance textile materials are reviewed, and examples of their applications in the engineering design of protective clothing are given, together with a brief introduction to the legal requirements for protective clothing. As a case study, the functions of high performance textile materials used in cut resistant products, firefighters' clothing, chemical protective clothing and materials for achieving thermophysiological comfort in protective clothing are discussed.

Key words: protective clothing, chemical protective clothing (CPC), fire fighter's clothing, textile materials, high performance fibres, membrane, protection, activated carbon, comfort, thermophysiological comfort.

3.1 Introduction

Protective clothing is a major type of personal protective equipment (PPE), defined in European Directive 89/686/EEC[1] as 'any device or appliance designed to be worn or held by an individual for protection against one or more health and safety hazards'. While PPE exists for every part of the human body, including head protection (helmets, hearing protection, masks), respiratory protection (gas masks, filter masks), arm/shoulder protection, hand protection (gloves), leg and foot protection (shoes, knee pads, hip pads), protective clothing mainly refers to garments specifically designed for body protection (trousers, coat, vest, suits such as turnout gear, bomb disposal suits, motorcycle suits and body armour) and to be worn in workplace in relation to the wearer's occupational job, sports, martial arts and combat.

Protective clothing is used in various industries for one of the following three purposes:

1. to protect the wearer in the workplace, sports, martial arts and combat;

2. to protect products (e.g., food, pharmaceutical/medical products, electronic products in clean rooms) processed in the workplace from damage/contamination/infection by environmental hazards and people (e.g., food, electronic products, cosmetics);
3. to protect both people and products/customers/patient (e.g., pharmaceutical, medical products and patients in hospitals).

Protective clothing can be grouped into the following two categories:

1. specialist protective workwear: specifically designed for the purposes of providing job-related occupational safety and health to the wearers, their products/customer/patient, or both from environmental hazards and contamination in various industries, e.g., chemical engineering, food, pharmaceutical, electronics, civil engineering, etc.
2. protective clothing for law enforcement personnel: specifically designed for law enforcement personnel to protect them from such hazards as bullets, shrapnel, stabbing, fire and flame, chemical and biological agents, radiative contamination, etc. This includes firefighters' clothing, stab- and bullet-proof clothing, chemical, biological, and radiological (CBR) protective clothing, etc.

The performance of protective clothing in the European market is regulated by the basic safety requirements set out in European Regulation (EC) Directive 89/686/EEC (or CE marking regulation) and associated with about 300 EN standards for EC Marking,[2] and the uses and selections of the clothing are regulated by various national/international directives, regulations and health and safety policies (e.g., in the EU/UK, PPE (EC Directive) in Work Regulations 1992, PPE Regulations 2002, The Noise at Work Regulations 1989, The Construction (Head Protection) Regulations 1989, Control of Substances Hazardous to Health Regulations 2002 (COSHH), etc., and also 29 CFR Part 1910 (USA)).

In the United States, PPE is regulated by the Occupational Safety and Health Act of 1970 (OSH Act) and the Occupational Safety and Health Administration (OSHA) with 29 CFR standards for the selection of chemical protective clothing (CPC). In addition, the National Fire Protection Association (NFPA) has a series of performance-based standards[3] for protective clothing used in various industries, for example: NFPA 1976 and NFPA 1977 standards on protective clothing for firefighting; NFPA 1991, NFPA 1992, NFPA 1994; and NFPA 1999 on CPC (Todd, 2001). Based on the requirements of the clothing performance set in those standards, the structures and properties of the textile materials used for protective clothing are tested and examined by using corresponding methods defined in American Standard Testing of Materials (ASTM) standards.[4,5]

When the protective clothing is designed to protect products from contamination during the manufacture process of food, cosmetic, pharmaceutical, medical and electronic products, it is covered by specific regulations to maintain the hygiene and safety of the products. For example, any materials intended to come into contact with food in its finished state must conform to: EC 1935/2004 and EC 2023/2006, and guided by: EC good manufacture practice (GMP) in food manufacture; Food Hygiene Directive (93/43/EEC); EU Regulation 852/2004 (Article 5); and Hazard Analysis and Critical Control Point (HACCP). In the USA, such legislation includes the Code of Federal Legislation (CFR): 21 CFR 174 ~ 21 CFR 190 (USA).

The required functions for protective clothing include protection (e.g., heat, cold, electricity, cutting, stabbing, etc.), comfort (thermal and tactile comfort properties such as moisture transport resistance, smoothness and softness), mobility (heaviness, flexibility and conformability), durability to withstand harsh cleaning, washing, ageing and hazardous attack conditions (protection properties and strength properties after washing, flex cracking and ageing), and others (environmental impact, toxicity, etc.). In some special cases, such as in the food, drink, pharmaceutical, cosmetic, medical, electronic manufacture and healthcare industries, the protective clothing requires further properties, in order to not only protect the wearer from environmental hazards but also to protect the product from contamination (e.g., linting, pilling, dissolved monomer) or cross-infection by people (i.e., the wearer) and the workwear itself.

While the performance of protective clothing depends on both the structural design of the clothing and the properties of component fabrics, the latter is essential. In the engineering design of specific fabrics to meet special performance requirements of clothing set in relevant regulations (e.g., CE Marking scheme, NFPA standards), the composition of fibres, membrane, coating and accessory materials (such as seals, tapes and zips) used in the garment, together with the microstructure of fabrics, yarns and fibres employed to construct the clothing, need to correspond to the hazardous environment, the human physiological comfort, durability and other special requirements (e.g., product hygiene) in order to fulfil the functions, properties and performance of the protective clothing.

The types of textile materials used in protective clothing are determined by the performance requirements of the clothing, which come from the analysis of the specific human–clothing–environment system. In this chapter, the application of high performance textile materials in the firefighters' clothing, body armour and chemical protective clothing are reviewed, and the effects of the properties of those new materials on the overall performance of those PPE products will be briefly reviewed.

3.2 Requirements for protective clothing

There are four requirements related to the application of any clothing system:

1. aesthetic properties
2. functional performance
3. clothing comfort
4. human mobility.

While aesthetic properties of clothing are important for every clothing system including fashion, sportswear, casual wear, outdoor clothing or protective clothing, the emphasis for protective clothing is laid on functional performance, thermal comfort and human mobility.

In the protective clothing–human body–environment system, the interface between clothing and human body forms a micro-environment that affects human comfort; the interface between clothing and environment acts as a barrier to prevent the human body from exposure to harmful hazards. In a summary, the protective clothing performs three functions:

1. Protection: protects the wearer from environmental hazards, which is the primary purpose of protective clothing.
2. Clothing comfort: provides human comfort including tactile comfort and thermal comfort.
3. Mobility: maintains human mobility while the wearer carries out certain tasks.

A further, fourth, requirement of protective clothing is its durability: the capability of the clothing system to maintain the three primary functions (protection, comfort and mobility) under various environmental conditions after repeated use.

Protective clothing is designed to deal with specific environmental hazards. Those hazards are frequently grouped according to their physical forms (e.g., solid, particles, liquids, gases, mist) and properties. Some typical examples of environmental hazards are:

- Mechanical hazards: abrasion, cuts, punctures, slashes, high speed bullet impacts, low speed impacts, explosion fragment, etc.
- Thermal hazards: flame, fire, heat (infrared heat, radiation heat, hot gases, steam, hot liquid), ultraviolet (UV) exposure and cold (wind, ice, snow, deep sea water, exterritorial weather, liquid gases, cold surface and solid materials) hazards.

- Water: rain, snow, seawater, wastewater, ice, water vapour, etc.
- Chemical hazards: oil, petrol, solvent, paint, alkali, acids, salt, heavy metals, toxic, flammable, explosive and erosive chemicals.
- Biochemical and biological hazards: bacteria, virus, blood, body fluids, drugs, etc.
- Electrical hazards: high voltage, high current electricity, lightning, electrical spark and arc.
- Electromagnetic wave hazards: infrared, UV, laser, invisible light, sunlight, high intensity light, SW, MW, LW, UH, UVW.
- Acoustic hazards: low and high frequency noise, ultrasound, explosive waves, thunder.
- Radioactive hazards: X-rays, beta rays, gamma rays, protons and neutrons, negative electron rays.
- Industrial hazards: hot metal liquids, welding arc and spark, saw, squeeze.

Therefore, specific protective clothing is engineered for use in defined environments to protect the wearer from identified hazards. Examples of individual protective clothing include the body armour system (e.g., bulletproof clothing, stab-proof clothing, cut-proof gloves, etc.), firefighters' clothing, and Chemical-biological-radiological protective clothing (CBR).

Different types of protective clothing need to withstand attack from specific environmental hazards and thus require the constituent textile materials to have high performance in specific properties. For example, a body armour system requires the textile materials used to have superior mechanical properties; firefighters' clothing requires its component materials to have exceptional fire resistance properties; and chemical protective clothing requires the materials to be resistant to the attacks of chemical agents.

All of those clothing systems require the guarantee of clothing comfort in their environments of application. Clothing comfort includes tactile comfort and thermal comfort. Tactile comfort is one of the key requirements to the clothing/equipment next to skin, and thermal comfort is required for all clothing systems. Thermal homeostasis of the body is a result of the balance between heat production and heat dissipation. The primary factors influencing this thermal balance are energy metabolism, clothing thermal properties and ambient climatic conditions. Several models or thermal indices[6] (such as Predicted Mean Vote (PMV), Predicted Percentage Dissatisfied (PPD), and Draught Rate (DR), taking into account all of these factors, have been developed to compute the thermal interactions and their effect on the body. Protective clothing serves the purpose of eliminating or reducing the effects of environmental stress factors.

High performance textiles include fibres, yarns, fabrics and composite structures made from functional organic polymers and inorganic materials and have exceptional functionality and properties that are unmatched by conventional textile materials, they are widely used in protective clothing systems. The examples of required special properties of textile materials used in protective clothing include mechanical properties (tensile strength and modulus, abrasion, puncture and flex durability), thermal and electrical insulation properties, resistance to heat, flame and chemical agents, liquid absorption and diffusion properties, and optical and radioactive absorption properties. Such properties are frequently integrated into a single protective clothing system to achieve required protection and comfort performance during wear.

In the following sections, performance textiles used in three types of protective clothing – body armour systems, firefighter's clothing and chemical protective clothing to address mechanical, heat and chemical hazards, respectively – are discussed.

3.3 High performance textile fibres

High performance textile materials having exceptional mechanical, fire resistance and chemical resistance, are available in the form of fibre, film, membrane and liquid and are frequently used for protective clothing. The high performance fibres are made from a wide source including inorganic materials such as ceramic fibres, carbon fibre, stainless steel and aluminium fibres[14]; and including organic materials such as aramid (including p-aramid (poly-para-phenylene terephthalamide) and m-aramid (poly(meta-phenyleneisophthalamide))), polyimide (PI), polybenzimidazole (PBI), polyethylene-2,6-naphthalate (PEN), p-phenylene-2,6–benzobisoxazole (PBO), ultra-high molecule weight polyethylene (UHMW-PE), polyarylate (LCP), polyphenylene sulphide (PPS), polyetheretherketone (PEEK), novoloid (cured phenol-aldehyde), melamine, poly(tetrafluoroethylene) (PTFE), and fire resistant natural (Zirpro wool) and regenerated (FR viscose) fibres. Performance textile fibres having exceptional mechanical properties, fire resistance and chemical resistance are summarized in Table 3.1.

High performance polymer fibres having ultra-high mechanical properties were achieved mainly by orientating molecule chains in fibres. There are two main routes for the preparation of high performance polymer fibres; one is based on flexible macromolecules, while the other is based on rigid-rod macromolecules. In the case of flexible chain polymers, such as polyethylene, solid-state drawing (or gel spinning) is used to obtain the necessary level of chain extension needed to exploit the intrinsic properties of the polymer chain. In the case of rigid-rod macromolecules, the chain extension is already built in by the chemist, and alignment in the extended chain configuration is achieved during fibre spinning.

Table 3.1 The properties of high performance fibres[7–13]

Types of fibres	Composition	Commercial products	Protection from mechanical hazards (stab, ballistic, puncture, chainsaw, etc.)	Protection from flame, fire and thermal hazards	Protection from chemical and biological hazards
Aramid fibres	p-aramid (poly-para-phenylene terephthalamide)	Kevlar (DuPont), Twaron (Acordis), and Technora (Teijin)	++ Tenacity: 194.3~229.6 cN/tex Modulus: 4061.8~9713 cN/tex	++ LOI: 25–28 T*: 190°C	+
	m-aramid (poly(meta-phenyleneisophthalamide))	Nomex, Kermel and Corex	++ Tenacity: 33.6~63.6 cN/tex Modulus: 106.0~123.6 cN/tex	+ LOI: 30 T*: 205°C	+
Polyamide imide (PAI) fibre and Polyimide	Polyimide (PI)	Torlon (Solvay) P-84® (Evonik)[8]	+ Tenacity: 37.1 cN/tex	++ LOI: 38 T*: 260°C	+ Poor in alkali
PBI fibres	Polybenzimidazole	PBI (Celanese)	++ Tenacity: 23.8 cN/tex Modulus: 282.6 cN/tex	++ LOI 41 T*: 250°C	+
PEN, PBO and PIPD fibres	PEN (polyethylene-2,6-naphthalate) PBO (p-phenylene-2,6—benzobisoxazole)	Zylon (Toyobo), M5 (Akzo Nobel) Novel yarn	++ (excellent abrasion) Tenacity: 370.9 cN/tex Modulus: 11479 cN/tex	++ LOI: 68 T*: 288~315°C	++ Fair in acid
UHMW-PE fibres[9,10]	Ultra-high Molecule Weight Polyethylene, Gel spun	Dyneema (DSM), Spectra (Honeywell), Tekmilon (Mitsui)	++ abrasion Tenacity: 264.9 cN/tex Modulus: 12362 cN/tex	– LOI 41 T*: 121°C	++
LCP fibre[11]	Polyarylate, liquid crystal polymer fibre	Vectran (Kuraray)	++ (excellent abrasion) Tenacity: 247.2~256.1 cN/tex modulus: 61.8~88.3 cN/tex	++ LOI 37 T*: >392°C	++ Good in solvent
PPS fibres	Polyphenylene sulphide, Crystalline thermoplastic fibre	Ryton® (Amoco), Procon® (Evonik), Toray PPS® (Toray)	+ Tenacity: 30.9~39.7 cN/tex modulus: 61.8~88.3 cN/tex	+ LOI: 34 T*: 288~315°C	++
PEEK fibre	Polyetheretherketone, Semi-crystalline thermoplastic fibre	ZYEX, ZEUS	++ Tenacity: 17.7 cN/tex modulus: 70.6 cN/tex	++ LOI: 95 T*: 260°C	++

(*Continued*)

Table 3.1 (*Continued*)

Types of fibres	Composition	Commercial products	Protection from mechanical hazards (stab, ballistic, puncture, chainsaw, etc.)	Protection from flame, fire and thermal hazards	Protection from chemical and biological hazards
Novoloid fibre	Cured phenol-aldehyde	Kynol	+ (poor abrasion) Tenacity: 11.5 cN/tex Modulus : 97.1 cN/tex	++ LOI: 33 T*: 96°C	++ Poor in concentrated acid
Melamine-based fibres	Melamine	BASOFIL (BASF)	+ Tenacity: 17.7 cN/tex Modulus : 97.1 cN/tex	++ LOI: 32 T*: 204°C	++ Poor in concentrated acid
Fire resistant natural and regenerated fibre	Modified cellulosic	Lenzin viscose FR DFG (Daiwabo), Avilon (Avilon)	–	++	–
	Modified wool	Zirpro wool	–	++	–
Inorganic fibre	Carbon fibre: pitch	/	++ (poor abrasion) Tenacity: 123.6~264.9 cN/tex Modulus: 8830-51656 cN/tex	++ T*: 300–538°C	++ Poor in concentrated acid
	PAN	/	++ Tenacity: 158.9~618.1 cN/tex Modulus : 14481.2-33995.5 cN/tex	T*: 300–538°C	++ Poor in concentrated acid
	Ceramic fibre	Ceramic	++	++	+
	Metal (stainless steel, aluminium)	Stainless steel, aluminium	++	++	–

Two typical examples of those high performance fibres, whose properties are significantly influenced by their chemical and physical structures, are explained below.

3.3.1 Aramid fibres

The properties of aramid (or aromatic polyamide) fibre is determined by both its chemical and physical microstructure. Its excellent thermal stability is due to its aramid ring, and aromatic/amide bonds have 20% higher dissociation energies than their aliphatic counterparts. The excellent physical properties are also due to its highly anisotropic (in comparison, graphite fibres are also anisotropic, while glass fibres are isotropic), crystalline and extended chain structure for p-aramid (i.e. there is no chain folding in the rod form of the para-aramid molecules).[15] The phenyl rings of adjacent molecule chains stack on top of each other very neatly to make the polymer highly crystalline after the extrusion process; such structures give para-aramid fibre outstanding high strength, high modulus fibre, relative insensitivity to temperatures up to T_g (~280°C) and dimensional stability.

There are three types of major commercial aramid fibres in the market: para-aramid (or p-aramid, poly-(p-phenylene terephthalamide), PPTA), meta-aramid (or m-aramid, poly-(meta-phenylene isophthalamides), MPIA) and co-polyamides fibres.

The commercial brands of p-aramid fibres include Kevlar® (DuPont) and Twaron® (Akzo), made from p-phenylene diamine and terephthaloyl in *N*-methyl-pyrrolidone/calcium chloride solution. The series of Kevlar® fibres, including Kevlar® 29, Kevlar® 49 and Kevlar® 149, are designed with progressively higher orientation of polymer molecules and crystallinity, and thus have higher modulus and tenacity (up to 260 cN/tex) in the above order; in addition, such a highly orientated microstructure results in highly anisotropic physical properties; for instance, Kevlar® fibres have a negative coefficient of expansion along the fibre axis –2 to –6 um/m/°C and +60 um/m/°C laterally; they are also stronger and stiffer in the axial direction than in the transverse direction (see Table 3.2).

Commercially available meta-aramid fibres include Nomex® (DuPont), Teijinconex® (Teijin), Newstar® (Yantai Tayho), X-Fiper® (SRO) and Kermel® (Kermel). m-aramid fibre is produced by condensation reaction from the monomers m-phenylenediamine and isophthaloyl chloride in an amide solvent. It has at least 85% of the amide linkages attached to two aromatic rings in its long chain of polyamides. Unlike Kevlar, Nomex polymer molecule chains cannot align during filament formation, and thus has

Table 3.2 Properties of different grades of Kevlar® fibres[16,17]

Kevlar grade	Density (g/cm^3)	Tensile modulus (cN/Text)	Tensile strength (cN/Tex)	Tensile elongation (%)
29	1.44	8567.5	371.7	4.0
49	1.44	13522.2	371.7–423.0	2.8
149	1.47	19199.4	351.0	2.0

poorer strength. However, it has excellent thermal, chemical, electrical and radiation resistance for a polymer material.

The Nomex® family products consists of staple fibres, continuous filament yarns, paper, woven, hydroentangled and coated fabrics, and they are used in: firefighter garments; insulation in fire resistant thermal protective apparel; and apparel fabrics to protect against flash fire and electric arc exposure. Examples of DuPont™ Nomex® fibres suitable for protective clothing are summarized in Table 3.3 below.

The two typical example of copolyamide type of aramid fibres are Technora®[21] (Teijin) and Kermel® (Kermel) fibres.[22] Technora® fibres is produced from reactions between a mixture of p-phenylenediamine (PPD) and 3,4′-diaminodiphenylether (3,4′-ODA) with terephthaloyl chloride (TCl),[23] and Kermel fibre is a polyamide imide (PAI) fibre, which has the advantage of having little shrinkage in hot water (= 0%) and hot air (0.2%).[22]

3.3.2 Poly(p-Phenylene-2, 6–Benzobisoxazole) (PBO or Zylon) fibre

The high performance fibre consisting of rigid-rod chain molecules of poly (p-phenylene-2, 6–benzobisoxazole) (PBO), was developed by Toyobo Co. Ltd and branded as Zylon.®[24] The chemical structure of PBO provides high thermal stability and outstanding mechanical properties. PBO has strength and modulus almost double that of p-aramid fibres (e.g., Kevlar®, Twaron®) and ultra-high molecular weight polyethylene fibres (e.g., Dyneema®, Spectra®). It also has a higher decomposition temperature (650°C) than the p-aramid fibres. The limiting oxygen index (LOI) is 68, which is the highest among organic super fibres.

There are two types of Zylon fibres: Zylon HM (high modulus) has greater tensile modulus and less moisture regain than Zylon AS (as spun). The properties of Zylon AS and HM fibres are shown in Table 3.4.[15]

While all the high performance fibres are susceptible to chemical, moisture, light and UV degradation, the degradation of PBO fibres in bulletproof vests has raised concerns and lawsuits in the commercial market.[26] It was

Table 3.3 Properties of some Nomex fibres[18–20]

Types of Nomex® fibres	N301	450	455 (Nomex III)	462 (Nomex IIIA)	430	
					1200 Denier	1600 Denier
Density (kg/m^3)	/	1370	/	/	1380	
Tensile strength, ultimate (MPa)	340	349	310	310	607	595
Elongation at break (%)	19	22	21	21	30.5	31
Tenacity (N/tex)	0.247	0.256	0.229	0.229	0.441	0.432
Coefficient of linear expansion (CTE), (μm/m°C) at 20.0°C	18.0	18.0	18.0	18.0	18.0	
Specific heat capacity (J/g-°C)	0.290	0.300	0.260	0.260	0.300	
Thermal conductivity (W/m*K)	/	/	/	/	0.250	
Shrinkage (%), in water at 100°C	/	<= 4.0	0.50	0.5	1.3	1.1
Moisture absorption at equilibrium (%)	4.5	4.5	4.5	4.5	4.5	
Water absorption (%)	8.3	8.2	8.3	8.3	4.0	
Note	Staple fibres of 1.5 Denier, 38–51 mm; a blend of 95% m-aramid and 5% Kevlar.	100% meta-aramid staple fibres of 1.5 Denier, 38–51 mm; higher crystallinity, greater strength and less dyeable than Type 455 and 462 fibres.	A blend of 95% m-aramid and 5% Kevlar fibres; lower crystallinity and easier to be dyed; for thermal protective apparel	A blend of 93% m-aramid fibre, 5% Kevlar and 2% antistatic fibre P-140, used for thermal protective apparel	100% m-aramid filament yarn; High crystallinity; higher strength and chemical resistance than staple spun yarns Suitable for firefighters' turnout gear shells and liners, coated fabrics and electrical insulation	

Table 3.4 Properties of Zylon® (PBO) fibres[24]

Type		Regular	High modulus
		AS	HM
Filament decitex	dTex	1.7	1.7
Density	g/cm^3	1.54	1.56
Moisture regain (65% RH)	%	2.0	0.6
Tensile strength	GPa	5.8	5.8
Tensile modulus	GPa	180	270
Elongation at break	%	3.5	2.5
Melting temperature	°C	None	None
Decomposition temperature in air	°C	650	650
Coefficient of thermal expansion	ppm	/	–6
Limiting oxygen index	LOI	68	68

reported that the strength of Zylon® fibre, one type of PBO fibre, decreases under conditions of high temperature, high humidity, and exposure to UV and visible light.[27] Scientific literature has revealed that the oxazole ring, the five-member ring that appears within the chemical structure of PBO, has characteristics that cause it to be susceptible to degradation due to moisture and light exposure.[28] It was found[18], during the ballistic and mechanical properties testing on 103 used Zylon®-containing body armours provided by law enforcement agencies across the United States by National Institute of Justice (NIJ) of US Department of Justice, that 60 of these used armours (58%) were penetrated by at least one round during a six-shot test series. Of the armours that were not penetrated, 91% had backface deformations in excess of that allowed by the NIJ standard–0101.04[29] for new armour. Only four of the used Zylon®-containing armours met all performance criteria expected under the NIJ standard for new body armour compliance. It was concluded that, although these results did not conclusively prove that all Zylon®-containing body armour models have performance problems, the results showed that used Zylon®-containing body armour might not provide the intended level of ballistic resistance.[9]

3.4 Conventional and high performance fibre blends for protective clothing

High performance fibres are usually mixed or interwoven in the fabrics, or used as layered fabric structures so as to have a combination of multiple protection functions in protective clothing. Most of those high performance fibres, except those converted from natural fibre sources, are stiff and do not have moisture management properties, and some of them have poor durability; therefore, the fabrics made from them are frequently used together with conventional fibres to achieve required durability and clothing comfort properties.

Conventional and high performance fibres are frequently used together to produce fabrics for protective clothing. Fabrics made from traditional synthetic fibres, such as polyethylene, polypropylene, polyester, polyamide and polyurethane, are still widely used in protective clothing, especially specialist workwear, for the purpose of balanced performance and cost effectiveness; for example, polyamide fibres (e.g., nylon) combined with polyurethane coating, are widely used as a durable layer[30] in many protective clothing to compensate poor abrasion resistance of some high performance fibres (i.e., m-aramid fibres) to achieve superior abrasion resistance properties of the clothing. Microfibres, bi-component fibres, hollow fibres, fibres/filaments having non-circular cross-sections made from those conventional synthetic polymers are also frequently used in the protective clothing system for protection, clothing comfort and durability purposes. Natural fibres such as cotton, wool, down[31] (see Fig. 3.1a) and fibres regenerated from natural polymers such as viscose rayon (including their special variants, e.g., FR modified, nanoparticles coated[32] (see Fig. 3.1b)), are also frequently employed to achieve protection (e.g., thermal and cold protection) and clothing comfort (both tactile and thermophysiological comfort) properties. Fabrics made from those conventional fibres are usually in a mixture form, and additionally blended with other special performance fibres for antistatic, fire resistance and chemical resistance. Those fabrics usually require special treatment and finishing, such as coating (see Fig. 3.1c), lamination and other special finishing (water/oil repellent, moisture permeable, softness, antibacterial, insecticide, etc.) to achieve multiple protection functions.

When different high performance fibres are used together, or incorporated with conventional fibres in protective clothing, the design of yarn and fabric structure is crucial to achieve targeted performance of durability, protection, comfort and mobility of the clothing. For example, composite cord threads, using metal filaments as core and other high performance

3.1 Microstructures of three different textile materials for special functions: (a) down fibre (Fuller *et al.*, 2013), (b) nanoparticles coated fabric (Du *et al.*, 2013), and (c) polymeric coated fabric.

fibres as sheath, are used in special fabrics to provide flexibility as well as cut protection,[33] ballistic-proof and stab-proof protections from gunshots, shrapnel and knives stabs are reported.[34] While the metal wire blunts the sharp edge of knives and shrapnel, the super-strong, high performance fibres (such as UHMW-PE fibres) absorb the impact energy to stop the damaged knife and minimize trauma. Strong polyester filament sheath–core composite yarns are also created to realize both strength and comfort properties. Fabric weaves are also important. For example, basket weaves[35] and ripstop structures are well known for their superior mechanical strength[36]; it was found from research[37] that the specific energy absorption increases with the increase of the fabric tightness and fabric areal density for fabrics made from p-aramid, PBO and UHMW-PE fibres. New fabric structures, such as fabrics made from metal wire high performance fibre composite yarns, spacer fabric, 3D fabric, multidirectional weave fabrics, honeycomb composite, auxetic structure, and biomimetic inspired materials and their uses in protective clothing have also been investigated.

Woven and non-woven fabrics made from synthetic filaments (e.g., polyester, polyamide, polyethylene and polypropylene) are frequently used in making reusable and disposable protective workwear worn by people working in food, medical, and pharmaceutical industries and the clean room environments of electronic industries. These fabrics not only protect people from many hazards, such as biological and chemical substances, steam, temperature extremes, sharp-edged tools, harsh cleaning solutions, but they also prevent contaminants whether carried by workers (e.g., electrostatics, bacteria, hair and skin flakes) or within workwear itself (e.g., monomer, filler, additives, dyestuff, lint, dust from polymer, fibre fabric and coating materials of the garments), from entering the controlled environment and making contact with equipment or product. The materials used in those products are required to be both lint-free and able to stand degradation during the harsh sterilization (either chemical or radiation) and laundering (for reusable clothing) process.

Protective clothing has remained as functionality-driven products equipped with hybrid materials, smart technology and nanotechnology. For example, novel energy absorption materials (e.g., shear-thickening fluid, magnetic shear-thickening materials, air inflatable structures) have been used in protective clothing to make flexible and lightweight fabric without compromising protection and comfort properties; new materials and structures including nanoparticle-fabric composites (see Fig. 3.1b), phase change materials (PCM), moisture/temperature responsive materials,[38 39] electric conductive yarns and carbon nanotubes[40] are also used for thermal comfort; some of these materials are able to sense the wearers' physiological conditions,[41] their posture and activity[42] and outside environment,[43] as well as responsive action.[44]

3.5 Cut resistant and energy absorption materials

Examples of composite materials and hybrid structures having unique cut resistance and energy absorption properties in protective clothing and personal protection equipment (helmet, knee protection panel, gloves, protection curtains, protection tent and shield, etc.) are discussed in this section.

3.5.1 Cut resistant textile materials

Composite materials and hybrid structures are usually employed to provide multiple functions in addition to cut resistance requirements. Composite cord threads using stainless steel filaments as core, and other high performance fibres as sheath, are used in cut resistance fabrics to provide both flexibility and cut protection.[45] A flexible body armour vest[46] made from a combination of Dyneema high performance polyethylene fibre from Royal DSM (Dutch State Mines), and Steelskin steel cord material from Bekaert, offering both ballistic-proof and stab-proof protections from gunshots, shrapnel and knives stabs has been reported.

Results on stress–strain relationship and deformation characteristics during cutting are reported[47] for three types of high strength yarns, Zylon, Kevlar and Spectra; it is found that Zylon fabric has the highest specific energy absorption in high speed fragment impact testing, and the specific energy absorption increases with the increase of the fabric tightness and fabric areal density.

Traditional protective clothing for chainsaw operators includes trousers and chaps that generally have a nylon, polyester or denim outer fabric and inner inserts comprising layers of ballistic nylon or high tenacity polyester fabric; or, in higher-quality clothing, layers of aramid fabric such as Kevlar® or ultra-high molecular weight polyethylene (UHMW-PE) such as Dyneema®. The cut resistant protective layers provide passive protection to the wearer and are designed to jam the chainsaw drive sprocket, thus stopping the saw before it can cut through the trouser leg and cause injury.

3.5.2 Abrasion resistant high performance textiles

Some high performance fibres (e.g., p-aramid and m-aramid fibres) have poor abrasion resistance; in this case, they are usually blended with fibres having superior abrasion resistance properties, combined with polyurethane coating, or layered with nylon fabrics. Fabric containing nylon filaments has been widely used as a durable layer for military[48] and civil PPE[49] to protect military personnel, chainsaw operators and motorcyclists. For example, the rigorous requirements of abrasion, impact, cut and burst

Table 3.5 The tear and abrasion properties between Cordura and Kevlar fabrics[65]

Fabric/material	Pounds of force until fabric tears	Abrasion cycles on pavement until fabric fails
Cotton jeans	4.5 pounds to tear	50 cycles to failure
70 Denier standard nylon	4.5 pounds to tear	165 cycles to failure
500 Denier polyester	8 pounds to tear	180 cycles to failure
200 Denier standard nylon	7.5 pounds to tear	275 cycles to failure
500 Denier Cordura	22 pounds to tear	710 cycles to failure
620 Denier Cordura	35 pounds to tear	1200 cycles to failure
Competition grade leather	80–110 pounds to tear	1200–1700 cycles to failure
1000 Denier Cordura	110 pounds to tear	1780 cycles to failure
Air mesh Kevlar	1260 pounds to tear	970 cycles to failure
Stretch Kevlar blend	420 pounds to tear	1800 cycles to failure

resistance for the protective clothing and equipment used in motorcycling are specified in a series of EU standards[50–58]. For example, in CE EN 13595, the material used in motorcycle protective clothing must have abrasion resistance of between 4 and 7 seconds for use over the high impact areas of the body (Zone 1 and 2) defined by Woods.[59] It was reported that a denim fabric of 200 gsm will last just over half (0.6) a second[60] in these testing conditions. New EN standards, including EN 1621–3 for chest protectors and EN 1621–4 for inflatable (air bag)-type protectors, are in the process of development.[61]

Nylon fibre is famous for its durability, including tensile strength, and tear and abrasion resistance properties, and thus is used widely for protective clothing, rather than ballistics protection. A typical example of nylon fabric used in protective clothing is Cordura® Ballistic nylon fabrics,[62,63] which is a thick, tough, synthetic nylon fabric having a construction of 2 × 2 basket weaves. Cordura fabric consisting of 1000 and 500 D nylon filaments is used in motorcycle jackets and workwear clothing for durability.[64] Comparison of the tear and abrasion properties between Cordura and Kevlar fabrics is shown in Table 3.5.

3.5.3 Materials for making flexible and lightweight body armour for impact absorption

In conventional protective clothing system, foams and other soft textile materials are employed to absorb impact energy, in combination with metal, composite or ceramic rigid plates used as inserts to spread the impact force over a larger area and thus reduce the impact force propagation to the human body. However, the rigid plate might reduce the energy absorption effect of the protection system. High performance materials are therefore sought to make flexible and lightweight body armour without compromising energy absorption. Those materials employ either new concept energy absorption materials (e.g., shear-thickening fluid, magnetic shear-thickening materials, air inflatable structures) or use new fabric structures, such as fabrics made from metal wire composite yarns, spacer fabrics, honeycomb composite and auxetic materials in integration.

Shear-thickening material (such as dimethyl siloxane hydroxy-terminated polymer) is a type of material which viscously flows at low rates of deformation but, at a significantly elevated rate of deformation such would be induced by impact, undergoes a substantial increase in viscosity with rate of change deformation and the material becomes substantially stiff or rigid. The mechanical energy is distributed throughout the matrix of a shear-thickening material, thus significantly reducing the effect of impact.

The shear-thickening materials can be combined with high strength materials to add additional mechanical properties to maximize the energy absorption effect, and can be employed with other energy absorption structures (such as honeycomb, spacer fabrics, etc.) to distribute the impact force. An example of this shear-thickening material is Dow Corning's Active Protection System (ASP); such a product may be based on three-dimensional spacer fabrics, or on an expanded thermoplastic rubber impregnated with shear-thickening silicones to offer high levels of protection.[66] An energy absorbing pad[67] used in protective clothing and a bone fracture prevention garment[68] employing shear-thickening material have been reported.

When shear-thickening impregnated products are incorporated in protective clothing assembly, it is usually used as an outer layer over the shock absorbing structure to receive the initial impact force;[69] it becomes rigid on impact, diffusing the impact forces over a greater area of underlying support material to improve the energy absorption. Also, an additional outer layer of ballistic woven fabric made from p-aramid fibres (e.g., Kevlar) or carbon nano-tube-based material, with or without additional shear-thickening fluid imbedded in the fibres, may be applied over the layered impact-shock-absorbing material, to provide protection against penetration of sharpened objects.

Shock absorbing liquid (SALi) Technology[70]-based cushions consist of small, compressible elastic capsules blended with an 'incompressible' matrix fluid for protection from violent impacts, crash, vibrations and shock waves. It is suggested that the cushions offer four damage mitigation mechanisms[71–74] in the impact energy absorption process:

1. The capsules absorb energy when they suffer bulk compression during an impact.
2. Viscous damping converts impact energy into heat when the matrix fluid swirls round the compressing capsules. This reduces the 'kickback' when

the capsules spring back into shape in the milliseconds following the impact.
3. The hydraulic characteristics of the matrix fluid distribute impact loading over the surface being protected; and
4. Shock waves are scattered at the capsule–fluid interfaces.

3.6 Clothing for firefighters

Structural firefighter protective clothing[75,76] is designed to protect firefighters from heat and other environmental hazards, such as abrasive objects and unpredicted chemical substances.[77] The turnout gear (jacket and trousers only)[69] typically consists of at least three functional layers, including an:

1. outer shell (OS)
2. moisture barrier (MB)
3. thermal liner (TL).

The TL layer primarily provides thermal protection, and is usually engineered as a highly porous padding or non-woven insulating structure fixed between an inner layer of woven fabric, which reduces the frictions between the turnout gear and the clothing in contact with the skin. The air trapped in the non-woven insulating padding between the inner layer of woven fabric and MB provides thermal protection. While a thicker TL provides better heat protection, it might be heavier and less breathable and thus less thermally comfortable.

The MB is typically the middle layer between the OS and TL; it is expected to prevent water, high temperature water vapour, chemicals and other pathogens from the environment passing through the firefighters' clothing. It is usually a PTFE permeable film barrier laminated to a thin polyaramid woven or non-woven backing substrate.

New materials and designs have been developed for heat protective clothing. For example, improved thermal insulation can be provided by non-wovens made with thin hollowed fibres, and can be made thermo-adaptive with two-way shape memory alloys such as nickel–titanium.[78] Better thermoregulation inside the garment is sought with PCM,[79] either encapsulated[80] or incorporated in a matrix.[81] Other solutions use external power, e.g., for liquid coolant circulation[82] or with Peltier cells embedded in the textile.[83]

3.6.1 Outer shell (OS) materials

The OS is the outermost layer that provides both resistance to fire and prevents water ingress, cut, stab, tear and abrasion hazards which might be present at the fire scene. It might provide 25–30% of the total thermal protection

Table 3.6 The composition, structure and properties of some OS fabrics[93]

OS brand	PBI gold	PBI MATRIX	Advance ultra	Advance	Gemini™	Millenia™ XT	Nomex® IIIA	BASOFIL® (NorFab)	NFPA 1971 requirements
Materials	40% PBI™, 60% Kevlar®	40% PBI™, 60% Kevlar®	60% Kevlar® 20% Meta-aramid 20% PBO	60% Kevlar® 40% Nomex	60% Kevlar® 40% PBI reinforced with Para-aramid filament networks	60% Para-aramid 40% PBO	93% Nomex®, 5% Kevlar®, 2% Antistatic	40% BASOFIL®, 60% Kevlar®	
Fabric weight (oz./yd2)	7.5	7.5	7.5	7.0	7.5	7.5		7.5	
Weave structure	Plain weave Ripstop	Plain weave reinforced with Kevlar® filament	Ripstop	Ripstop	Modified plain	Ripstop	Plain weave	Ripstop weave	
Thermal stability	Superior	Superior	Very good	Very good			Basic	Excellent TPP value	
Tensile strength (lbs) ASTM D 5034 (After 10 launderings (warp × fill))	Very good	Excellent	320 × 310 (Excellent)	270 × 250 (Very good)	230 × 220	575 × 565 (Superior)	Very good	Good	140.0 min
Tensile strength (lbs) (after 17.5)	/	/	110 × 100	15 × 10	75 × 50	150 × 135		/	
TPP exposure (2.0 cal/cm²•s) (warp × fill))									

(*Continued*)

Table 3.6 (*Continued*)

OS brand	PBI gold	PBI MATRIX	Advance ultra	Advance	Gemini™	Millenia™ XT	Nomex® IIIA	BASOFIL® (NorFab)	NFPA 1971 requirements
Trapezoidal tear strength (lbs) ASTM D 5587* (After 10 launderings (warp × weft))	Very good	Excellent	40 × 35 (Excellent)	35 × 30 (Very good)	35 × 35	80 × 70 (Superior)	Very good	Good	22.0 min
Abrasion resistance	Very good	Excellent	Excellent	Very good		Superior	Good		
Water repellency		Durable water repellent SST finish							

of the gear. A typical OS has woven structure with a ripstop construction to prevent tear propagation. OS is usually made of p-polyaramid, m-polyaramid, PBI, poly(melamine-formaldehyde) fibres, and or a composite structure of these fibres.

The high strength para-aramid fibre Kevlar® has an extremely high level of tear and cut resistance, as well as good heat resistance. However, unlike meta-aramid fibres, para-aramid fibres have a few properties that can be classified as restrictive for fire protection clothing. When exposed to flames, para-aramid fibres absorb hardly any heat energy, so a large proportion of the energy can penetrate the fabric. Fabrics with a high content of para-aramid fibres are generally relatively heavy, feel rigid to touch and are uncomfortable to wear because of their high tensile strength. Furthermore, a fabric made of para-aramid is less UV stable and less abrasion resistant than a meta-aramid fabric, and has a tendency to pilling.

The heat-resistant meta-aramid fibre Nomex® can withstand exposure to extreme heat for over 8 seconds, and does not melt or drip, even at high temperatures. It offers excellent heat insulation by absorbing the heat energy when exposed to a heat source. It also boasts high UV resistance, abrasion resistance and durability, and is soft to the touch.

While the tensile strength and cut resistance of Kevlar® para-aramid fibre are key additional components for increasing the performance of flame resistant fabrics, it is shown, from recent research results[84] based on a number of fabrics with varying contents of para-aramid fibres subjected to a series of tests, that too high a content of para-aramid may negatively impact on the performance of the flame- and heat-resistant fabric. In comparison with fabrics with a 60% content of para-aramid fibres, fabrics with a para-aramid content of between 5% and 23% achieved better values in terms of thermal protection (heat penetration resistance), as Nomex® fibres absorb a large part of the heat energy when exposed to heat; it also has better abrasion resistance and UV resistance, as well as washing behaviour. In contrast, fabric containing 60% para-aramid exhibited a significantly higher level of tear resistance in its original condition than fabrics containing less para-aramid.[85]

Other reactive OS materials for flame and heat protection are also engineered to provide minimal insulation during normal wear, but maximum insulation when heat threat impinges. For example, intumescent treatments are normally in the form of thin and low insulation coatings on a lining fabric[85], when activated by excessive heat or flame, the formulation swells instantly to form an inert insulative char, protecting the wearer.

PBI fibre originated from a collaborative project between NASA and Celanese for the purpose of spacesuits and space vehicles. PBI has excellent thermal resistant properties, it has one of the greatest LOI (41%), is resistant to ignition and hydrolysis, and neither burns in air nor melts, nor drips. PBI fabric can withstand arc flash, flash fire and other dangers associated with firefighting.[86] However, PBI has relatively low tensile strength, and does not resist acids[87] or UV radiation exposure.[88,89] PBI fibre degrades in a moist environment, and thus the fibre used in firefighters' clothing loses its mechanical properties in a firefighting environment having higher relative humidity;[90] it was reported[91] that PBI fibres lost 30% of their tensile strength after exposure to the environment at 50°C, and RH 60% for 84 days, and 60°C, and RH 37% for 73 days.

Currently, there are four forms of PBI fabrics available for firefighters' gear in the market[92]:

- PBI Gold
- PBI Matrix
- PBI TriGuard
- PBI BaseGuard.

PBI Gold, which is available as PBI Gold knits, PBI Gold Ripstop and PBI Gold Twill fabrics, has been widely used as an OS in firefighters clothing (e.g., Bristol Uniform) since 1978; it blends 40% of thermal resistant PBI fibres with 60% of p-polyaramid fibres, resulting in a fabric which is not easily to shrink, become brittle, or break open under extreme heat and flame exposure. PBI Matrix, which was introduced into the market in 2003, employs a durable matrix of high strength aramid filaments woven into the PBI Gold fabric to enhance and reinforce its resistance to wear and tear while retaining its superior flame and heat protection.

PBI TriGuard fabric, a blend of three types of high performance fibres including PBI, Lenzing FR and MicroTwaron, is designed for flame protection, comfort and durability. It is certified for use in Wildland Firefighter's clothing, special operations and motorsports applications, as well as the petrochemical, gas utility and electric utility industries. PBI TriGuard and PBI Gold knits are now in use at several major motorsport racetracks.

PBI BaseGuard is a type of lightweight, soft and flexible knitted fabric designed to be a next-to-skin base layer in firefighters' clothing systems to provide fire protection along with moisture management and durability. PBI fibres are also used in military flame protection, automotive braking systems and fire blocking layers in aircraft seats. A summary of some OS fabrics used in the turnout gear of structural firefighters' clothing is shown in Table 3.6.

3.6.2 Moisture barrier (MB)

A summary of some commercially available MB materials[93–95] used in firefighters' clothing is shown in Table 3.7.

The PTFE or expandable polytetrafluoroethylene (ePTFE)-based, waterproof, breathable fabrics dominate the MB in firefighters' clothing. One of the most significant fabrics is the Goretex-based bi-component polymeric film made of PTFE. The Goretex film is claimed to contain more than 1.3 billions of micropores of size around 0.2 μm in an area of one centimetre square. However, modern MB materials are mostly bi-component membranes comprising ePTFE porous structure embedded in a continuous polyurethane film to prevent the skin of the wearer being burned from the hot water vapour penetrating through the MB.

PTFE/ePTFE barrier membrane is frequently bonded to a variety of substrates, such as Kevlar fabrics, to form a multiple layer laminate to provide the membrane with superior mechanical properties.

New-generation waterproof and breathable membranes include Crosstech™ membrane,[96] which combines microporous expanded PTFE dispersed in a monolithic element, and Stedfast's Stedair® membrane,[97] which is based on laminated layers of various hydrophilic and oleophobic properties. 3D innovative functional barrier materials (e.g., Goretex Airlock)[98] to provide still-air space between the MB and OS fabrics for firefighting clothing enable the clothing to combine the functions of protection, comfort and heat insulation. Its aim is to eliminate the conventional, bulky, thermal insulation layer and substitute it with a protective still-air cushion. Dots consisting of foamed silicone are discontinuously applied to a fibre substrate and anchored within the microporous Goretex membrane. They measure a few millimetres in height, creating a defined air cushion between the adjacent flame-retardant face fabric and the inner lining.

Temperature-dependent permeability to moisture could even be provided by polyurethane-based shape memory polymer (SMP).[99] Better comfort is also provided by moving humidity and sweat away from the surface of the skin, either with hydrophilic linings[100] or by the use of channelled cross-section fibres with reinforced wicking properties.

3.6.3 Thermal liner (TL) materials

Besides the high performance fibres used in the OS layer, flame-retardant-treated natural and regenerated fibres such as wool and cellulosic fibres might also be used to make the thermal insulation liner. Some commercially available TL materials for firefighters' clothing are shown in Table 3.8.

Wool is the only natural fibre having inherent flame-retardant properties, and wool fabric has been used as a fire resistant layer in various products.[102] Due to its relatively high nitrogen content (16%) and strong disulphide bond network in the microstructure of the wool fibre, it has a high ignition

Table 3.7 Examples of some commercially available MB materials for firefighters' clothing[22,94,95]

MB brand	CROSSTECH® 2-Layer	CROSSTECH® 3-Layer	GORE RT7100™	CROSSTECH® S/R Fabric
Substrate	Nomex® fabric woven from spun fibres (pyjama check)	Top: Nomex® fabric woven from spun fibres (pyjama check) Bottom: Nomex® fabric woven from spun fibres	Nomex® non-woven spunlace	Nomex® fabric
Film	Enhanced bi-component ePTFE membrane	Enhanced bi-component ePTFE membrane	Bi-component ePTFE membrane	Enhanced bi-component ePTFE technology
Fabric weight (oz./yd2)	5	7.0	4.3	4.9
Overall thermal performance	The highest THL[1]/TPP[2] combinations	Best durability/THL/TPP combinations	Lightweight and flexible	Flame- and heat-resistant Meets NFPA 1951 and NFPA 1999
Liquid resistant	Best	Best	Better	
Breathability	Breathable		Breathable	Most breathable
Durability	Very good abrasion resistance	Most durable flex and abrasion resistance		Most durable in abrasion, cuts, and punctures

temperature, at 570–600°C, and an LOI[103] of 25~26% with low heat of combustion (4.9 kcal/g) and low heat release.[104]

There are durable flame-retardant treatments,[105,106] based on different fire retardant mechanisms developed for wool and its blending fabrics, including the treatments based on phosphorous compounds[107–109] and intumescent agents.[110] The treatment based on phosphorous compounds tend to lower the thermal decomposition temperature of wool fabrics, allowing the volatile fuel to escape before the ignition temperature is reached, and the intumescent agent[111] combines the attributes of flame retardancy with the formation of a high thermal resistance insulating char layer, and thus enhances the natural flame resistance and char formation properties of wool. However, wool fabrics treated by using both phosphorous

CROSSTECH® EMS Fabric	Goretex Airlock	STEDAIR® 3000 ePTFE	Stedair® 4000	NFPA 1971 requirements
Texturized polyester fabric	Nomex® fabric with silicone dots spacer	DuPont's E89	3.2 oz./yd2 woven Nomex containing 2% carbon fibres	
Enhanced bi-component ePTFE technology	Enhanced bi-component ePTFE technology	ePTFE membrane from BHA Technologies	Bi-component ePTFE matrix in a continuous hydrophilic and oleophobic polymer layer.	
3.5				
Lightweight and flexible Meets NFPA 1999		THL[3]> 240 TPP > 45 NFPA 1971 CGSB 155.1 and BNQ	THL[4]> 220 TPP > 45 NFPA 1971, NFPA 1999 EMS, NFPA 1951 USAR	THL_{min}: 130 watts/m^2 TPP_{min}: 35 Cal/cm^2
Most breathable Durable in abrasion, flexing, and tear resistance				

compounds and intumescent agents have a limited market so far. Flame-retardant treatments for wool fibres, by introducing halogen donors[112] into wool treatment, although highly effective in interfering free radical process, have unavoidable associated environmental problems and their uses are thus restricted.

Zirpro wool (or wool with Zirpro treatment)[113–118] has been the most popular wool fabric used in firefighters' clothing so far; it is developed by Benisek using a treatment with either hexafluorotitanate or hexafluorozirconate. Zirpro treatments are based on the exhaustion of negatively charged zirconium or titanium salts onto positively charged wool, resulting in the deposition of about 3% of flame retardant inside the fibre, and causing the formation of intumescent char during the burning process.[119] Zirpro wool

Table 3.8 Examples of some commercially available TL materials for firefighters' clothing[101]

MB brand	QUANTsUM3D™ (2-layer)	QUANTUM3D™ (1-layer)	CALDURA® SL2	CALDURA® SL3
Face cloth	Nomex® blend of filament and spun fibres	Nomex® blend of filament and spun fibres	Nomex® fabric woven from a blend of filament and spun fibres	Nomex® blend of filament and spun fibres
Batting	Two layers of Nomex®, Kevlar® blend spunlace (1 layer 2.3 oz./yd2 with 3D waffle design and one layer 1.5 oz./yd^2)	One layer of Nomex®, Kevlar® blend spunlace (2.3 oz./yd^2 with 3D waffle design)	Two layers of Nomex® E-89™ (1 layer each of 2.3 with 3D waffle design and 1.5 oz./yd^2)	Three layers of Nomex® E-89™ (each layer is 1.5 oz./yd^2)
Total weight (oz./yd^2)	7.7	6.2	7.6	8.3
Donning/ doffing properties	Excellent	Excellent	Excellent	Excellent
Flexibility	Most flexible, thin and lightweight	Most flexible, thin and lightweight	Lightweight and flexible	The most flexible choice for this much insulation
Thermal protection per unit weight	More thermal protection			
Wickability and quick drying	Excellent	Excellent	Very good	Very good

has an LOI of 27~33%[104], without any decomposition temperature found. Benisek and Craven[120] also showed that the combination of Zirpro and tetrabromophthalate has a synergistic effect in fire resistance, although tetrabromophthalic acid or its salts tend to smoke on burning and are not durable to washing.

It was claimed[121] that Zirpro wool fabric assemblies showed a relatively longer time to reach pain (first degree burn) and blister (second degree burn) thresholds, as well as the longer pain alarm time – the time available to the wearer to withdraw from the flame heat source before serious injuries occur. It is also claimed that,[121] compared with some other fibres including

CALDURA® NP	ARALITE® SL2/SL3/NP	FIREFLITE™ (NorFab)	SYNERGY (NorFab)	Q-8™
Nomex® fabric woven from a blend of filament and spun fibres	Nomex® fabric woven from spun fibres	Calendared Nomex® fabric (for improved slide)	Calendared Nomex® fabric (for improved slide)	50% Meta-aramid, 50% FR Modacrylic
50% Kevlar®, 50% Nomex® fibre batt (blend of virgin and reprocessed fibre)	Two or three layers of Nomex® E-89™ (1 layer each of 2.3 and 1.5 oz./yd^2) 50% Kevlar®, 50% Nomex® fibre batt (blend of virgin and reprocessed fibre)	Two layers spunlace blended from BASOFIL® fibres, Nomex® fibres and wool	Two layers of BASOFIL®/Nomex® blend spunlace	Reprocessed Aramid fibre, FR Rayon
7.6	7.2/8.0/7.2	7.2	7.2	8.0
Excellent	/	/	/	/
Lightweight and flexible	Lightweight and flexible	Lightweight and flexible	Lightweight and flexible	
Good	Very good/good/good	Superior	Quick drying	

aramid and novoloid fibres, Zirpro wool fabric assemblies have one of the lowest residual heat transfers after a limited flame exposure to the pain threshold and thus reduce the possibility of second degree burns.

When used in firefighters' clothing, it was found[121] that a woven Zirpro wool fabric of high density over a bulky knitted Zirpro wool underwear fabric offered significantly better protection than a single layer of a woven or knitted fabric or a double layer of a woven fabric of the same total weight.

However, the Zirpro wool treatment process and Zirpro wool products have certain problems. Firstly, both the zirconium and titanium hexafluorozirconate salts used in Zirpro wool treatment have associated potential environmental problems, and might be affected by government legislation, and directives such as REACH[122]; in addition, Zirpro wool fibres tend to be brittle, which might lead to problems in yarn spinning and fabric weaving. Therefore, research into developing new chemical treatments utilizing synergic effects of phosphorous and nitrogen elements to achieve wool fibres of improved flame-retardant properties has been reported.[123]

The use of a phosphorus additive[124] in viscose rayon (e.g., Clariant 5060) developed by Sandoz and Lenzing AG, and silicic acid-containing rayon fibres[125] developed by Sateri Fibres (formerly Kemira), are two typical fire resistant viscose rayon fibres potentially used as a blend in firefighters' clothing. Remarkably, the phosphorus additives can be added to the highly alkaline cellulose xanthate 'dope' before fibre spinning. They survive the acidic coagulating bath as well as alkaline laundering.

Visil[126] and other similar viscose fibres[127] are inherently flame-retardant silicic acid-containing viscose rayon fibres used as a blend component. Visil[33] is made by wet spinning of alkaline cellulose xanthate (viscose) containing a sodium silicate (equivalent to about 30–33% SiO_2) with some aluminosilicate component. During fire combustion, flame retards by both endothermic water release and char formation.

Halogen-containing fibres, such as modacrylic fibres[128] (e.g., Saran fibres[129]), are also used as flame-retardant components in blends of fire protection equipment.[130,131] Modacrylic fibres are typically copolymers of vinyl chloride or vinylidene dichloride and acrylonitrile. However, while modacrylic fibres are non-flammable and do not melt or drip, they shrink rapidly when exposed to fire, and thus are rarely used in firefighters' clothing.

PPS fibres (e.g., Torcon (Toray), Procon (Evonik) and Diofort (Diolen)) have inherent fire resistance, chemical resistance and high tenacity properties, and are widely used in high temperature filtration applications and chemical protective clothing.

3.7 Chemical protective clothing (CPC)

The purpose of CPC is to protect military personnel and law enforcement officers from the contamination of various chemical hazards in the battlefield, chemical industrial production operation, and in the event of industrial accident, riot and counter terrorism activity.[132] The hazardous chemicals and chemical warfare agents dealt by the existing CPC are either gaseous or liquid;[133] they include blister agents/vesicants (e.g., sulfur mustard (mustard gas), nitrogen mustard, phosgene oxime and Lewisite), blood agents (arsine, cyanide), nerve agents (e.g., GA, GB, GD, and VX), riot control agents/tear gas (e.g., CA, CN, CS, PS, CR, phosgene) and other chemicals including caustics/acids (e.g., hydrogen fluoride), alcohol and solvents.

There are two types of CPC: impermeable, and permeable which includes semi-permeable and selectively permeable CPC available in the market.[134] Impermeable garments (e.g., Tychem[135]) resist the chemical agents from penetration through the clothing materials to provide an effective barrier against chemical penetration and diffusion through the fabric of the clothing, but it prevents the perspiration released by human body from evaporation through the clothing materials, to produce additional thermal stress to the wearers[136] and it is thus suitable for wearing only for short periods of time (e.g., from 30 min to a couple of hours).

It is important to note that no single barrier material used in this type of clothing will protect against all chemicals, and that no material is totally impermeable to all chemical hazards, thus specific barrier materials are designed to protect against particular chemicals based on the chemical resistance performance upon contact with the chemicals. Barrier material is required to demonstrate:

- no penetration,
- no significant degradation,
- a designated breakthrough time greater than the duration of the task and
- a low permeation rate upon contact with the chemicals.

The penetration of chemicals occurs because of leakage through seams, zippers, pinholes and other imperfections in the clothing; degradation (e.g., softening, swelling, shrinkage, dissolving, hardening and embrittlement) is the physical deterioration of a material due to contact with a chemical; permeation rate refers to the rate at which a specific chemical diffuses through a barrier material at the molecular level; breakthrough time is the total elapsed time from the initial contact of the chemical on the material's exterior to the time of detection of the chemical on the inside surface. Both the permeation rate and break through time are affected by such factors as the type of chemical, chemical concentration, material thickness, humidity, temperature and pressure. A CPC would ideally have a low permeation rate and long break through time.

The permeable CPC containing a permeable or semi-permeable barrier OS and certain amount of activated carbon-based air-permeable interliners[137] (such as Mark IV NBC Suit (UK), Paul Boyé's CBRN Protective Suit (France),[138] and JSLIST overgarment (USA)[139]) were therefore developed to provide protection over a longer period of wearing (up to 24 h).[140]

A permeable CPC usually consists of three layers – barrier layer, sorptive layer and inner layer. The barrier layer (OS) is a liquid repellent layer giving protection from liquid chemicals. The sorptive layer is the key component of protective clothing in absorbing liquid and airborne chemicals. The inner layer provides comfort to the wearer. In permeable CPC, when the activated carbon materials become saturated with the contaminant, they need to be replaced, regenerated or reactivated (usually by heat). While the Mark IV NBC Suit was developed over 30 years ago, one of the targets for the future CPC garments[141] is to enable the wearers to operate longer in a chemically and biologically contaminated environment comfortably, safely and effectively, by using new technologies such as selectively permeable membranes (SPM), to reduce its weight to about a third to a half that of existing CPC systems, and to design clothing for different weather and contamination environments. Self-detoxifying materials including reactive nanoparticles and catalysts have been investigated to enhance the effective protection of CBR protective clothing.[142]

While there are various materials used in CPC,[143] the two key functional elements employed in CPC are discussed in this section:

1. the barrier fabric and
2. the chemical vapour absorbent interliner materials.

3.7.1 Chemical barrier materials

Chemical barrier fabrics can be permeable, semi-permeable or impermeable.[144] Permeable barrier OSs are traditionally made of woven fabric, which is permeable to gases, liquid, aerosol, mist and vapour from water and chemical agents; they might be finished to provide liquid repellency and waterproofness.

There are two types of semi-permeable membrane barrier materials – porous membrane and non-porous monolithic membrane. Different types of porous membrane might contain different sizes of pores, ranging from macropores (pores of micrometres or tens of nanometres in size), micropores (nanosized pores), to ultrapores (molecular sieving pores), and they will allow gases, particulates and vapours of different sizes to pass through the pores. Examples of such products are special non-woven fabrics and fabrics containing porous polymeric membranes (e.g., Microgard® series fabrics, Crosstech® fabric, Chempak® fabric and Gore® chemical splash fabric from Goretex; Tyvek ® and ProShield® from DuPont; and some electrospun nanofibrous membranes[145]), many polymers[146] including expanded PTFE, sulfonated tetrafluoroethylene (Nafion),[147] sulfonated poly(styrene-isobutylene-styrene) (S-SIBS),[148] sulfonated poly(styrene-ethylene-*ran*-butylene-styrene) (S-SEBS) and sulfonated poly(phenylene sulphide) (S-PPS)[149] are employed to make such porous membranes. Non-porous monolithic membranes are solution–diffusion membranes and they usually contain different types of polymers (e.g., polyurethane). This type of permeable membrane also includes SPM, which has a selective solution–diffusion process for certain liquids, gases and vapours. A wide range of polymer materials could be used to make SPMs, depending on the targeted chemical hazards to be handled. These polymers include cellulose, cellolusic acetate, polyallylamine, polyallylamine and polyvinyl alcohol, and many of them are already widely used in the separation industry.

Materials such as butyl, halogenated butyl rubber, neoprene and some sulfonated polymer such as polyphenylene sulfide (PPS), are widely used to make impermeable barrier membrane. One type of impermeable chemical protective garment (e.g., Tychem series clothing[135]) is made of a single impermeable laminated fabric to protect against the diffusion of chemical hazards. The laminated fabric consists of multiple sheets of different materials to protect against the penetration of chemical liquids.[150]

Smart protective clothing having a responsive barrier and self-decontaminating functions[151] was also investigated. It was found that acyclic amide, such as acrylamide grafted textiles, can be activated by chlorination to form N-halamine, which can demonstrate effective bacterial killing. A similar N-halamine structure was also shown to be able to detoxify some carbamate pesticides upon contact. Since crosslinked acrylamide can function as an absorbent to form a hydrogel, it is highly likely that both a smart response and a biocidal/detoxifying function can be combined without compromising the comfort of the modified clothing. It is expected that the grafted network responds by swelling in contact with the biological fluids/pesticides. This causes insidious fluids to be blocked physically, and the absorbed fluid will thus be decontaminated *in situ* to perform a smart dual protective function. Flexible composite membranes for selective permeability potentially used as smart barrier materials for permeable CPC, have been reported.[152,153]

3.7.2 Activated carbon-based interliners

Both chemical and biological agents in either gaseous or liquid form can be adsorbed into adsorbent materials via adsorption, which is a dynamic adsorption–desorption process for forming a monolayer of gaseous or liquid molecules (i.e., adsorbates) on the surface of solid adsorbents such as activated carbon via van der Waals forces. During adsorption, part of the adsorbate molecules in fluid state are transferred on to the solid surface of adsorbents (i.e., adsorption phase). Meanwhile, part of the adsorbate molecules captured on the adsorbent surface are released again to the fluid state (i.e., desorption phase). When the rates of adsorption and desorption become equal, an adsorption equilibrium (or so-called adsorption isotherm) is obtained. Adsorbent materials are usually porous materials. Their adsorption capacities depend on the type of pores and the total surface area available for the access of adsorbate molecules. For the adsorption of gaseous, the amount of adsorbates adsorbed is a function of gas phase concentration inside the pores and is measured by equilibrium partial pressure $p = p_o$ at constant temperature.

The adsorbent materials having adsorption properties include silica gel and activated alumina, activated carbon, zeolites (molecular sieves) and some synthetic polymers. Silica gel and activated alumina are used primarily in the dehydration of gas streams. Zeolites (molecular sieves)[154] are alumina–silicate crystalline structures with uniformly sized pores (also called windows) throughout the structure. They are used to capture or separate gases on the basis of molecular size and shape. Polymeric adsorbents[155] (for example, Dowex™ Optipore and Amberlite[156] polymeric resins from Dow Chemical Company, Treversorb[157] from Chemra company, Germany) are formed by crosslinking long chain polymers that have a variety of functional groups. These polymeric materials have a rigid microporous structure, with surface areas up to more than 1000 m^2/g. Also, sol-gel derived carbon xerogel coated on cotton and plasma-treated polypropylene fabrics are also being proposed for protection against volatile organic compounds.[158] However, among all those adsorbent materials, only activated carbons are widely used in protective clothing.

Activated carbon material is a versatile material, from granular to fibre form, with desirable properties such as large surface area, adsorptivity, high thermal conductivity, fire resistance, robustness and wide availability. Both activated carbon granules and activated carbon fibres are employed in permeable protective clothing to absorb/adsorb chemicals.[159,160]

The activated carbon granules were coated in substrates (e.g., polyurethane foam and non-woven pads) for use as an interliner in the clothing system; however, the coated polyurethane foam usually has low air and water vapour permeability, and it is reported that needle-punched fabrics coated with activated carbon have better thermal comfort properties.[161] The coated activated carbon material is combined with other textile materials in permeable CPC to provide two main functions:

1. to absorb chemical agents to provide the wearer protection against chemical hazards and
2. to permit water vapour produced from the human body to pass through the clothing materials to provide clothing thermophysiological comfort.

Activated carbon fibres[162,163] have larger surface area and contain more open or accessible micropores (less than 2 nm) and mesopores (2–50 nm) but fewer non-adsorbing macropores (wider than 50 nm) than activated carbon granules. While the adsorption capacity of activated carbon is closely related to its pore size and both micropores and mesopores determine adsorption activities, the large pore volumes (up to 1.6 cm^3/g) and surface areas (up to 2500 m^2/g) associated with these mesopores and micropores result in the activated carbon fibres[164] that have high capacities for adsorbing chemical liquids and its vapour[165] in CPC.[166,167] There is also a special class of active carbons having small pore size, called carbon molecular sieves (CMS),[168] which have a sharp distribution in a range of micropores around 0.2~0.5 nm but with a smaller pore volume (e.g., < 0.2 mL/g) than other activated carbons. Those CMS activated carbons might also be interesting adsorbent materials for protective clothing.

It was found that the diameter of the pores in which adsorbate is adsorbed ranged from 1.7~21 times of the diameter of the adsorbate, and the most effective adsorptive pores were between 1.7~6 times of the adsorbate sizes,[169] in which there is only one molecule of adsorbate adsorbed in these pores. It was proposed that the energy after adsorption would increase and result in a repulsion between adsorbates if the pore size is smaller than 1.7 times of the adsorbate sizes. For the pores smaller than 3 times of the adsorbate sizes, the adsorbates molecule receives an attractive force from the surrounding wall of the pores and thus is tightly adsorbed, while the tightly adsorption of the absorbate by such strong attractive force in those smaller pores might create a problem for the solvent regeneration of exhausted activated carbon materials. The attractive force that the adsorbate molecule receives in the pores greater than 3 times of the adsorbate sizes is just from one side of the surrounding wall of the pores and it is thus relatively weaker.[170]

The use of activated carbon in fibre form (~10 μm in diameter) in protective clothing has many advantages over granular and powder forms. For example, it has been shown that adsorption of methylene blue from solution at ambient temperature in rayon-based activated carbon fibres is two orders of magnitude faster than in a granular active carbon (~100 μm in diameter) and one order of magnitude faster than in active carbon powder (~1000 μm)[171,172]; also, the adsorption capacities were relatively higher,[171,172] due to the lack of non-adsorbing macropore spaces in the fibre compared with powder and granular.

The production process of activated carbon fibre[164] consists of the development of amorphous carbon fibres[172] at around 1000°C from fibre precursors, followed by physical activation either by steam at 800–1000°C or chemical activation.[173] The organic fibre precursors include polyvinylidene chloride (**PVDC**) fibres (Saran),[166,174,175] phenolic fibres (Kynol)[176–183], poly(acrylamide),[184] polyacrylonitrile (PAN)[185] and rayon fibres[186]; activated carbon fibres can also be made from vapour-grown fibres, nanotubes and others.[187] Most of the activated carbon fibres used in protective clothing are made from rayon,[188] pitch[163] or PAN.[166] For activated carbon used in CPC, it usually needs an additive coating for the adsorption of toxic chemicals[176–186, 189] and dynamic detoxification of the chemicals adsorbed on the fibres. For example, a rayon-based activated carbon fibre (ACF) cloth is impregnated with organo-metallic compounds such as copper (II) tartrate for the adsorption of hydrogen cyanide gas.[189] While rayon-based activated carbon fibre is still used in current CPC, the activated carbon fibres made from other precursors, such as PAN, pitch and phenolic-resin, appear to be easier and/or more economical to make, as well as generally having greater surfaces areas and other associated properties.[190]

However, activated carbon material has inherent disadvantages:

- The rayon-based activated carbon fibres are quite expensive to make; while the production of activated carbon fibre at lower cost by activating phenolic-resin carbon coatings on glass fibre substrates has been proposed,[191] this has had no success in commercial scale production yet.
- The poor mechanical properties, especially poor abrasion resistance and tensile strength,[192–194] is still an unresolved problem.
- It only adsorbs nerve and mustard agents; various coating resins need to be developed to absorb different individual types of hazardous chemical agents.
- Its lack of thermophysiological comfort properties because it has little capability of absorbing large amounts of moisture from the human body.
- If used in a mask, it might affect the wearer's breathing, because moist activated carbon will adsorb oxygen from the environment giving rise to breathing difficulties.
- It has other inherent disadvantages, including greater density, brittleness and less mechanical strength than traditional textile materials.
- The disposal of the suits after usage, because activated carbon is classified as a hazardous material once it becomes saturated. The used wear is normally sent back to the supplier or manufacturer, where it would be essentially decontaminated by bleaching and then disposed of by incineration or landfill.

3.7.3 Evaluation of protective clothing

Methods for the determination of protective clothing have been standardized for both protective function and clothing comfort for different types of protective clothing by various international, national and industrial organizations including International Standard Organisation (ISO), European Standardisation Organisations (CEN), British Standard (BS) and ASTM/AATCC.

3.8 Materials to improve thermo-physiological comfort of protective clothing

One of crucial requirements on protective clothing is that, while not compromise the ultimate protection of wearer, they need to address the thermal burden caused by the metabolic heat and moisture produced by human body enclosed in the protective clothing ensembles. Various high performance materials and functional textile structures are incorporated into protective clothing to reduce these thermo-physiological burdens, and a few examples of such functional materials and textile structures are discussed in this section.

3.8.1 Thermal burden in protective clothing

Clothing comfort relates to human feelings of satisfaction when the clothing is seen, touched, or worn by the user. Thermal stress is a particular concern for all protective clothing.[195] The three characteristics of thermal stress in protective clothing are:

1. human body metabolic heat,
2. the weight of protective clothing and
3. impermeability of the fabrics to water vapour.

These are discussed below.

The quantity of heat produced by a human being depends very much on physical activity, and can vary from 100 W while resting to over 1000 W during maximum physical performance.[196] Particularly during the cooler seasons (approx. 0°C), the recommended thermal insulation is defined in order to ensure that the body is sufficiently warm when resting.

Some protective clothing is very heavy (e.g., body armour, NBR clothing). Currently, the majority of available bulletproof vests are very heavy and bulky: the weight of such a bulletproof vest for US police uses ranges from 2.5 kg for US NIJ level III protection up to 10 kg when it includes ceramic plates for US NIJ level IV protections;[197] for some special military purposes, the weight of body armour could be up to 20 kg. It is found that the weight of clothing is linearly related to the metabolic energy consumed by the human body,[198,199] and the proposed relationship between metabolic energy consumed during loaded walking, *Mw*, and the load carried by the walker, L, by Givoni and Goldman[200] is as follows,

$$Mw = T(W + L)\left(2.3 + 0.32(V - 2.5)^{1.65} + G(0.2 + 0.07(V - 2.5))\right)$$

Mw = metabolic energy consumed during walking (Watts), W = body mass (kg), L = load mass (kg), T = terrain factor, V = velocity or walking rate (m/s), G = slope or grade (%)

Pandolf *et al.*[201] expanded the work of Givoni and Goldman[184] to develop an equation to predict the energy cost of load carriage:

$$Mw = 1.5W + 2.0(W + L)\left(\frac{L}{W}\right)^2 + T(W + L)(1.5V^2 + 0.35VG)$$

where T = terrain factor (1.0 = black top road; 1.1 = dirt road; 1.2 = light brush; 1.5 = heavy brush; 1.8 = swampy bog; 2.1 = loose sand; snow, dependent on depth of depression ($T = 1.30 + 0.082*D$, where D = depression depth in cm))[202]

As the thermal burden is proportional to the clothing weight, it thus requires more metabolic energy to be transported away from human body within the microclimate of the heavier protective clothing in order to maintain wearer comfort. Many different technologies, rarely used in traditional clothing systems, have been invented to achieve thermophysiological comfort in a protective clothing system; one example is to place humidity absorbing modules containing hygroscopic endothermic salts[203] within the protective clothing to reduce the relative humidity, which increases the wearer's comfort and allows more sweat to evaporate, leading to additional cooling.

In CPC, especially for clothing against the penetration of hazards in the form of gaseous, liquid and mist, the thermal burden usually arises from the lack of breathability of the clothing due to the impermeability of the fabrics to water vapour. A serious challenge associated with body armour and CPC, apart from its excessive weight, is the heat caused by lack of effective perspiration, eliminating the body's natural ability to dissipate and dispose of metabolic heat. Increasing heat stress exhausts the human body within a short time, even under normal conditions, let alone the extreme heat encountered in the Middle East and Central Asia.

Both smart materials and structures can be used to achieve physiological comfort of protective clothing. In this chapter, three groups of typical examples of incorporating smart textile technologies into protective clothing to achieve clothing comfort are discussed.

1. Typical examples of using smart materials to achieve clothing comfort are: heat storage and thermo-regulating materials including PCM,[204,205] temperature sensitive materials[206] including SMP/alloy[207]; smart polyurethane film for moisture transport and evaporative cooling technologies.
2. Typical examples of using smart structure to achieve clothing comfort are Hainsworth® TI-Technology, and clothing having a forced ventilation system.

3. A range of active and proactive systems used in protective clothing to control the temperature and humidity of the clothing microclimate, such as an active cooling system using ice, water and Peltier plate are discussed.

3.8.2 Phase change materials (PCMs)

PCMs[208] are the materials possessing the ability to absorb, store and release latent heat through undergoing one of the four types of phase transformations (solid–solid, solid–liquid, liquid–gas and solid–gas phase transformations) in a nearly isothermal process within a certain range of environmental temperature. During this process, the latent heat can be stored or released from PCMs going from one physical state (e.g., solid) to another (e.g., liquid) or vice versa.

Solid–liquid PCM is among the most favoured, due to its relatively large energy storage with acceptable volume change (≤ 10%) over small temperature variations. Traditionally solid–solid PCMs, changing their crystalline structure from one lattice configuration to another, have the lowest thermal storage capability and are seldom used in commercial applications; however, this type of PCM has some invaluable advantages, such as simplicity of use, low erosion, no leakage, no contaminant, long lifespan with little change in appearance and no need of nucleation to prevent supercoiling. Therefore the latest development of solid–solid PCMs makes this type of PCM able to store and release large amounts of heat, comparable to the most effective solid–liquid PCMs at a fixed and well-defined temperature.[209,210] Currently, this range of phase changing temperature remains between 25°C (77°F) and +180°C (356°F). Solid–gas and liquid–gas transformations have relatively higher latent heat, but can hardly be used in smart textiles due to the complex storing technique requirement with pressurized container.

In general, PCMs can be grouped into three categories: organic, inorganic and eutectics materials. Organic materials used as PCMs include aliphatic compounds and polymers with long chain molecules composed primarily of carbon and hydrogen (e.g., paraffin waxes (PW) (or n-alkanes), oils, fatty acids and polyglycols (PEGs)); they tend to exhibit high orders of crystallinity when freezing and mostly change phase above 0°C (32°F). Currently crystalline alkyl hydrocarbons are used exclusively for textile applications, due to their large latent heat, good thermal and chemical stability, low vapour pressure and self-nucleating behaviour.[211] Inorganic PCMs include salt hydrates, metallic and ice – typical inorganic PCMs are Glauber's salt. Eutectic materials are mixtures of chemical compounds or elements (e.g., solutions of salts in water) that have a single chemical composition that solidifies at a lower temperature (eutectic temperature) than any other composition made up of the same ingredients. It is known that the phase change temperature of salt hydrate-based eutectics could be below 0°C

Table 3.9 Latent heat of adsorption, emission and crystallization temperature of selected waxes[213–219] and strearates

Hydrocarbons	No of C atoms	Latent heat of adsorption (ΔH) in J/g	Latent heat of emission ($-\Delta H$) in J/g	Melting point	Crystallization temperature (T_c, °C)	References
n-Hexadecane	16	235.2	236.6		12.2	B. Pause[216]
n-Heptadecane	17	176.4	182.6		16.5	B. Pause[216]
n-Octadecane	18	244.8	246.4		22.0	B. Pause[216]
n-Nonadecane	19	177.6	182.6		26.4	B. Pause[216]
n-Eiscosane	20	242	230		30.4	Zhang XX[182]
Butyl stearate		120		19	21	D. Feldman *et al.*[215]
Vinyl stearate		122		27	29	D. Feldman *et al.*[215]
Isopropyl stearate		142		14	18	D. Feldman *et al.*[215]

Table 3.10 Latent heat and melting point of selected salt hydrates[220]

Material	Melting point (°C)	Heat of fusion (kJ/kg)	Latent heat (MJ/m³)
$MgCl_2.6H_2O$	117	169	242
$Mg(NO3)_2.6H_2O$	89	163	252
$CH_3COONa.3H_2O$	58	226	287
$MgCl_2.6H_2O/Mg(NO_3)_2.6H_2O$	58	132	201
$Na_2HPO_4.12H_2O$	34	265	379
$Na_2SO_4.10H_2O$	32	251	335
$Na_2CO_3.10H_2O$	32	233	340
Waxes	28–4	220–245	170–195
Polyethylene glycols	28–15	146–155	165–175
$CaCl_2.6H_2O$	27	191	298
Glauber's salt ($Na_2SO_4{\cdot}10H_2O$) + additives	24–4	Wide range	Wide range
$CaCl_2.6H_2O/CaBr_2.6H_2O$	15	140	249
Water	0	335	335
Range of water/salt eutectics	0–64	Wide range	Wide range

(32°F) or above 0°C (32°F) depending on the formula of the mixture; some of those water-based eutectics can change phase at temperatures up to 117°C (242.6°F).[212] (See Tables 3.9 and 3.10).

The PCMs having the most suitable phase change temperature for clothing are organic compounds such as paraffins or linear alkyl hydrocarbon, non-paraffinic materials (e.g., hydrocarbon alcohol, hydrocarbon acid, polyethylene or polytetramethylene glycol and aliphatic polyester) and inorganic compounds such as hydrated inorganic salts, eutectics and polyhydric alcohol–water solution.[202]

PCM microcapsules[221,222] were initially developed in textile structures and clothing in the early 1980s by NASA to provide improved thermal protection in textile and clothing having smart acclimatizing properties for astronauts' spacesuits against the extreme temperature fluctuations in outer space. When used to make functional fabrics for clothing, PCM can be incorporated into various fibres,[223–225] non-wovens,[228] PU foams[226] and coated on fabrics[227] in original form, micro-capsulated PCMs[21,22] and nano-capsulated PCMs.[228]

Outlast fibre[229] is a bi-component fibre, including both staple fibres and filaments, containing patented tiny microcapsules (c.1–3 μm in diameter) of PCMs; Outlast Thermocules, substances similar to paraffin, are stored. The new fibre with temperature management is claimed to align particularly well with underwear and other products worn next to skin such as socks, T-shirts, shirts and trousers. Outlast technology used in firefighters' clothing[230] has been made from 60% MAC Protex and 40% CV Outlast. In this bi-component yarn structure, the flame-retardant Protex fibre stops the flame from spreading by producing minute amounts of inert non-combustible gases which 'seal' the fabric surface from oxygen; after removal of the flame source, flame spread is stopped. The Outlast fibre in the new fabrics provides a uniquely higher level of comfort, temperature regulation and moisture management not normally associated with protective clothing.

Beginn[231] reported on an ultra-high molecular weight polyethylene (UHMW-PE)/PW composite gel structure and examined its thermal and mechanical properties. It was reported that these gel composites were superior to microencapsulated paraffin PCMs with respect to heat capacity. On exceeding the melting temperature of the paraffin, the solid compounds reversibly transformed to soft, but shape persistent, gels. The UHMW-PE/PW gels can store and release much greater enthalpies, up to 200 J/g at the melting temperature of the PW; even with the cheap paraffin wax PW 42, the melting enthalpy could be up to 120 J/g, while commercial polyurethane PCM foams[232] only carry a latent heat of 60–65 J/g. In the gel state, the liquid alkane can leave the gel when in direct contact with porous materials, and above the melting transition of the paraffin compound in the composites, the capillary forces prevent the liquid wax from free flow. However, while it was believed that the paraffin wax crystallizes without destruction of the UHME-PE gel network, the gel's mechanical strength was found to be slightly lower than that of standard UHMW-PE polymers.

Another example of phase change technology used in protective clothing utilizes evaporative cooling technology. Cooling fibres[233] introduce water-retaining fibres into the fabric structure, sandwiched between breathable outer fabrics, e.g., cotton and Nomex; the inner layer containing cooling fibres conducts heat and moisture away from the body through evaporation of water.

Fabrics[234] incorporating PCMs were used in protective clothing[235] for clothing comfort. When the protective clothing is in use, PCMs absorb human body metabolic heat in the clothing microclimate during its melting process to give a cooling effect to the wearer and release the stored energy to the environment during a reverse solidifying process, to give the wearer a warming effect[236] to achieve the desired thermo-regulating functions in protective clothing.[237,238] Thus, this type of performance textile structure containing PCMs can function as a heat buffer to store a large quantity of heat from environment and human body as a thermal insulation layer; it can also release the absorbed latent heat during large fluctuation of environmental temperature through conduction, convection or radiation.

3.8.3 Smart structure for both protection and thermal comfort of firefighter clothing

Outer textile materials for firefighting clothing have frequently been manufactured from 100% meta-aramid or PAI, blends of meta-aramid and para-aramid fibres, or by use of core spun yarns or staple mixtures with poly-paraphenylene terephthalamide copolymer or fibres comprising para-aramid cores with meta-aramid or PAI covers. The combination of these fibres in the fabric enhances mechanical properties and durability of the product. However, meta-aramid and PAI fibres shrink, consolidate and thicken when exposed to a high temperature heat source. The presence of para-aramid or polyphenylene terephthalamide copolymer, in either the fibre blend or as a core, can be used to prevent fibre shrinkage and consequent breaking open of the garment. However, the inclusion of para-aramid fibre in the blend has been found to be insufficient in tightly woven fabrics to prevent breaking open. Consequently, there is a need of improved textile materials for the manufacture of firefighting garments.

Hainsworth® TI-TECHNOLOGY[239–241] use a double layer spacer fabric construction that combines the high performance of Nomex® and Kevlar® fibres in an intelligent way by keeping them as separate layers to maximize the benefits of each fibre. Nomex® is on the face of the fabric, with the high strength of Kevlar® protected from the effects of UV degradation and abrasion on the back. When the fabric assembly is hit by extreme heat, the Nomex® layer consolidates, while the Kevlar® layer remains unchanged. The different thermal shrinkage of the fibres means that the two layers move and react against each other. The dynamic nature of the fabric means that the open, breathable construction required for everyday activities moves to form air pockets between the two layers, thereby increasing thermal protection only when it is most needed.

3.8.4 Active cooling system in protective clothing for clothing comfort

Various smart textiles have been used to actively control clothing comfort for protective clothing. They include materials to sense the wearers' physiological condition,[41] their posture and activity[42] and outside environment,[43] as well as responsive action.[44] Conductive yarns can be produced by coating with conductive polymers or by embedded conductive fillers such as carbon nanotube.[40] Electrically active structures can then be formed, e.g., through specially patterned knitting[38] and temperature responsive materials for clothing.[39] However, numerous challenges remain, in particular with contactless sensors, interconnects, electronic reliability, data and power transmission lines, and shielding.[242]

Personal cooling systems sustain a microclimate circulating dry, fresh air, liquid, ice or wax to draw and absorb heat from wearer's body. One example[243] is a lightweight 'spacer vest', designed to assist natural cooling through perspiration, which is blocked by the thermal insulation of the body armour. The Interceptor Body Armour (IBA) system is distanced from the wearer's skin surface to increase evaporative cooling around the torso. Improvement of up to 20% in evaporative cooling potential is achieved when wearing the spacer vest, in comparison with wearing standard IBA.

An Israeli company, Rabintex, developed a lightweight cooling system known as Breeze,[244] utilizing a lightweight battery powered ventilation system that re-establishes the natural cooling by perspiration. By feeding fresh air beneath the body armour, the Breeze vest facilitates cooling by wicking moisture from the body and cooling through the natural phase change process.

A more advanced cooling system is under development for the future force warrior (FFW) Program.[245] This system supports the soldier operating the standard FFW suite, as well as full NBC protected gear. The 3.5 lb system provides microclimate control by circulating cooling water. In the UK, another microclimate suite is currently under development to provide 100 W of cooling for operating for 4 h for the British Army.

Cooling can also be achieved by applying thermal regulation elements into the body armour. Such a concept is described in a patent registered by John Hopkins University's Applied Physics Lab (APL). This concept eliminates entirely the need for a liquid coolant or PCMs, by using high thermal-conductivity channels embedded into the body armour itself. Heat conductors are applied over a moisture wicking layer, pulling moisture from the body and dispersing the moisture to the outer surface where it can evaporate.[245]

The Body Ventilation System (BVS)[246] developed by GlobalSecure Safety Corporation uses a battery powered blower to circulate air through an air distribution device worn under the soldier's body armour. As fresh air is circulated beneath the IBA and battle dress uniform (BDU), natural body cooling is retained, resulting in improved physiology.

The microclimate cooling system (MCS)[247] is provided as part of US Army Air Warrior Program, for helicopter crews operating in the hot climate of Southwest Asia. The microclimate cooling garment is connected to a condenser unit which chills water and pumps it through small tubes embedded in the vest; it is worn as an undergarment beneath the body and closing.

Protective suits for firefighters and others working in extremely hot conditions with a built-in cooling systems[248] has been developed to provide a high quality thermal and moisture management layer based on a three-dimensional warp-knitted fabric coated with a water binding polymer. Its structure mimics the physical mechanisms for thermoregulation in the human body. A three-dimensional textile structure is used for the thermal and moisture management layer with a cooling tubing system inserted into cavities in the three-dimensional textile structure. The tubes have liquid circulating through them, and remove heat in a similar manner to that of blood vessels in the body.

3.9 References

1. Council Directive 89/686/EEC of 21 December 1989 on the approximation of the laws of the Member States relating to personal protective equipment (amended), http://ec.europa.eu/enterprise/sectors/mechanical/documents/legislation/personal-protective-equipment/.
2. http://ec.europa.eu/enterprise/policies/european-standards/harmonised-standards/personal-protective-equipment/index_en.htm.
3. National Fire Protection Association (NFPA), http://www.nfpa.org/codes-and-standards/document-information-pages.
4. ASTM F903-10, *Standard Test Method for Resistance of Materials Used in Protective Clothing to Penetration by Liquids: Procedure C.*
5. ASTM F73-07, *Standard Test Method for Permeation of Liquids and Gases through Protective Clothing Materials under Conditions of Continuous Contact.*
6. BS EN ISO 7730:(2005), Ergonomics of the thermal environment, Analytical determination and interpretation of thermal comfort using calculation of the PMV and PPD indices and local thermal comfort criteria.
7. Smith B., *High Performance and High Temperature Resistant Fibers*, http://www.intexa.com/downloads/hightemp.pdf.
8. http://www.swicofil.com/high_performance_fiber_comparison.html.
9. http://www.fiber-line.com/websites/implementatie/mediadepot/348bf1bbaf2.pdf.
10. http://www.p84.com/product/p84/en/Pages/default.aspx.

11. http://www51.honeywell.com/sm/afc/products-details/fiber.html#Product Literature1.
12. www.dyneema.com/.
13. *www.vectranfiber.com/.*
14. Shishoo R (2002), Recent development in materials for use in protective clothing, *International Journal of Clothing Science and Technology*, **14**(3/4), pp. 201–215.
15. Yang HH (1989), *Aromatic High Strength Fibers*, New York, J. Wiley, c.
16. http://www.swicofil.com/high_performance_fiber_comparison.html
17. Mark JE, *Oxford Polymer data handbook*, 2009, Oxford University Press, Oxford, UK
18. Technical Guide for NOMEX® Brand Fiber, Technical Guide for NOMEX® Brand Fiber, http://www.nakedwhiz.com/gasketsafety/nomextechnicalguide.pdf.
19. Protection Against Fire Hazards with Nomex® III A, http://www.dupont.com/products-and-services/personal-protective-equipment/thermal-protective/brands/nomex/products/nomex-iii-a.html.
20. http://www.matweb.com/search/QuickText.aspx?SearchText=nomex.
21. US Patent 4075172, 1974.
22. Polyamide imid, http://www.swicofil.com/products/223polyamideimide.html.
23. Ozawa S (1987), A new approach to high modulus, high tenacity fibers, *Polymer Journal*, **19**(1), pp. 119–125.
24. http://www.toyobo-global.com/seihin/kc/pbo/menu/fra_menu_en.htm.
25. http://www.p84.com/product/p84/en/Pages/default.aspx
26. Gonzale AR, Schofield RB and Hart SV (2005), *Third Status Report to the Attorney General on Body Armor Safety Initiative Testing and Activities*, National Institute of Justice, Office of Justice Programs, U.S. Department of Justice, 24 August 2005.
27. Technical Information Bulletin, 'PBO Fiber Zylon', Toyobo Co. Ltd., revised 2001.
28. Holmes GA, Rice K and Snyder CR (2006), Review Ballistic fibers: A review of the thermal, ultraviolet and hydrolytic stability of the benzoxazole ring structure, *J Mater Sci.*, **41**, pp. 4105–4116.
29. *NIJ Standard–0101.04*, National Institute of Justice, Office of Justice Programs, U.S. Department of Justice.
30. MIL-C-12369F (1974), Military Specification: Cloth, Ballistic, Nylon.
31. Fuller M, Mao N and Taylor M (2013), The internal structure and tensile properties of goose and duck down feathers, *Proceedings of 12th Asian Textile Conference (ATC-12)*, Shanghai, China, 23–26 October 2013.
32. Du M, Mao N and Russell SJ (2013), The influence of sol-gel processing parameters on the porous structure of flexible silicone aerogel made from methyltrimethoxysilane (MTMS), *Proceedings of 14th National & 1st International Recent Developments, Textile Technology and Chemistry Symposium*, Bursa, Turkey, 08–10 May 2013.
33. Zhu R and Prickett LJ, US Patent 6,534,175, Cut resistant fabric, 2003.
34. Dual Protection Vests Made of Steel Reinforced Fibers, *Defence Update*, 1, 2004, http://defense-update.com/products/d/Dyneema.htm.
35. http://www.cordura.com/en/fabric-technology/ballistic-fabric.html.

36. Review of evidence relating to the effect of protective clothing for motorcyclists, http://info.wirral.nhs.uk/document_uploads/evidence-reviews/Revieweveffectofprotectclothingmotcycls_1e577.pdf.
37. Office of Aviation Research (2002), Washington, DC 20591, *Improved Barriers to Turbine Engine Fragments*, Interim Report IV, June 2002.
38. Dias T (2007), Electrically active knitted structures, In: *4th International Avantex Symposium for Innovative Apparel Textiles,* Frankfurt am Maine, Germany: Messe Frankfurt Exhibition.
39. Crespy D and Rossi RM (2007), Temperature-responsive polymers with LCST in the physiological range and their applications in textiles, *Polymer International*, **56**(12), pp. 1461–1468.
40. Devaux E, Koncar V, Kim B, Campagne C, Roux C, Rochery M and Saihi D (2007) Processing and characterization of conductive yarns by coating or bulk treatment for smart textile applications, *T I Meas Control.*, **29**(3–4), pp. 355–376.
41. Horter H, Linti C, Göppinger B, Loy S and Planck H. (2007) Garment with sensors, electronics and mobile energy supply. In: *4th International Avantex Symposium for Innovative Apparel Textiles*. Frankfurt am Maine, Germany: Messe Frankfurt Exhibition.
42. Tognetti A (2007), Sensing fabrics for body posture and gesture classification. In: *4th International Avantex Symposium for Innovative Apparel Textiles.* Frankfurt am Maine, Germany: Messe Frankfurt Exhibition.
43. Hertleer C and Van Langenhove L (2008), Interactive PPE and embedded electronics. In: *1st International Conference on Personal Protective Equipment: for More (Than) Safety*. Zwijnaarde, Belgium; CENTEXBEL.
44. Janssen D. (2008) Responsive materials for PPE, In: *1st International Conference on Personal Protective Equipment: for More (Than) Safety*. Zwijnaarde, Belgium; CENTEXBEL.
45. Zhu R and Prickett LJ (2003), USP6,534,175, *Cut Resistant Fabric.*
46. Dual Protection Vests Made of Steel Reinforced Fibers, Defence Update, 1 2004, http://defense-update.com/products/d/Dyneema.htm.
47. Improved Barriers to Turbine Engine Fragments, Interim Report IV, June 2002, Office of Aviation Research, Washington, DC 20591.
48. MIL-C-12369F (1974), Military Specification: Cloth, Ballistic, Nylon.
49. EN 381, PPE Product standards concerning chain saw protection.
50. EN 13595-1:2002 – Protective clothing for professional motorcycle riders. Jackets, trousers and one piece or divided suits.
51. EN 13595-2:2002 – Protective clothing for professional motorcycle riders. Jackets, trousers and one piece or divided suits. Test method for determination of impact abrasion resistance.
52. EN 13595-3:2002 – Protective clothing for professional motorcycle riders. Jackets, trousers and one piece or divided suits. Test method for determination of burst strength.
53. EN 13595-4:2002 – Protective clothing for professional motorcycle riders. Jackets, trousers and one piece or divided suits. Test methods for the determination of impact cut resistance.
54. EN 13634:2002 – Protective footwear for professional motorcycle riders. Requirements and test methods.
55. EN 13594:2002 – Protective gloves for professional motorcycle riders. Requirements and test methods.

56. EN 1621-1:1998 – Motorcyclists' protective clothing against mechanical impact. Requirements and test methods for impact protectors.
57. EN 1621-2:2003 – Motorcyclists' protective clothing against mechanical impact. Motorcyclists back protectors. Requirements and test methods.
58. EN 1938:1999 – Personal eye protection. Goggles for motorcycle and moped users.
59. Woods RI (1996), Specification of motorcyclists' protective clothing designed to reduce road surface impact injuries. *Performance of Protective Clothing: Fifth Volume*, ASTM STP 1237, JS Johnson and SZ Mansdorf (Eds.), American Society for Testing and Materials, Philadelphia, pp. 3–22.
60. SATRA (2002), Motorcycle Impact Abrasion Testing, Unpublished report, DTI Battran Project No: 712000, SATRA Technology Centre, SATRA House, Rockingham Road, Kettering, Northamptonshire, NN16 9JH, United Kingdom.
61. http://www.roadsafety.mccofnsw.org.au/a/88.html.
62. http://www.cordura.com/en/fabric-technology/ballistic-fabric.html.
63. http://www.thetechnicalcenter.com/search/fabric-search.cfm.
64. Review of evidence relating to the effect of protective clothing for motorcyclists, http://info.wirral.nhs.uk/document_uploads/evidence-reviews/Revieweveffectofprotectclothingmotcycls_1e577.pdf.
65. Holmes GA, Rice K and Snyder CR (2006), Review Ballistic fibers: A review of the thermal, ultraviolet and hydrolytic stability of the benzoxazole ring structure, *J Mater Sci.*, **41**, pp. 4105–4116.
66. http://www.dowcorning.com/content/deflexion/deflexionmaterials/deflexion-literature.aspx.
67. Palmer RM and Green PC (2007), Energy absorbing materials, GB-A-2349798.
68. Hayes WC, McMahon TA and Robinovitch SN (1997), US Patent 5599290, Bone fracture prevention garment and method.
69. Ferguson JR (2012), US Patent 8,087,101, Impact shock absorbing material.
70. Courtney WA (1997), Device incorporating elastic fluids and viscous damping, World Intellectual Property Organisation, WO 97/25551.
71. Courtney WA (1998), Improved impact absorber with viscous damping, *World Intellectual Property Organisation*, PCT/GB98/03594.
72. Courtney WA (1998), Impact absorbent building structures, *Britis Intellectual Property Office, GB9805887.8.*
73. Courtney WA (2009), Impact energy absorbing device incorporating bunching capsules *British Intellectual Property Office*, GB0907996 3.
74. Courtney WA (1998), Preliminary investigations into the mechanical properties and potential applications of a novel shock absorbing liquid, *MPhil Thesis,* Manchester School of Engineering, University of Manchester.
75. EN469:2005.
76. NFPA 1971:2007, Standard on protective ensembles for structural fire fighting and proximity fire fighting. National Fire Protection Association.
77. NFPA1851 (2008), Standard on selection, care, and maintenance of protective ensembles for structural fire fighting and proximity fire fighting. National Fire Protection Association.
78. Lee J, Kim E, Yoo S, Baek B and Hwang S (2007), Development of an intelligent turnout gear for dynamic thermal protection using two-way shape memory

alloy. In: *4th International Avantex Symposium for Innovative Apparel Textiles*, Frankfurt am Maine, Germany: Messe Frankfurt Exhibition.
79. Pause B (2006), New cooling undergarment for protective garment systems. In: *Proceedings of the 3rd European Conference on Protective Clothing and Nokobetef 8* [CDROM]. Gdynia, Poland: CIOP-PIB.
80. Shin Y, Yoo D-I and Son K (2005) Development of thermoregulating textile materials with microencapsulated phase change materials (PCM). II. Preparation and application of PCM microcapsules, *J Appl Polym Sci.*, **96**(6), pp. 2005–2010.
81. Meister F, Bauer R, Melle J and Gersching D. (2007) Smart duotherm®: the thermoregulating cellulose fibre with large heat storage capacity. In: *4th International Avantex Symposium for Innovative Apparel Textiles*, Frankfurt, Germany.
82. Wang J, Dionne P and Makris A (2006), Influence of different parameters on cooling efficiency of liquid circulating garments. In: *Proceedings of the 3rd European Conference on Protective Clothing and Nokobetef 8* [CD-ROM]. Gdynia, Poland: CIOP-PIB.
83. Aubouy L (2007), Natural thermoregulatory system for smart textile applications. In: *4th International Avantex Symposium for Innovative Apparel Textiles*. Frankfurt, Germany.
84. http://www.dpp-europe.com/Too-much-para-aramid-may-reduce.html?lang=en, 24/05/2011.
85. Kozlowski R, Wesolek D, Wladyka-Przybylak M, Duquesne S, Vannier A, Bourbigot S and Delobel R (2007), Intumescent flame-retardant treatments for flexible barriers, *Materials Science*, **97**(1), pp. 39–61.
86. http://www.sti.nasa.gov/tto/Spinoff2008/ps_3.html.
87. Sugama T (2004), Hydrothermal degradation of polybenzimidazole coating, *Materials Letters*, **58**(7–8), pp. 1307–1312.
88. Zhang H, Zhang J, Chen J, Hao X, Wang S, Feng and Guo Y (2006) Effects of solar UV irradiation on the tensile properties and structure of PPTA fiber, *Polymer Degradation and Stability*, **91**(11), pp. 2761–2767.
89. Tincher W, Carter W and Gentry D (1977), Protection of Nomex from Ultraviolet Degradation, http://oai.dtic.mil/oai/oai?verb=getRecord&metadataPrefix=html&identifier=ADA041494.
90. Davis R, Chin J, Lin C and Petit S (2010), Accelerated weathering of polyaramid and polybenzimidazole firefighter protective clothing fabrics Original Research Article, *Polymer Degradation and Stability*, **95**(9), pp. 1642–1654.
91. Chin J, Forster A, Clerici C, Sung L, Oudina M and Rice K. (2007) Temperature and humidity aging of poly(p-phenylene-2,6-benzobisoxazole) fibers: Chemical and physical characterization, *Polymer Degradation and Stability*, **92**(7), pp. 1234–1246.
92. http://www.pbiproducts.com/en/pbi_fiber/.
93. Structural Outer Shell Materials, http://www.globeturnoutgear.com/turnout-gear/materials/outer-shell
94. http://www.goreprotectivefabrics.com/remote/Satellite/Structural-Fire-Fighting-Turnout-Gear/Structural-Fire-Fighting-Turnout-Gear-Select-Your-Fabric.
95. http://stedfast.com/english/fabrics.
96. http://www.goreprotectivefabrics.com/remote/Satellite/Structural-Fire-Fighting-Turnout-Gear/Structural-Fire-Fighting-Turnout-Gear-Select-Your-Fabric.

97. http://www.stedfast.com/.
98. AIRLOCK® Spacer Technology, http://www.gore-workwear.co.uk/remote/Satellite/Innovations/AIRLOCK-Insulation.
99. Hayashi S and Ishikawa N. (1993) High moisture permeability polyurethane for textile applications, *J Coated Fabrics*, **23**, pp. 74–83.
100. Bartels VT (2007), Hydrophilic linings to enhance the liquid sweat transport through water tight clothing. In: *Proceeding of 4th International Avantex Symposium for Innovative Apparel Textiles*, Frankfurt, Germany.
101. http://www.tencate.com/TenCate/Protective_fabrics_USA/documents/ EmergencyResponse/ThermalBarriers/THERMAL_LINERS_COMPARISON.pdf
102. Owens Jr JW and Burns JJ (2004), US Patent Application 2004/0166353, Wool-based textile of flame resistant character and articles formed therefrom.
103. Horrock AR (1986), A review of flame-retardant finishing of textile, *Journal of the Society of Dyers and Colourists*, **16**, pp. 62–101.
104. Flame Resistance of Wool, CSIRO, http://www.csiro.au/files/files/p9z9.pdf.
105. Dermeik E, Braun R, Lemmer KH and Lung M (2006), WO 2006/102962, Process for the flame-retardant treatment of fiber materials.
106. Dermeik E, Braun R, Lemmer KH and Lung M, WO 2006/105833, Process for the flame-retardant treatment of fiber products.
107. Hawkes JA, Lewis DM, Webb P and LU Y, WO/2010/082063, Methods of treating wool, 2010.
108. Zubkova NS, Butylkina NG, NA and Khalturinsky AA (2005) Berlin, US6863846, Combustion retardant for polymeric materials.
109. Zubkova NS and Butylkina NG, US Patent 6995201, Flame retardant for polymeric materials.
110. Eskind LG, Pliskin RV, Satin R and Satin SP (2008), US Patent 2008/0302544 Fire barrier fabric and related fire protective systems.
111. Eskind LG, RV. Pliskin, R Satin and SP Satin (2008), US Patent 2008/0302544 Fire barrier fabric and related fire protective systems.
112. Hanekom EC and Barkhuysen FA (1975), US Patent 3994681 Oxidation of wool and like keratin.
113. Benisek L (1971), Use of titanium complexes to improve the natural flame retardancy of wool, *Journal of the Society of Dyers and Colourists*, **87**, pp. 277–278.
114. Benisek L (1972), New aspects of flame protection using wool, *Textile Manufacturer* (Manchester), **99**: 36–39.
115. Benisek L (1974), Improvement of the natural flame retardants of wool part I: metal-complex applications, *Journal of Textile Institute*, **65**, pp. 102–108.
116. Benisek L (1974), GB 1372694 Flame-proofing textiles.
117. Benisek L (1975), GB1379752, Flame-resist treatment of polyamide fibres.
118. Benisek L (1982), US Patent 4447242 Textile Finishing.
119. Horrocks AR (2000), Char formation in flame-retarded wool fibres, Part 1 effect of intumescent on thermogravimetric behaviour, *Journal of Fire and Materials*, **24**, pp. 151–157.
120. Benisek L and Craven PC (1983), Evaluation of metal complexes and tetrabromophthalic acid as flame retardants for wool, *Textile Research Journal*, **53**, pp. 438–442.

121. Benisek L, Edmondson GK and Phillips WA (1979), Protective clothing: Evaluation of zirpro wool and other fabrics, *Fire and Materials*, **3**(3), pp. 156–166.
122. REACH – Registration, Evaluation, Authorisation and Restriction of Chemicals, http://ec.europa.eu/enterprise/sectors/chemicals/reach/index_en.htm.
123. Zhu J and Mao N (2014), Wool keratin polymer having improved thermal properties, *Proceedings of 14th AUTEX World Textile Conference (to be published)*, Bursa, Turkey, 26–28 May 2014.
124. Mauric C and Wolf R (1980), US Patent 4,220,472.
125. Paren A and Vapaaoksa P (1995), US Patent 5,417,752.
126. Bajaj P (2000). Heat and flame protection, In: Horrocks AR and Anand SC (eds), *Handbook of Technical Textiles*, pp 223–263, Cambridge, England, Woodhead Publishing Limited.
127. Fushitani S and Nakano M (2007), WO 2007/023777, Flameproof rayon fiber and process for production thereof.
128. Frushour BG and Knorr RS (2006), Acrylic fibers, In Lewin M (ed.), *Handbook of Fiber Chemistry*, 3rd edition, pp. 811–974, Woodhead Publishing Limited.
129. www.asahi-kasei.co.jp/sarannet/en/seihin_sarannet01.html.
130. Zhu R, Guckert D and Lovasic SL (2006), US Patent 7065950, Modacrylic/aramid fiber blends for arc and flame protection.
131. K Ichibori, T Matsumoto and Y Kanbara (1996), US Patent, 5503916, Flame-retarded clothing.
132. http://www.navysbir.com/n09_3/N093-160.htm.
133. http://www.bt.cdc.gov/chemical/overview.asp#categories.
134. http://www.itpltd.com/en/literature.
135. http://www2.dupont.com/Personal_Protection/en_US/tech_info/literature_tychem.html.
136. Marzi WB (1989), Development of a new impermeable NBC protective suit for German civil defence, *Proceedings of the Third International Symposium on Protection Against Chemical Warfare Agents*, Stockholm, Sweden 11–16 June 1989, pp. 21–24, Swedish Defence Research Establishment, UMEA.
137. Military Specification MIL-S-43926, Suit, Chemical Protective, http://mil-spec.tpub.com/MIL-S/.
138. http://www.paulboye.com/products_nbcf_1.html.
139. Military Specification MIL-DTL-32102, JSLIST Coat and Trouser, Chemical Protective, 3 April 2002.
140. Katz MG (1989), A new approach to heat stress relief in chemical protective clothing, *Proceedings of the Third International Symposium on Protection Against Chemical Warfare Agents*, Stockholm, Sweden 11–16 June 1989, pp. 25–31, Swedish Defence Research Establishment, UMEA.
141. Military Medical/NBC Technology (2001), *NBC Threat Specialist Q&A – Detection, Protection Rank High on the Army's Medical Technology Agenda*, **5**(3), pp. 20–24.
142. proceedings.ndia.org/dod_cb/Tuesday_PM/Wilusz.pdf.
143. Dolez PI and Vu-Khanh T (2009), Recent developments and needs in materials used for personal protective equipment and their testing, *International Journal of Occupational Safety and Ergonomics (JOSE)*, **15**(4), pp. 347–362.
144. Ramkumar SS, Sata U and Hussain M (2008), Chapter 8. Personnel protective fabric technologies for chemical countermeasures, In *Advances In Biological*

and Chemical Terrorism Countermeasures, Ed by Kendall RJ, Presley SM, Austin GP, and Smith PN, Woodhead Publishing Limited, http://www.crcnetbase.com/doi/abs/10.1201/9781420076554.ch8.

145. Sundarrajan S, Venkatesan A and Ramakrishna S (2009), Fabrication of nanostructured self-detoxifying nanofiber membranes that contain active polymeric functional groups, *Macromolecular Rapid Communications*, **30**(20), pp. 1769–1774.
146. Sloan JM, Napadensky E, Crawford DM and Elabd YA (2005), Nanostructured polymer membranes for chemical protective clothing, *Proceedings for the Army Science Conference (24th)*, 29 November–2 December 2005, Orlando, Florida, USA.
147. Vishnyakov AM and Neimark AV, Molecular Modeling of Nafion Permselective Membranes, www.dtic.mil/cgi-bin/GetTRDoc?AD=ada431689.
148. Napadensky E and Elabd YA, Breathability and Selectivity of Selected Materials for Protective Clothing, www.dtic.mil/cgi-bin/GetTRDoc?AD=ada425206.
149. Fuller TJ and Dobulis BT (2005), US Patent 7601449, sulfonated poly(phenylene sulfide) films as polyelectrolyte membranes.
150. www.airtecsafety.com/graph/shared/adv/GSSM-AirTec-Flyer.pdf.
151. www.inda.org/papers/enhancements2005/Sun.pdf.
152. www.iaea.org/inis/collection/NCLCollectionStore/_Public/.../27037866.pdf.
153. www.testsymptomsathome.com/PDF.../Smart_strip_brochure.p.
154. Ruthven DM (1988), Zeolites as selective adsorbents. *Chemical Engineering Progress*, **84**(2), 42–50.
155. Goltz HR, Jones KC and Tegen MH (1994), High Surface Area Polymeric Adsorbents for VOC Capture and On-Site Regeneration. *Paper Presented at the Air & Waste Management Association, 87th Annual Meeting and Exhibition*, Cincinnati, OH, June 1994.
156. http://www.dowwaterandprocess.com/products/adsorb.htm.
157. www.chemra.com/treversorb-polymeric-adsorbents.html.
158. Cireli A, Mutlu M, Kutlu B, Onar N, Kayatekin I and Celik E. (2006) Sol-gel coatings of plasma modified polypropylene fabric for gas defence. In: *Proceedings of the 3rd European Conference on Protective Clothing (ECPC) and Nokobetef 8*, Gdynia, Poland: CIOP-PIB; 10–12 May 2006.
159. Langston RD (1992), CBW protective clothing, US Patent 5112666.
160. Alexandroff EE (1986), Saratoga: carbon pellet technology in chemical warfare protective clothing, *Proceedings of the 2nd International Symposium on Protection Against Chemical Warfare Agents*, Stockholm, Sweden 15–19 June 1986, pp. 67–76, National Defence Research Establishment, Umea, Sweden.
161. Suri M, Rastogi D and Khanna K (2002), 'Development of protective clothing for pesticide industry: part II', *Indian Journal of Fibre & Textile Research*, **27**(9), pp. 259–265.
162. Rouquerol J, Avnir D, Fairbridge CW, Everett DH, Haynes JH, Pemicone N, Ramsay JDF, Sing KSW and Unger KK (1994), Recommendations for the characterization of porous solids, *Pure Applied Chemistry*, **66**(8), p. 1739.
163. Mays T (1999), Active carbon fibres, In *Carbon Materials for Advanced Technologies* Burchell T (Ed.), Elsevier Science Ltd, Oxford, UK, pp. 95–125.
164. Suzuki M (1994), Active carbon fibre: fundamentals and applications, *Carbon*, **32**, pp. 577–586.

165. Abbott WF (1962), US Patent 3053775, Process for carbonizing cellulosic textile materials.
166. Boucher EA, Cooper RN and Everett DH (1970), Preparation and studies of Saran-carbon fibres, *Carbon*, **8**(5), pp. 597–605.
167. Kaiser R, Kulczyk A, Rich D, Ronald JW, Minicucci J and MacIver B (2007). Effect of pore size distribution of commercial activated carbon fabrics on the adsorption of CWA simulants from the liquid phase, *Industrial & Engineering Chemistry Research*, **46**(19), p. 6126.
168. Inagaki M (2000), *New Carbons: Control of Structure & Functions*, Elsevier Science Ltd, Oxford, UK, pp. 126–146.
169. Branton P and Bradley RH, Effects of active carbon pore size distributions on adsorption of toxic organic compounds, *Adsorption*, 2011, 17: 293–301.
170. Zhang X, Wang Z and Gu X. (1991), Simple combination of biodegradation and carbon adsorption, *Water Research*, **25**(2), pp. 165–172.
171. Shmidt JL, Pimenov AV, Liebeman AI and Cheh HY (1997), Kinetics of adsorption with granular, powdered and fibrous activated carbon, *Separation Science and Technology*, **32**(13), pp. 2105–2114.
172. Bansal RC (1988) *Active Carbon*, New York, Dekker.
173. Derbyshier F, Jagtoyen M and Thwaites M (1995), Porosity in carbons, In Patrick J W (ed.), *Characterization and Applications*, Edward Arnold, London, p. 227.
174. Adams LB, Boucher EA and Everett DH (1970), Adsorption of organic vapours by Saran-carbon fibers and powders, *Carbon*, **8**(6), pp. 761–772.
175. Adams LB and Boucher EA (1978), Activation of Saran carbon fibers and powders with carbon dioxide, *Carbon*, **16**(1), pp. 75–76.
176. Arons GN and Macnair RN (1972), Activated carbon fiber and fabric achieved by pyrolysis and activation of phenolic precursors, *Textile Research Journal*, **42**(1), pp. 60–63.
177. Arons GN, Macnair RN and Erikson RL (1973), Sorption characteristics of activated carbon fabric, *Textile Research Journal*, **43**(9), pp. 539–543.
178. Arons GN and Macnair RN, Coffin LG and Hogan HD (1974), Sorptive textile systems containing activated carbon fibers, *Textile Research Journal*, **44**(11), pp. 874–883.
179. Arons GN and Macnair RN (1975), Activated carbon fabric prepared by pyrolysis and activation of phenolic fabric, *Textile Research Journal*, **45**(1), pp. 9–10.
180. Macnair RN and Arons GN (1978), Sorptive textile systems containing activated carbon fibers, In *Carbon Adsorption Handbook*, Cheremisinoff PN and Ellerbusch F (ed.), Ann Arbor Science, Ann Arbor, pp. 819–859.
181. Economy J and Lin RY (1973), German Patent 2246572.
182. Lin RY and Economy J (1973), Preparation and properties of activated carbon fibers derived from phenolic resin precursor, *Applied Polymer Symposium*, **21**, pp. 143–152.
183. Economy J and Lin RY (1976), Adsorption characteristics of activated carbon fibers, *Applied Polymer Symposium*, **29**, pp. 199–211.
184. Bailey A and Maggs FAP (1972), GBP 1301101.
185. Bailey A, Maggs FAP and Williams JH (1973), GBP 1310011.
186. Bailey A, Goodyear DJ, Maggs FAP, Schwabe PH and Williams JH (1974), GBP 1376888.

187. Burchell TD (1999), *Carbon Materials for Advanced Technologies*, Elsevier Science Ltd, Oxford, UK.
188. http://www.chemvironcarbon.com/en/activated-carbon-cloth/what-is-zorflex.
189. Gurudatt K, Tripathi VS and Sen AK (1997), Adsorbent carbon fabrics: New generation armour for toxic chemicals, *Defence Science Journal*, **47**(2), pp. 239–250.
190. Ilk MR, Jovanic PB, Radosevic PB and Rajakovic, LV (1995), Sorption of hydrogen cyanide onto activated carbon cloth impregnated with metallo-organic compounds, *Separation Science and Technology*, **30**(13), pp. 2707–2729.
191. Economy J, Daley M and Mangun C (1996), *Activated Carbon Fibers – Past, Present and Future*, ACS Preprints (Fuel Chemistry Division), **41**(1), pp. 321–325.
192. Capon A, Maggs FAP and Robins GA (1980), The mechanical properties of activated charcoal cloth, *Journal of Physics D: Applied Physics*, **13**(4), pp. 897–907.
193. Ko TH, Chiranairadul P, Lu CK and Lin CH (1992), The effects of activation by carbon dioxide on the mechanical properties and structure of PAN-based activated carbon fibers, *Carbon*, **30**(4), pp. 647–655.
194. Alcaiiiz-Monge J, Cazorla-Amoros D, Linares-Solano A, Yoshida S and Oya A (1994), Effect of the activating gas on tensile strength and pore structure of pitch-based carbon fibers, *Carbon*, **32**(7), pp. 1277–1283.
195. http://defense-update.com/features/du-2–07/infantry_armor_cooling.htm.
196. Holmes DA (2000), Performance characteristics of waterproof breathable fabrics, *Journal of Coated Fabrics*, **29**(4), pp. 306–316.
197. http://www.engardebodyarmor.com/tactical.htm.
198. Knapik J, Loads carried by soldiers: Historical Physiological, Biomechanical and Medical Aspects, *TECHNICAL REPORT T19/89*, U.S. Army Research Institute of Environmental Medicine, Exercise Physiology Division, Natick, MA 017M0-5007, June 1989.
199. Soule RG and Goldman RF (1969), Energy cost of loads carried on the head, hands or feet, *Journal of Applied Physiology*, **27**, pp. 687–690.
200. Givoni B and Goldman RF (1971), Predicting metabolic energy cost, *Journal of Applied Physiology*, **30**, pp. 429–433.
201. Pandolf KB, Givoni B and Goldman RF (1977), Predicting energy expenditure with loads while standing or walking very slowly, *Journal of Applied Physiology*, **43**, pp. 577–581.
202. Pandolf KB, Haisman MF and Goldman RF (1977), Metabolic energy expenditure and terrain coefficients for walking on snow, *Ergonomics*, **19**, pp. 683–690.
203. http://www.prospie.eu/info/.
204. Zhang X (2001), Heat-storage and thermo-regulated textiles and clothing, In *Smart Fibres, Fabrics and Clothing*, Tao X (ed.), Woodhead Publishing Limited, Cambridge, pp. 34–57.
205. Shim H, McCullough EA and Jones BW (2001), Using phase change materials in clothing, *Textile Research Journal*, **71**(6), pp. 495–520.
206. Bajaj P (2001), Thermally sensitive materials, In *Smart Fibres, Fabrics and Clothing*, Tao X (ed.), Cambridge, Woodhead Publishing Limited, pp. 58–82.

207. Chung H and Cho G (2004), Thermal properties and physiological responses of vapor-permeable water-repellent fabrics treated with microcapsule-containing PCMs, *Textile Research Journal*, **74**, p. 571.
208. Mondal S (2008), Phase change materials for smart textiles – an overview, *Applied Thermal Engineering,* **28**(11–12), pp. 1536–1550.
209. Zhou X (2009), Preparation and characterization of PEG/MDI/PVA copolymer as solid–solid phase change heat storage material, *Journal of Applied Polymer Science*, **113**(3), pp. 2041–2045.
210. http://www.pcmproducts.net/Solid_Solid_PCMs.htm.
211. Zalba B, Marín JM, Cabeza LF and Mehling M (2003), Review on thermal energy storage with phase change: materials, heat transfer analysis and applications, *Applied Thermal Engineering,* **23**(3), pp. 251–283.
212. http://www.pcmproducts.net/Phase_Change_Material_Products.htm.
213. Zuckerman JL, Pushaw RJ, Perry BT and Wyner DM (2003), Fabric coating containing energy absorbing phase change material and method of manufacturing same, US Patent 6514362.
214. Mondal S (2008), Phase change materials for smart textiles – An overview, *Applied Thermal Engineering*, **28**(11–12), pp. 1536–1550
215. Feldman D, Shapiro MM and Banu D (1986), Organic phase change materials for thermal energy storage, Solar Energy Materials, **13**(1), pp. 1–10.
216. B. Pause (2001), Building conditioning technique using phase change materials, US Patent 6230444.
217. Gonzale AR, Schofield RB and Hart SV (2005), *Third Status Report to the Attorney General on Body Armor Safety Initiative Testing and Activities*, National Institute of Justice, Office of Justice Programs, U.S. Department of Justice, August 24.
218. Zhang XX, Fan YF, Tao XM and Yick KL (2005), Crystallization and prevention of supercooling of microencapsulated n-alkanes, *Journal of Colloid Interface Science*, **281**(2), pp. 299–306.
219. *NIJ Standard–0101.04*, National Institute of Justice, Office of Justice Programs, U.S. Department of Justice
220. *www.vectranfiber.com/*
221. Nelson G (2001), Microencapsulation in textile finishing, *Review of Progress in Coloration,* **31**, pp. 57–64.
222. Fabien S (2011), The manufacture of microencapsulated thermal energy storage compounds suitable for smart textiles, In Marco Aurélio dos Santos Bernardes (eds), *Developments in Heat Transfer*, Rijeka, Croatia, InTech.
223. Zhang XX, Wang XC, Tao XM and Yick KL (2006) Structures and properties of wet spun thermo-regulated polyacrylonitrile-vinylidene chloride fibers, *Textile Research Journal*, **76**(5), pp. 351–359.
224. Zhang XX, Wang XC, Zhang H, Niu JJ and Yin RB (2003), Effect of phase change material content on properties of heat-storage and thermo-regulated fibres nonwoven, *Indian Journal of Fibre and Textile Research,* **28**(3), pp. 265–269.
225. Zhang XX, Wang XC, Tao XM and Yick KL (2005), Energy storage polymer/MicroPCMs blended chips and thermo-regulated fibers, *Journal of Materials Science*, **40**(14), pp. 3729–3734.

226. N Sariera and E Onder (2007), Thermal characteristics of polyurethane foams incorporated with phase change materials, *Thermochimica Acta*, **454**(2), pp. 90–98.
227. Sánchez-Silva L, Sánchez P and Rodríguez JF (2011), Effective method of microcapsules production for smart fabrics, In Marco Aurélio dos Santos Bernardes (eds), *Developments in Heat Transfer*, Rijeka, Croatia, InTech.
228. Choi K and Cho G (2011), Physical and mechanical properties of thermostatic fabrics treated with nanoencapsulated phase change materials, *Journal of Applied Polymer Science*, **121**(6), pp. 3238–3245.
229. http://www.innovationintextiles.com/articles/902.php.
230. http://www.innovationintextiles.com/outlast-and-protex-launch-new-development/.
231. Beginn U (2003), Applicability of frozen gels from ultra high molecular weight polyethylene and paraffin waxes as shape persistent solid/liquid phase change materials, *Macromolecular Materials and Engineering*, **288**(3), pp. 245–251.
232. Ye H and Ge XS (2000), Preparation of polyethylene–paraffin compound as a form-stable solid-liquid phase change material, *Solar Energy Materials and Solar Cells*, **64**(1), p. 37.
233. Stull JO (2000), Cooler fabrics for protective apparel, *Industrial Fabric Products Review*, March 2000, **76**(11), pp. 62–68.
234. Bryant YG and Colvin DP (1992), Fabric with reversible enhanced thermal properties, US Patent 5366801.
235. Colvin DP and Bryant YG (1993), Micro-climate cooling garment, US Patent 5415222.
236. Gao C, Kuklane K and Holmér I (2008), The heating effect of phase change material (PCM) vests on a thermal manikin in a subzero environment, *7th International Thermal Manikin and Modelling Meeting* – University of Coimbra, September 2008, USA.
237. B. Pause (2003), Nonwoven protective garments with thermo-regulating properties, *Journal of Industrial Textiles,* **33**(2), pp. 93–99.
238. http://defense-update.com/features/du-2–07/infantry_armor_cooling_PCM.htm.
239. http://www.hainsworth.co.uk/technical-and-industrial-textiles/firefighters-ppe-fabric.
240. Hainsworth T and Walker D (2004), US Patent 6,699,802.
241. Hainsworth T and Walker D (2005), US Patent 6,955,193.
242. Linz T (2007), Technologies for integrating electronics in textiles, In: *4th International Avantex Symposium for Innovative Apparel Textiles*. Frankfurt am Maine, Germany: Messe Frankfurt Exhibition.
243. http://defense-update.com/products/i/IBA.htm.
244. http://defense-update.com/products/b/breeze.htm.
245. http://defense-update.com/products/f/ffw-atd.htm.
246. http://defense-update.com/products/b/bvs.htm.
247. http://defense-update.com/products/m/mcs-cooling.htm.
248. Performance Apparel Markets, 1st quarter 2006.

4
High performance textiles for heat and fire protection

A. R. HORROCKS, University of Bolton, UK

DOI: 10.1533/9780857099075.144

Abstract: The principles of textiles requiring the properties of heat and fire protection are introduced within the context of the fundamental physical and chemical characteristics of component fibres and textile fabrics coupled with the underlying fire science requirements to understand the main textile design issues required. The large range of high performance fibres available for fire resistant end-uses is discussed in terms of generic chemical differences including the need for thermoset, totally aromatic or inorganic (ceramic) structures. The need for and types of fire resistant coatings and treatments are also introduced. Of especial relevance are the evolving technologies for conferring heat and fire resistance and these include the use of intumescent treatments and the potential usefulness of nanotechnologies, which has received much attention at the research level during the last 10 years. Typical applications of heat and fire resistant textiles are those where regulations usually demand their use and these include protective clothing, contract and domestic furnishing fabrics, fire barrier textiles (e.g. theatre fire curtains, insulative textiles) and military applications and, in particular, many textiles used in mass transport systems, such as commercial airlines, luxury cruise liners and the current generation of fast trains. The chapter concludes with a brief discussion of two projects undertaken in the author's laboratories that illustrates two recent challenges posed by the needs of the commercial aerospace industry. The first outlines how decorative textiles comprising very expensive and aesthetically valued fabrics might be rendered fire resistant to the levels required by the regulations and yet preserve their exotic characteristics. The second shows how the development of textile composites for use as an acoustic and fire resistant barrier was successfully undertaken to meet the stringent and new international regulatory requirements for commercial airliner fuselage fire requirements.

Key words: protective textiles, fire resistance, heat resistance, aramid, thermoset, ceramic, intumescent, nanotechnology.

4.1 Introduction

The need to develop flame resistant textiles and related materials, such as composite building structures based on wood and other plant derivatives, has been realised for thousands of years with reported treatments in Roman times being examples. While the Roman civilisation was aware of the need for fire regulations,[1] their use of asbestos[2] in woven materials may be seen perhaps as the first development of a high performance material to resist fire and heat. In fact, the Romans referred to asbestos as 'linum vivum' or 'immortal linen' because it was used to wrap the bodies of the dead before cremation so that the ashes could be preserved. While today we design textiles capable of protecting our living bodies against damage, it is ironic that the Romans were interested in using heat protective asbestos to enhance the preservation of the dead! Over one thousand years later, discussions at the Royal Society in 1685 regarding its remarkable resistance to fire referred to asbestos as 'incombustible linen' (sic).[3]

In more recent times, use of chemical treatments has been made in order to reduce the flammability hazard associated with the more flammable textiles such as cotton and linen. The significant patent of Wyld[4] in 1735 describing a finishing treatment for cellulosic textiles using alum, ferrous sulphate and borax, and Gay-Lussac's first systematic study of the use of flame retardants in 1821[5] have formed the basis for the more recent and modern approaches to developing heat and fire resistant textiles.

Thus the foundations were laid for the generation of the present range of what might be termed today as high performance, heat and fire resistant textiles. Protection may be offered to a wearer not only by means of clothing that insulates the human body against the effects of an external heat source, but also by use of barrier textile structures. These may be present as interlinings in clothing assemblies and around seating fillings in domestic and contract furnishings. They may also be included as a fire protective component in a building or other confined space, such as an aircraft interior. Thus, all fire and heat resistant textiles function by having a reduced tendency to ignite in the first instance, to burn slowly and ideally self-extinguish once ignited, and they must protect and prevent underlying materials from thermal damage or ignition. Examples of these include protective clothing for industrial personnel and firefighters, interior furnishings including carpets and barrier fabrics specifically designed to contain potential fire hazards such as furnishings in public transport, theatre fire curtains, laggings for pipework, textile structures for panel, structural beam, fire door and bulkhead protection and intumescent packing and sealing materials.

4.1.1 Fire statistics and exemplar regulations

The need for fire protection is usually driven by regulations, some of which may relate to overarching legislation. As this is a subject beyond the scope of this chapter, it is sufficient to provide just a few examples by way of illustration. Noteworthy in the UK is the legislation[6] introduced in 1988 that required all domestic soft furnishings and furniture (apart from curtains and drapes) to pass the so-called cigarette and match tests defined in BS 5852:Part 1:1979. While the details of these and similar procedures have been described elsewhere,[7] the essence of the regulations requires that the outer fabric, when tested over an actual sample of combustion-modified filling material used in the final product, is shown to resist a cigarette ignition source. However, when the same fabric is tested over a sample of flammable, unmodified polyurethane foam filling, the assembly must resist ignition to a simulated match flame in contact with the fabric surface for 20 s. In this way, the furnishing fabric is being assessed for its ability to protect an underlying flammable material from a defined ignition source as well as its being part of a totally ignition-resistant composite. These regulations have given rise to a whole new family of flame retardant, protective fabrics for use in the UK furnishing industry. Furthermore, and more importantly from the regulatory point of view, it has been estimated that their imposition saved on average about 140 lives in the UK each year, and that this has been a significant cause of the reduction of fire deaths in dwellings, which have reduced from a figure close to 700 per annum during the 1980s to less than 400 in recent years.[8]

In terms of regulatory requirements for protection for commercial transport, all commercial airlines, land public and maritime transport must abide by stringent textile flammability requirements for internal furnishings applied at national or international levels. While every textile present must usually pass a basic ignition resistance requirement based on a simulated match flame, there is the need to provide protection to underlying fillings in upholstered furnishings and bedding by incorporation of fire barrier properties within the construction. These fire barrier or blocker fabrics are usually tested under conditions that realistically attempt to replicate their application thermal hazards. A prime example is the accepted US Federal Aviation Administration (FAA) kerosene burner test, used to test commercial aircraft seating, Federal Aviation Regulations (FAR) 25.863(c).[9] Here a full seat mock-up is exposed to a burner with a heat flux of 115 kW/m^2 for 120 s and fabric/seat assemblies must suffer a weight loss of less than 10% and meet specified maximum and average burn length criteria while not sustaining afterburning and smouldering for more than 5 min. Fabrics that pass the tests may have a variety of structures and weights, depending on whether aluminised or not and the choice of fibres present. For instance, fabrics comprising flame-retarded wool, oxidised acrylics, aramid, arimid and glass in various blends and in woven or non-woven structures with area densities from 250 to 400 gsm are typically used.

In other cases, where a blocker is part of a larger system, such as part of a building or transport system, then the test will be related to the related building, motor vehicle, rail vehicle, aircraft or ship fire testing regime.[10] While building fire tests relating to national standards requirements, for aircraft and ships there are overarching agencies responsible for these tests with international standing, such as the US FAA and the International Maritime Organisation (IMO).[10]

It is probably evident from the above introductory discussion, and important to realise, that the whole issue of heat and fire protection with respect to textiles is a complex area involving knowledge of elements of fire science, flame retardant treatments, development of heat and fire resistant fibres and derived textile structures and the inter-relationships between regulations, applications and markets. The bibliography lists a number of prime sources of information in these areas, and for an overview of textiles for fire and heat protection the reader should refer in particular to a recent article by this author.[11] In this chapter, the focus will be only on high performance textiles in which fire and heat protection are essential requirements.

4.2 Principles of textile heat and fire protection

Heat and fire protection afforded by any material is environmental and time dependent. Levels of protection are therefore relative rather than absolute and so, for example, should be qualified by time of exposure to a heat source having a defined character (e.g. radiant versus flame), intensity and temperature as well as a measure of access to oxygen. Fire primarily involves the presence of a flame, which is the location of an intense chemical reaction in which a gaseous fuel is undergoing oxidation with oxygen from the surrounding air. This reaction zone gives out energy sufficient to cause emission of light, and sufficient heat to yield temperatures typically, for burning textiles, in the range 600–1000°C (see Table 4.1). Most fires that cause loss of human life occur in confined spaces, such as dwellings where oxygen is initially consumed, and so the local levels in the atmosphere may reduce below the normal 21 vol % level. Such a vitiated atmosphere may lead to reduced heat flux rates and even self-extinction of flames, and fire doors in buildings are present to reduce the rate of spread of a fire and its intensity via vitiation. However, to create a worst possible situation, most fire tests assume that oxygen is freely available and so fire testing is undertaken to represent or model a real-fire hazard in terms of intensity and for a time that is assumed to allow a reasonable reaction and escape time. Typically, a fire resistant door may be expected to resist a given heat flux for at least 30 min, while resistance times as long as 2 and 4 h are required for fire curtains or doors used to seal off areas within buildings during a fire (as defined, for example, in British Standard BS EN 1634-1 *(Fire resistance tests for door and shutter assemblies. Fire doors and shutters)*). Thus, any textile protective element of such a system will be chosen depending upon the magnitude of the applied heat flux and the duration that resistance is required. Typically, heat fluxes are chosen to replicate known hazards; those associated with a domestic room at the time of flashover, when all the contents are fully combusting, is equivalent to about 50 kW/m^2, whereas a larger room such as a public theatre may generate up to 100 kW/m^2 at flashover. Flashover is the condition achieved in a confined space when the heat flux emitted by a fire is sufficient to cause the almost immediate ignition of all combustible items in that space. It is at such a moment that fires often become out of control, and their hazard to life reaches a maximum. Fuel or pool fires may generate as much as 150 kW/m^2 heat intensity or higher.

Heat alone is often less damaging to a textile material than flame in the shorter time intervals, because of the absence of associated flame chemistry that will promote more rapid ignition of a previously cool surface. Radiant energy impinging up on a surface must first raise it sufficiently to initiate the chemistry associated with ignition and combustion. At heat fluxes typically of 25 kW/m^2 or less, heat is normally insufficient to cause ignition of a polymeric material such as a textile but may promote physical transitions such as softening and melting as well as the onset of degradation or pyrolysis (see below). However, such transitions may cause deformation and weakening of a fabric at best, and complete disintegration at worst. Consequently, heat resistant textiles are those designed to resist the effects of lower heat fluxes for considerable times and may be associated with service temperatures of up to 250°C, typical of those experienced by some hot gas filters in industrial processes. Higher heat resistance is only possible if component fibres do not thermally degrade and burn at temperatures typically above 500°C, and ceramic fibres including glass are typical of this group. Consequently, insulating textile media used to surround the combustion chambers within jet engines, and thus protect surrounding aircraft structures from the intense heat generated, comprise ceramic fibres such as silica that can withstand temperatures as high as 1100°C or so for many hours, if not months.

In the case of protective clothing, such as that designed for fire fighters for example, there are a number of issues that must be considered when designing textiles and tests for assessing their suitability. Of prime importance are the intensity of the hazard (e.g. 50 kW/m^2 in a typical domestic environment versus 100 kW/m^2 in an industrial or public environment) and the maximum time for which protection must be sustained. The latter, in the case of a firefighter, is determined by the time that the wearer may work prior to succumbing to heat fatigue in a low heat flux environment (e.g. 20–30 min) and the need to breathe unvitiated air (using breathing apparatus) at a reasonable (< 50°C) temperature in a high risk area, which may be of the order of less than 1 min. In the case of a racing car driver, the maximum escape time from a burning vehicle may be assumed to be 30 s or so before asphyxiation and/or lung damage ensues. Thus, in designing protective clothing there is no 'one size fits all' and the choices of fibres, yarn and fabric structure, as well as garment design, are determined by the end-use hazard levels. In developing test methods for protective textiles, individual components may be assessed for their potential ignition resistance, flame retardancy and heat transfer, as exemplified by the suite of tests within the international testing standard for firefighters' composite clothing, BS EN 469:2005. However, recently real-fire simulations using instrumented manikins have become well established and standardised; for example, in BS ISO 13506:2008 (an option within BS EN 469), whole garment assemblies may be exposed to heat fluxes, typically of 84 kW/m^2 (200 cal/cm^2) for 8 s and extent of burn intensity to an underlying simulated human torso in terms of first, second and third degree burn percentage torso areas being defined.[11]

Because of the relative character of the term 'fire resistance', 'flame retardancy' is often preferred and the whole science and technology of reducing the flammability of textiles is usually referred to as such. Thus the use of flame retardant additives and treatments enables the ease of ignition and rates of burning of textile structures at fibre, yarn, fabric and final product levels to be reduced.[12] Flame retardant agencies never prevent a previously flammable textile from degrading and forming some combustible fuel; they merely interfere with the so-called combustion cycle (see below) in a manner that more or less suppresses the tendency of the material to maintain a self-propagating flame after an ignition source has been removed. In the presence of an igniting source there is sufficient energy to initiate and maintain the burning process, albeit at a reduced rate. Once the source is removed, the flame retardant agent is usually sufficient to cause eventual extinction of any remaining flames and afterglow.

4.2.1 Fundamental physics and chemistry of heat and fire protection

Thermal protection relates to the ability of textiles to resist conductive, convective or radiant thermal energy, or a combination of these. As stated above, for example, a flame constitutes a convective oxidative chemical reaction zone in which the energy is contained within the extremely hot gas molecules and particulates, including smoke. When a flame impinges upon a textile surface, not only is it subjected to the high temperature of the flame (see Table 4.1),[11] but also to the chemical intermediates of the reaction zone, which may increase the likelihood of its degradation and ignition. Thus the selection of textile fabrics exposed to flame hazards, for example a firefighter's tunic, should take account of both the temperature and reactivity of an impinging flame source.

Radiant thermal energy, however, is absorbed by the molecular structure of the surface fibres and any coatings present, raising the temperature, which, if high enough (typically > 300°C), promotes thermal degradation (or pyrolysis) and even ignition. In a real fire, it is usually the radiant heat present that causes it to spread rapidly, by heating up and igniting materials some distance away from the flame itself, culminating in an eventual flashover if present in a confined space. Therefore, heat and fire resistant textiles having the highest performance requirements are often rendered as heat reflective as possible by the use of metallised coatings such as aluminium, and even stainless steel.

Finally, conductive heat protection is required for textiles that may come into direct contact with a heat source other than a flame. Major threats here include those met by metal industrial workers, who risk contact with hot metal tool handles and molten metal splashes. For this reason, not only must the thermal insulating characteristics of the textile be paramount, but also surfaces which minimise contact, for instance by resisting wetting by molten metals, must be considered. However, in many thermally hazardous environments, a combination of conduction, convection (or flame) and radiation may be operating in concert, and usually the last two are associated with flame sources in particular.

Once heat is absorbed by a textile fibre, it may promote physical or chemical change, or both. All organic fibre-forming polymers will eventually thermally degrade at or above a threshold temperature, often defined as the pyrolysis temperature, T_p, often influenced by the presence of oxygen in the air. Typical values of T_p are listed in Table 4.1 for the more common and selected high performance fibres. Once heated above this temperature in air, ignition at a higher temperature, T_i, usually follows and these are also listed in Table 4.1.[11] However, flame temperatures are usually higher still, and where available, are included in the table for those fibres that burn in air.

Some fibres undergo physical change as an initial softening, followed by melting at temperatures below those required for thermal degradation. The former is usually defined by a second order temperature, T_g, and the latter by a melting temperature, T_m. For the conventional textile fibres such as polypropylene, polyester and nylons (or polyamides) 6 and 6.6, the relatively low temperatures at which these physical transitions occur means that textiles they comprise offer little or no protection against thermal energy. For high performance applications, it is essential that both these transitions are as high as possible so that thermoplastic effects are minimised and as Table 4.1 shows, for non-thermoplastic fibres, T_m values and even T_g values become greater than respective T_p values.

4.2.2 Fire science principles relevant to textiles

Fire scientists tend to classify materials as being either thermally thick or thermally thin. By thermally thin means that heat absorbed on one surface of the material will penetrate its thickness sufficiently rapidly, so that there will be no significant temperature gradient through the material depth.[13] Textiles in the form of single-layered fabric assemblies are usually dimensionally thin and hence could be classed as thermally thin materials.

Table 4.1 Thermal transitions and flame temperatures of common, higher performance fibres[11]

Fibre	T_g (°C) (softens)	T_m (°C) (melts)	T_p (°C) (pyrolysis)	T_c (°C) (ignition)	T_f, (°C)*	LOI, vol %
Wool	–	–	245	570–600	680, 825 (v)	25
Cotton	–	–	350	350	974 (h)	
Viscose	–	–	350	420	–	18.9
Nylon 6	50	215	431	450	–	20–21.5
Nylon 6.6	50	265	403	530	861 (h)	20–21.5
Polyester	80–90	255	420–447	480	649 (h), 820 (v)	20–21
Acrylic	100	> 220	290 (with decomposition)	> 250	910 (h), 1050 (v)	18.2
Polypropylene	–20	165	470	550	–	18.6
Modacrylic	< 80	> 240	273	690	–	29–30
PVC	< 80	> 180	> 180	450	–	37–39
Meta-aramid (e.g. Nomex)	275	375–430 (decomp)	425	> 500	–	29–30
Para-aramid (e.g. Kevlar)	340	560 (decomp)	> 590	> 550	–	29

Note: *Flame temperatures recorded only for fibres that burn in air; h is for horizontal and v is for vertically downward burning fabrics (Rebenfeld, Miller and Martin, 1979).

A specimen is considered to be thermally thick when the heat wave penetration depth is less than the physical depth such that increase in physical thickness does not influence the time to ignition under a given set of conditions.[14] The challenge for designers of protective textiles requiring the highest levels of fire and heat protection is one of maximising the thermal thickness of the final structure. This usually involves maximising actual thickness and hence mass of the final product, which then presents problems of excessive weight (and discomfort from poor moisture transmission and overheating in the case of clothing) and cost. In practice, however, it is the air entrapped within the yarn and fabric structures that provides the thermal insulating medium and so, while thickness of the overall fabric is crucial, use of lofty yarn and fabric structures to minimise area and specific densities is of paramount importance if overall excessive weight is to be avoided.[11]

4.3 Routes to achieving fire and heat protection

For the purpose of this discussion, high performance requirements will be arbitrarily defined as those in which a textile must resist a heat or fire hazard for a significant period and so protect against thermal threats yielding temperatures above 200°C for periods of greater than 60 s. Therefore, non-thermoplastic fibres only should be used, and these comprise two broad groupings, the first of which are flame-retarded versions of non-thermoplastic fibres based on naturally-occurring polymers. Thus, flame-retarded cotton, viscose and wool are included. The second group includes all fibres which are inherently heat resistant, with T_g values largely in excess of 200°C and which will also coincidentally have acceptable levels of flame retardancy and hence fire resistance. The former will yield textiles that resist defined ignition sources and, if they do ignite when the source is present, have reduced burning rates and times once the source has been removed.[12] Ideally, these fibres, when heated, convert to char and so the textile now provides a carbonaceous char replica of itself that is extremely flame resistant and offers a barrier to underlying surfaces.[15]

4.3.1 Heat and fire resistant organic polymeric fibres and blends

Heat resistant fibres, however, are those having chemical structures that are little changed physically or chemically by temperatures above the 200°C, and even the 300°C, levels and, in the case of ceramic fibres, above 1000°C. For textiles used in high temperature industrial processes, such as gas and liquid filtration, long term exposure to temperatures of about 100°C is often required, but not all these fibres are used in thermally protective applications.[16] However, in long term exposure thermally protective applications, we need to be able to define maximum service life temperatures, and these are listed in Table 4.2 for selected heat resistant fibres.

Flame resistance may be conveniently measured as a limiting oxygen index value,[17] which for textiles with values greater than 21 vol % (the percentage volume concentration of oxygen in air), indicates less vigorous burning than if less than 21. Fibres and textiles having limiting oxygen index (LOI) values > 26–28 vol % tend to be flame retardant in air and will pass simple vertical fabric strip tests and, if non-thermoplastic and char-forming, will offer excellent barrier properties. Those fibres having values in excess of 30 vol % are highly flame retardant and it is noteworthy that those listed in Table 4.2 all fit into these flame retardant categories. Bourbigot *et al.* have produced one of the few recent comparisons of the fire performance of a number of these high temperature and fire resistant polymer-based fibres based on cone calorimetric data.[18] If the data sets are expressed as fire growth index (FIGRA) values, which give a better measure of relative fire propagating behaviour, the increasing fire growth property is in the order:

PBO < Kynol ~ PPTA < Technora < Oxidised acrylic

Table 4.2 Maximum service lifetimes for heat resistant fibres in thermally protective textiles[16]

Fibre genus	Second order temperature (°C)	Melting temperature (°C)	Onset of decomposition (°C)	Maximum, continuous use temperature (°C)	LOI, vol %
Melamine -formaldehyde	NA	NA	370	190	32
Novoloid	NA	NA	> 150	150/air; 250/inert	30–34
m-aramid	275	375–430 (decomp)	425	150–200	28–31
p-aramid	340	560 (decomp)	> 590	180–300	29–31
Copolymeric p-aramid	–	–	500	200–250	25
Arimid (P84)	315	–	450	260	36–38
Aramid-arimid	< 315	–	380	NA	32
Semicarbon	NA	NA	NA	~200/air	55
PBI	> 400	NA	450/air; 1000/inert	~300 (est)	> 41
PBO	–	–	650; > 700/inert	200–250(env)	68
Glass	650–970	NA	850	> 300	NA
Silica-based (e.g. Quartzel®)	–	> 1700	–	1200	NA
Alumina-based (e.g. Saffil®)	–	> 2000	NA	1600	NA
Alumina-silica-based (e.g. Nextel®)	–	> 1800	NA	1260–1370	NA
Silicon carbide (e.g. Nicalon, Nippon Carbon)		2650–2950	NA	< 1800	NA
Basalt (e.g. Basaltex)	–	1300–1700	NA	650–850	NA

Notes: NA = not applicable; (decomp) = with decomposition; (est) = estimated value; (env) = depends on chemical inertness of the environment.

The decreasing order in terms of LOI as a fire measure is:

PBO > Oxidised acrylic > < PPTA ~ Kynol > Technora

Both measures used to order these fibre types indicate that phenylene benzobisoxazole (PBO) is the least flammable and hence most protective fibre, while the respective orders for the other fibres differ. However, given that cone calorimetry is recognised to be a reasonable simulation of a fire, then the FIGRA order is probably closer to defining the relative fire protective behaviour of these fibres.

Absent from this group is the important series of inorganic and ceramic fibres, which have no tendency to burn and so may be used in applications where high temperature and chemical reactivity of the environment combine to determine respective durabilities. These fibres include the various forms of glass, silica and alumina available, although more exotic ones like stainless steel, boron nitride and silicon carbide are available for specialist end-uses. Such extreme thermal protection as shown in Table 4.2 is a desirable feature in applications such as furnace linings or hot component insulation in car exhaust catalysts or around combustion chambers in jet engines where working temperatures and occasional flash temperatures are in excess of 500°C and even 1000°C in extreme circumstances. However, while fire resistance is an intrinsic feature of these inorganic fibres and textiles, their poor aesthetics limits their use to these extreme technical applications, although glass- or ceramic-cored, organic fibre-wrapped yarns may be used to improve this feature.

Each of these major fibre groups has been described elsewhere[11,16] and so their major features only are presented below. The main groupings of these inherently heat and fire resistant fibres may be divided into the following, namely, the thermosets, the aramids and arimids, the polybenzazoles, the semicarbons and inorganics. Not only are the generic chemistries similar within each grouping, but their properties and potential application suitabilities are similar.

Thermoset polymeric fibres

Typified by the melamine-formaldehyde fibre Basofil® (BASF) and the phenol-formaldehyde (or novoloid) fibre Kynol® (Kynol GmbH), they have in common that, when heated, they continue to polymerise, cross-link and thermally degrade to coherent char replicas. Derived chars have especially high flame and heat resistance, as a consequence of their high carbon content, although their relatively low strength prevents their being processed easily into yarns and so they are more often incorporated into non-woven fabric structures. In addition, their inherent colour (pink for Basofil® and gold for Kynol®) ensures that they are usually used as a barrier fabric and not in face fabrics, although the melamine-formaldehyde structure in Basofil® does allow the fibre to be dyed with small molecular disperse dyes. Respective thermal properties are listed in Table 4.2, which indicates very similar high levels of heat and flame resistance. Typical end-use applications of both Basofil ®and Kynol® in thermal protection include fire blocking, heat insulating barriers, and heat and flame protective apparel. Fibres may be blended typically with meta- and para-aramid fibres to improve tensile properties including strength and abrasion resistance in both non-woven felts and fleeces for fire blocking aircraft seat fabrics and firefighter's clothing, for example. Such fabrics may be aluminised to improve heat reflection and hence fire performance.

The aramid and arimid families

This group is perhaps the best known and exploited of all the inherently heat and flame resistant fibres developed since 1960, and all members of this group are typified by having thermal resistances in excess of 300°C for short-term exposures and high levels of inherent flame resistance (see Table 4.2).

The most commonly used thermally resistant aramids are based on a meta-chain structure as typified by the original Nomex® (Du Pont) fibre and more recent commercially available fibres, e.g. Conex® (Teijin), Apyeil® (Unitika) and Fenilon® (Russia), in addition to those having modified tensile properties (e.g. Inconex, Teijin) and antistatic properties (Apyeil-α, Unitika). Nowadays, these fibres have improved dyeing properties and are available in a full colour range. They have the advantage of acceptable 'nylon-like' tensile and physical properties, they have minimal thermoplastic characteristics with second order transition temperatures (T_g) of about 275°C and they have an ill-defined melting point accompanied by thermal degradation starting at 375°C (see Tables 4.1 and 4.2). They are ideal for use in protective clothing. Improvements in thermal performance in terms of increased char strength have been achieved by blending with small amounts of para-aramid fibres (e.g. Nomex III contains 5% Kevlar) and so are more suitable for applications where direct heat exposure is possible, e.g. firefighters' station uniforms, coveralls, jackets, trousers, gloves, flight suits or tank crew coveralls. Other variants include antistatic and moisture management properties (Nomex Comfort) and one specifically designed for firefighters' clothing (Nomex Outershell).

The para-aramids are typified by Kevlar® (DuPont) and Twaron® (Teijin) and are based on poly (para-phenylene terephthalamide) or PPTA. While having enhanced tensile strengths and moduli as a consequence of the extreme symmetry of their polymer chains and hence order or crystallinity, they also have improved thermal performance with a second order transition temperature of about 340°C and decomposing above 590°C (see Table 4.1). However, since thermal degradation is similar to that occurring in the meta-aramids, the LOI values are also similar at 30–31 vol %. Their higher cost, poorer textile processing properties and higher modulus ensure that use in applications such as protective textiles is limited to 100% contents only when performance demands are exceptional, and so more often than not they are used as minor blend components (e.g. 5% in Nomex III®).

A copolymeric derivative of the para-aramid fibres was introduced by Teijin in 1985 under the trade name Technora®. This fibre is based on the 1:1 copoly(terephthalamide) of 3,4′-diaminodiphenyl ether and p-phenylene-diamine[18] and is claimed to have a much higher chemical resistance than PPTA, as well as increased abrasion and steam resistance, which are useful properties in many protective applications. Technora® has a decomposition temperature of about 500°C and other properties are comparable with PPTA although its LOI value is slightly lower at 25 vol %.

Of a number of reported arimid fibres, only the example P84® introduced by Lenzing during the mid 1980s and now produced by Inspec Fibres (USA) has been commercially exploited. As Table 4.2 indicates, these fibres have superior thermal properties to aramid and so find use in applications including protective outerwear, underwear and gloves, either as 100% or blended with lower cost fibres such as flame retardant viscose. For instance, a 50/50 P84/Viscose FR (Lenzing) blend is available for knitted underwear with high moisture absorbency. Spun dyeing of P84 fibres enables their natural gold colour to be replaced by those often demanded by customers who may require more appropriate and bright safety colours.

The final member of this grouping is the poly (aramid-arimid) fibre, Kermel®, which was produced initially by Rhone-Poulenc of France in 1971 and is now produced by Rhodia Performance Fibres. Its overall properties are very similar to those of the meta-aramids, and in 1993 a so-called third generation Kermel was announced, claimed to have superior colouration properties. Typical of this group is its poor UV stability, and so it must be protected from intense radiation sources. It competes in protective clothing markets where again it is used as 100% or as blends with other fibres, including fire retardant (FR) viscose and wool. Composite yarns with high modulus aromatic fibres such as the poly (para-aramids) have yielded the modification Kermel HTA®, a yarn with a para-aramid core (35%) and a Kermel fibre wrapping (65%) to give improved abrasion resistance.

Polybenzoxazole group: polybenzimidazole and polybenzoxazole fibres

These fibre-forming polymers are so-called 'ladder polymers' and are essentially wholly aromatic polymer chains. The two examples commonly available commercially are the polybenzimidazole PBI® (Celanese) with the full chemical name poly (2,2′-(m-phenylene)-5,5′-bibenzimidazole), and the polybenzoxazole, Zylon® (Toyobo) with the full chemical name poly(paraphenylene benzobisoxazole) and generic acronym, PBO. Their similarity in polymer chain structures and high degree of chain rigidities gives both of these fibres their superior thermal properties, as shown in Table 4.2, with thermal degradation temperatures well in excess of 400°C and superior LOI values well over 40 vol %.

PBI® has been introduced to the commercial markets only during the last 20 years or so, in spite of its development during the early 1960s. The current PBI fibre is a sulphonated version of the above polymer, which improves shrinkage resistance at high temperature. Like many highly aromatic polymers it has an inherent colour – bronze – and cannot be dyed. The fibre is more often than not used as a blend, and one well-known blend is PBI Gold® in which a yarn is spun with both PBI and Kevlar in a 40/60 blend. This gives rise to gold-coloured fabrics with fire protective properties claimed to be superior even to those made from Nomex III®. This blend is now well established in firefighters' clothing, both in the USA and UK, for outer shells, underwear, hoods, socks and gloves. Other uses include industrial workwear, aluminised proximity clothing, military protective clothing and fire barrier/blocker applications but, because PBI is several times as expensive as the meta-aramids, this superior performance comes at a price.

Zylon® or PBO is a more recently developed fibre than PBI® and has outstanding tensile properties, as well as thermal and fire properties superior to any of the polymer-based fibres mentioned in this chapter (see Table 4.2). While there are at least two variants of the fibre, Zylon-AS and Zylon-HM, of which the latter has the higher modulus, both have the same thermal and burning parameter values. Principal examples of thermally protective textiles include heat protective clothing and aircraft fragment/heat barriers, where its price, similar to that of PBI, restricts its use to applications where strength, modulus and fire resistance are at a premium.

Semicarbon fibres

The semicarbon fibres include any in which the structure is essentially carbon while retaining acceptable textile properties, unlike true carbon fibres.[19] Within the group, the oxidised acrylics represent the sole commercial group and are produced following controlled, high temperature oxidation of acrylic fibres during the first stages of carbon fibre production. Since the early 1980s, a number of commercial versions were announced including Celiox (Celanese), Grafil O (Courtaulds), Pyron (Stackpole), Sigrafil O (Sigri Elektrographit, now SGL) and Panox (SGL UK Ltd., formerly R K Textiles), many of which are now obsolete. Current examples include Panox ® (SGL

Carbon Group), Pyromex® (Toho Rayon) and Lastan® (Asahi). Their low tenacity creates the problem of ease of processability for these weak fibres, although they can be spun into yarns by the woollen system. Thus they are produced as a continuous tow that is stretch-broken by conventional means for eventual conversion into coarse woollen-type yarns. The limiting oxygen index is typically about 55 vol % and so fabrics are extremely thermally resistant, giving off negligible smoke and toxic gases when subjected to even the most intense of flames. Unfortunately, the fibres are black and so are rarely used alone except in military and police coverall clothing where the colour is a bonus. Therefore, oxidised acrylic fibres are usually blended with other fibres, typically wool and aramid in order to dilute the colour and introduce other desirable textile properties. Because of their extreme fire resistance, and lower cost than PBI and PBO, they find applications as blends in anti-riot suits, tank suits, FR underwear, fire blockers for aircraft seats and heat resistant felts (insulation), hoods and gloves. When aluminised, they are very effective in fire entry/fire proximity suits.

Blends of high performance organic fibres

Blends of these fibres have been commercially exploited for a number of years, with aramid-FR viscose being perhaps one of the first attempts to reduce the cost of the final fabric while maintaining a high level of fire protection. As also mentioned in the foregoing sections, blends of meta-aramids with para-aramid or PPTA (e.g. Nomex III®, Du Pont; Kermel HTA®, Rhodia) and PBI and para-aramid (PBI Gold®, Celanese) have been developed to give a balance between the fire properties of both with the higher modulus and strength of the para-aramid. If any blends have been produced to generate synergies of fire resistance and other properties while, perhaps, improving other desirable features, very few includes cost. Bourbigot *et al.*[20] have recently reported a number of possibly synergistic blends made by mixing yarn by yarn of wool with PPTA improved the flame retardancy and generally improved thermal stability of the whole fabric. The suggested mechanism was that the molten char of wool coats adjacent para-aramid fibres, hindering the diffusion of oxygen to them and so negating their sensitivity to oxygen from the air and consequent thermal oxidative degradation. Subsequent research[21] has indicated that intimate blends of wool/PPTA show synergy when only 30% or greater PPTA is present, as opposed to 70% or more in the previously blended yarn results. Furthermore, synergy was noted in wool/Technora® blends, which can show reduced peaks of heat release rates with respect to 100% Technora®; some wool/PBO blends show similar encouraging results, suggesting that the possibility of enhanced fire performance at reduced cost and improved aesthetics is feasible commercially using these interesting blends.

4.3.2 Heat and fire resistant inorganic fibres

The best established are the family of glass fibres, which has been recently exhaustively reviewed[22] and continues to service many protective needs where heat and fire resistance are essential. The prime use has usually focussed on their reinforcing quality, and so the fibres, because of an inherent brittleness and poor textile general character, are used as reinforcing elements in flexible textile as well as rigid composite structures. Where glass fibre assemblies are used in non-reinforcing applications, they are usually contained to prevent fibre damage and loss of product coherence. Their use as filter media for high temperature gas and liquid filtration, as battery separators, and as fire and acoustic insulation in aircraft and other transport systems, are examples here. To respond to these different application needs, a wide range of glass compositions is available to suit many textile applications, and fibres made from various compositions have softening points in the range 650–970°C. When heated above 850°C, devitrification and partial formation of polycrystalline material occurs as the former glass fibres become more similar in character to ceramic materials. This devitrified form melts at 1225–1360°C, which is high enough to contain most fires for several hours. Recent examples of their use in high performance applications include their use in flexible roofing for sports and similar constructions – the recently developed roof of the Olympic Stadium in Berlin with a surface area of 42 000 m^2 is a recent example, in which the upper roof is made up of a highly tear-resistant fibreglass fabric coated with polytetrafluoroethylene (PTFE), to give a lifespan of at least 30 years, excellent fire resistance and a self-cleaning surface.[23]

Ceramic fibres on the other hand, have even poorer textile properties, are often more expensive, and are mostly used as refractory fibres as insulating and fire barrier materials for applications requiring resistance to temperatures of at least 1000°C for prolonged periods. They tend to have polycrystalline structures, hence their exceptional high temperature characteristics. These often very specialist fibres, reviewed recently,[24] are not often produced in appropriate fibrous dimensions for normal textile processing and are more usually available as non-woven or wetlaid webs. For instance, Saffil® (Saffil Ltd., UK), a polycrystalline alumina-based fibre, is available as a lofty, non-woven, wetlaid web or 'blanket' with a density of about 100 kg/m^2 and may be used in refractory and fire barrier applications at temperatures as high as 1600°C (see Table 4.2). On the other hand, the silica-based fibres, such as Quartzel® (Saint-Gobain, France), while having slightly inferior fire and heat performance (see Table 4.2), are available as continuous filament yarns, filament-based non-wovens and wetlaid papers. Thus, they may be knitted and woven to yield fabrics with applications in furnace insulation, combustion chamber insulation in aircraft, ablative composites for military and other markets, and hot corrosive gas and liquid filtration. Falling between these two extremes are the alumina-silica fibres, exemplified by the Nextel® range of products (3M, USA), which are determined by the alumina:silica ratio and rarely contain less than 60% alumina. For example, Nextel 610 comprises 99% (being similar to Saffil®), Nextel 720 comprises 85% and Nextel 312 comprises 62% Al_2O_3 to give a maximum user temperature range of 1260–1370°C. These again are available in yarn, fabric and non-woven forms for similar applications. Variations to the properties of these mixed oxide fibres, principally fibre physical characteristics such as flexibility or strength retention at high temperature with time, may be made by introducing other oxides such as boric oxide, B_2O_3 (Nextel 312 and 440), and zirconia, ZrO_2 (Nextel 650); although these do not enhance the maximum thermal working conditions.

There are a range of other inorganic fibres, such as basalt fibres (e.g. Basaltex®, Marureel, Belgium) from a naturally-occurring complex silica/alumina/other oxide basalt rock similar to glass in composition, used as an asbestos-replacement and available in filament and non-woven forms with claimed superiority to glass fibres in terms of temperature performance (see Table 4.2).

One non-oxide fibre that receives attention is the group based on silicon carbide (e.g. Nicalon®, Nippon Carbon), which is available as cf tow for use in high temperature rigid composites with either polymer-based resin or metal matrices such as tungsten.

4.3.3 Fire retardant and heat resistant treatments and coatings

As stated above, FR treatments applied to non-thermoplastic fibres of natural origin play important roles in high performance textiles requiring heat and fire resistance, provided that the chars formed on heating are sufficiently coherent and possess acceptable fire barrier properties. Durably finished flame retardant cotton (e.g. Proban CC®, Solvay formerly Rhodia), viscose (e.g. Lenzing FR®, Lenzing; Visil, Sateri) and wool (e.g. Zirpro®, formerly International Wool Secretariat, more recently Wool Development International Limited) fall into this category and find significant application in protective clothing and furnishing fabrics in transport. In the latter applications, wool and its flame-retarded analogue is often blended with an inherently fire resistant fibre. This whole subject has been dealt with in detail elsewhere and is outside the immediate remit of this chapter.[11,12,15] However, where and when relevant, attention will be drawn to these treated fibres.

Within the area of textiles, coatings and back-coatings are applied either to confer a level of flame retardancy alone or as an additional flame retardant property to another coating function such as water repellency, for example.

Back-coatings mainly feature in rendering furnishing fabrics flame retardant while preserving the aesthetics of the front face. These are sufficiently effective to protect underlying flammable fillings from an igniting source as discussed above and described elsewhere.[6,7,12] On the other hand, coatings used to confer fire and heat resistance are only well-developed for the paint and steel structure protection industries and usually incorporate an intumescent property. This area will be introduced in the following section.

4.4 Evolving technologies and future trends for heat and fire protection

A number of technologies have been considered and actively researched over the last ten years or so, although commercialisation of these potentially interesting areas has been slow.

4.4.1 Intumescent systems

Intumescence is a phenomenon where heat or other agency promotes formation of an expanded or swollen condition; in the case of a flame retardant, this is an expanded char.[15,25] Intumescent coatings when applied to textile fabrics, on heating, generate a foamed insulating layer on the fabric surface. This intumescent layer provides an extremely effective insulating barrier, and prevents any further ignition until the fire consumes the layer itself. The stable and thermally insulating char barrier created by the mechanism of intumescence is resistant to both radiant heat and flame. These intumescent coatings normally contain four major components, namely a 'carbonific' or char-former, a catalyst, a 'spumific' or gas-former and a binder. Other components that may be used include fillers, pigments, viscosity-modifiers and fibres. A typical intumescent coating[26] has the same surface appearance as that of a conventional coated textile, but on heating the surface of the binder melts and spreads, forming a barrier which inhibits exposure of the underlying fibres to atmospheric oxygen. The heat also activates the catalyst, which decomposes to yield an inorganic acid and chars the carbonific component present by a dehydration mechanism. Under the action of the heat, the specific compound generates non-flammable gases which are released and trapped within the viscous carbonaceous char, 'blowing' up the layer to form a thick thermally insulative foam – this solidifies as its fractional carbon content increases. It is usually the case that the generation of intumescence as part of the overall flame retardant property will also decrease the smoke and other toxic fire gas emissions.

The application of intumescent materials to textile materials has been reviewed[15,21] and is exemplified in the patent and research literature by the following examples:

- More conventional, flexible textile fabrics to which an intumescent composition is applied as a coating have been reported.[25,26] These coatings have been developed in order to produce seating barrier fabrics to resist US regulations[27] not dissimilar to those in the UK.[6] In the UK, Thor Chemicals have introduced the intumescent Flammentin IST as replacements for antimony-bromine systems in coating and back-coating formulations.[6] This product has since been discontinued, however a parallel development for technical non-wovens has been an intumescent coating based on modified carbon and developed by Schill & Seilacher.[6]
- Of a completely different character is the patent by Horrocks *et al.*[28] describing a novel range of intumescent-treated textiles that derive their unusually high heat barrier properties from the formation of a complex char that has higher-than-expected resistance to oxidation. These require the intumescent to be in intimate contact with the surfaces of flame-retarded fibres which also form char on heating and in which the respective char-forming mechanisms are similar and promote physical and chemical interaction. Exposure to heat therefore promotes simultaneous char formation of both intumescent and fibre to give a so-called 'char-bonded' structure. This integrated fibrous-intumescent char structure has a physical integrity superior to that of either charred fabric or intumescent alone and, because of decreased oxygen accessibility, demonstrates an unusually high resistance to oxidation when exposed to temperatures above 500°C, even as high as 1200°C. Furthermore, these composite structures show significantly decreased rates of heat release when subjected to heat fluxes of 35 kW m^{-2}, thus demonstrating additional significant fire barrier characteristics.[29] This work has shown that these intumescents, which are based on ammonium and melamine phosphate-containing intumescents applied in a resin binder, can raise the fire barrier properties of flame-retarded viscose and cotton fabrics to levels associated with such high performance fibres as aramids. Furthermore, charred residues are considerably stronger than those from flame-retarded fabrics alone.[30]
- Cazé *et al.* have also undertaken recent research into novel intumescents, and have described a means of producing an intumescent coating for polyester fabrics,[31] and more recently Bourbigot *et al.* have produced intumescent polypropylene fibres.[21] The latter demonstrated the effects of incorporating high concentrations and large particle sizes of intumescent components such as ammonium polyphosphate have on reducing resulting fibre and yarn tensile properties.

In all these cases, and others, a prime reason for poor commercial take up has been the inherent water solubility of one or other of the components of intumescent systems. Subsequent work by Horrocks *et al.*, however, demonstrated that durable intumescent modifications of cotton, polyamide and wool are possible by reacting phosphorylated pentaerythritol and 1,3-propanediol acid chlorides with the respective fibres to form covalent hydrolysis-resistant covalent bonding between the intumescent and respective fibre molecule.[32] A patent exists for this innovation.[33]

4.4.2 Potential impact of nanotechnologies

Horrocks *et al.* have recently reviewed the potential for the inclusion of otherwise inert nanoparticles, such as nanodispersed functionalised phyllosilicate clays, synthetic silicates (e.g. polyhedral silsesquioxanes (POSS) – see below) and carbon nanotubes, to increase tensile properties on the one hand with some improvements to fire behaviour on the other.[34,35] The claimed improvements occur at very low concentrations (e.g. 1–5% w/w) and, when present alone, influence only the peak heat release rate (PHRR) observed in cone calorimetry as observed by Bourbigot *et al.* for clay-containing polyamide 6 fibres and as nanoparticle-containing coatings applied to textiles like woven polyester.[36] While the introduction of nanoparticles often reduces ignition and extends burning times, and so offers little flame retardant activity, the selection of the nanoparticle, its functionalising group and resin (e.g. polyurethane resin containing POSS) has been shown to increase ignition time. In all cases when nanoparticles are present, an increase in char formation and its coherence is observed, thereby offering opportunities for developing improved heat barrier textiles. The mechanism by which the nanoparticles work is not clear, but it includes the formation of a heat resistant ceramic surface on the textile during pyrolysis of the surrounding polymer matrices (both fibre and resin in a coating) plus the generation of additional char by suppression of the volatile-forming reactions of polymer pyrolysis. The potential for, and effect of, a ceramic coating on the surfaces of a textile has also been demonstrated by Bourbigot *et al.*[21] by laying down a thin film of silica on polyamide 6 (PA6) and a clay-containing polyamide 6 nanocomposite (PA6nano) by reaction with 1.1.3.3-tetramethyldisiloxane monomer. LOI values for PA6 and PA6nano sample treated by cold remote nitrogen plasma rise from 21 and 23 vol % to 25 and 46 vol %, respectively, and are confirmed by cone calorimetry experiments where peaks of heat release rate compared to virgin PA6 are decreased by 30% and by 60%, respectively. It was thus concluded that the polysiloxane coating of about 10 μm thick on the textile surface reduced overall flammability.

This whole area is of increasing interest, with an expanding literature[21,34,35] to which the reader is directed. Suffice it to say for the purposes of this review that the potential for improving the fire performance of fibres and textiles is now considered to lie in the ability of some nanodispersed particles to interact additively or synergistically with conventional flame retardants, such that overall lower retardant concentrations are possible.[35] This gives the potential for developing flame retardant fibres having reduced cost on the one hand, and improved tensile properties and processing characteristics on the other. It must be pointed out that positive effects of introducing nanoparticles are not always seen, as noted for polypropylene to which the nanoparticles POSS and multiwalled carbon nanotubes were added[21] – no reduction in PHRRs was observed, although increased fibre moduli were recorded.

However, as yet, no commercialisation of any flame retardant fibres or textiles containing nanoparticle systems has been reported.

4.5 Applications of heat protective textiles

The need for, and application of, heat and fire protective textiles is both great and various with markets increasing as the improvements in performance are matched by the increasingly challenging requirements of regulators and customers for increased safety levels. In the latter case, these demanding customers are usually supported by regulations and national/international standards (e.g. aerospace, contract furnishings), codes of practice (e.g. barrier fabrics) or governmental demands (e.g. police and civil emergency forces and military applications). The major applications for high performance textiles in fire and heat protection fall into one or more of the following groups, many of which are covered by Scott,[37] Horrocks and Price[38] and Wilusz[39]:

- protective clothing including hot metal splash (e.g. workwear, firefighters clothing systems)
- contract furnishings (e.g. soft furnishing, curtains/drapes and carpets in hotels, public buildings, airport, etc.)
- fire barriers (e.g. for use in other applications or alone, as in theatre curtains, machinery and workplace protection, insulation for engines, etc.)
- land transport (e.g. engine and transmission components, structural elements and furnishings)
- maritime (as for land transport plus bulkhead fire insulation)
- aerospace (engine and fuselage fire insulation, all internal textiles, structural members)
- military applications (clothing for service personnel working in areas having defined fire hazard protection requirements as well as equipment components).

Selection of high performance fibres and textiles is often determined in the first instance by the expected service conditions in terms of continuous and

Table 4.3 FR fabrics/products for use up to 100°C continuous use

Application	Typical fibre or textile product/finish
Protective clothing: e.g. workwear Barrier fabrics: e.g. theatre curtains Furnishings and interior textiles	Pyrovatex®-treated FR cotton Proban®-treated FR cotton FR wool (e.g. Zirpro®) FR viscose (e.g. Lenzing FR, Visil®) FR polyester (e.g. Trevira CS®) FR acrylic (e.g. Kanecaron)

Table 4.4 High performance heat and fire resistant fabrics/products for use above 150°C continuous use

Application	Typical fibre or textile product
High performance protective clothing: e.g. firefighters' kit Defence and emergency textiles High performance barrier composites: aerospace, surface vessels, transport	Meta-aramids (e.g. Nomex®, Teijinconex®, Kermel ®) Para-aramids (e.g. Kevlar®, Twaron®) Arimid (e.g. P84®) Novoloid (e.g. Kynol®) PBI (e.g. PBI®) PBO (Zylon®) Semicarbon (Panox®) Carbon Ceramics (e.g. glass, Nextel®)

maximum user temperatures (see Table 4.2) and environments relating to a given application. As an arbitrary division across the gamut of all heat and fire protective textiles, many are used at ambient temperatures for most of their lives with a need to withstand a single high temperature or fire exposure, while others have to withstand elevated temperatures for considerable periods (if not the whole of their expected service lives). Table 4.3 summarises the applications of the more conventional flame retardant textiles, which may be defined as those which may be used at service temperatures of ambient up to about 100°C, while Table 4.4 outlines high performance fire and heat resistant ones that may be used continuously above 150°C. Both groups of products will be flame retardant, to the extent that they resist ignition by a small source, and will self-extinguish in air.

To examine each of the above areas in detail is beyond the scope or needs of this text; however, to illustrate the challenges involved, a brief examination of the needs of the commercial aerospace industry will be presented. According to Troitzsch,[40] within a modern high capacity jet such as the Boeing 747 there are about 4000 kg of plastics materials of which about half comprises glass- and carbon-fibre-reinforced composites. Within the other half are the textiles that are part of the aircraft itself including decorative features. In addition, there will be carpets, blankets and other textile-based

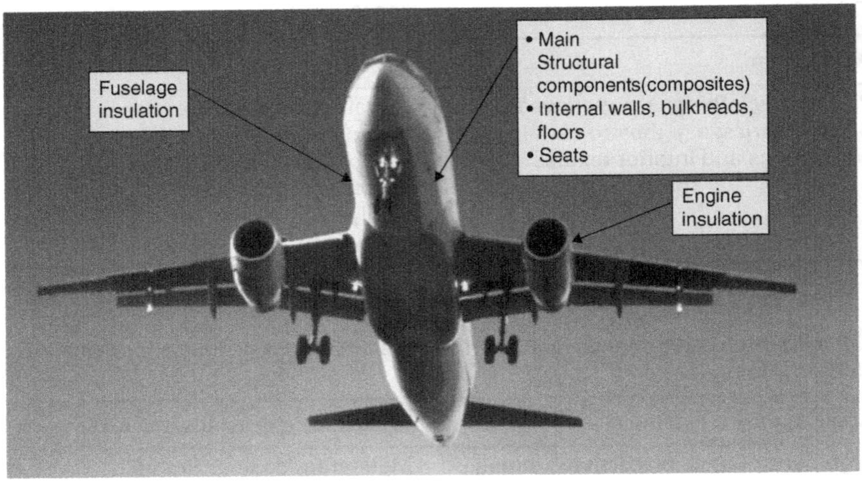

4.1 Components of a modern airliner which involve heat and fire resistant textiles.

equipment. Clearly, the potential fire load from these fibre-based elements will be substantial and in the new generations of airliner, composite use is increasing; for instance, the Airbus 380 airframe comprises over 25% composite, and the Boeing Dreamliner or 787 will have much higher levels. Figure 4.1 outlines the major contributions of heat and fire resistant textiles in making of a modern commercial airliner.

Most countries have adopted the various US Federal Aviation Regulations regarding commercial aircraft safety, and so all textiles, such as seating fabrics, carpets, curtains/drapes, blankets, etc., used anywhere in a commercial aircraft flying on national and international flights must pass a simple ignition test defined in the requirements given in FAR 25.853(b) using the test procedure defined in FAR Part 25 (Federal Aviation regulation). This is essentially a series of 'Bunsen burner/vertical, 45° or horizontal strip' ignition tests which assess whether or not a given material is self-extinguishing. For instance, vertical strip samples (75 × 305 mm) of textile materials used in blankets and seatings are subjected to a flame at the specimen bottom edge for 12 s. After its removal, the sample must experience a burn or damaged length ≤ 152 mm, an afterflame time ≤ 15 s and a flame time of any drippings ≤ 3 s. For textiles used in liners for cargo and baggage compartments, the 45° test is used with similar requirements except that no flame penetration through the fabric should occur. Thus, there is a need for textiles having varying degrees of heat and fire resistance dependent upon the level of perceived risk. In addition, there are more stringent requirements for areas having higher fire risk. Textiles which form a decorative or reinforcing element of structures within the passenger cabin must be tested as a composite or assembly for their ability to resist the spread of fire. Here, the heat release rate is the parameter that is to be measured using the Ohio State University (OSU) calorimeter according to the requirements of FAA specification FAR 25.853 Part IV Appendix F.[9,37,38] In this test, textiles used, say, as decorative coverings for wall panels are mounted on an appropriate wallboard material and are subjected to a heat flux of 35 kW m^{-2} yield. If the assembly, and hence the textile, is to pass the test, then the burning composite must emit a maximum heat flux output below 65 kW m^{-2} and an average over 2 min of less than 65 kW m^{-2}. For fabrics to pass this so-called 65/65 test, they must usually have LOI values in excess of 30 vol %, and so prime candidates are FR wool and aramids used individually or in blends, although other more exotic fibre examples may be used (see below).

Other high risk assemblies are seatings, and the external fabrics must be able to prevent ignition of internal filling materials. In specification FAR 25.853(c), a seat assembly is subjected to a kerosene burner having a heat flux of about 115 kW m^{-2} for 2 min. After extinction of the burner, the assembly must extinguish within 5 min, not burn beyond the seat dimensions, and the overall mass loss must be $\leq$ 10%. For seating assemblies to pass this test, it is usual to have an outer fabric (e.g. FR wool or FR wool/polyamide 6.6) that passes FAR 25.853(b) and an underlying fire blocking or barrier layer typically based on high performance fibres such as aramid, oxidised acrylic, glass or blends of these with either each other or with fibres such as FR wool.

Apart from the normally accepted textile products, heat and fire resistant textiles find use in engine insulation (e.g. ceramic structures around combustion chambers), reinforcements for composites (e.g. carbon fibre reinforcements for major structural elements), aramid honeycomb reinforcements for wall and floor structures, and fuselage acoustic and fire/heat insulation, each of which has its own fire performance requirements.[38,40] Associated with all these tests and materials or composites are toxic fire gas and smoke requirements, and so the choice of fibre and textile structures will be influenced by the need to pass the minimum emission standards for gases including carbon monoxide, nitrogen oxides, sulphur dioxide, hydrogen chloride and hydrogen cyanide.

Two case studies representing recent research in our own laboratories will illustrate the challenges posed by the need to develop novel high performance textiles having the required levels of heat and fire performance while maintaining the required textile properties of aesthetics, insulation, moisture transport, etc.

4.5.1 Development of exotic fibre-containing fabrics for interior décor in executive jets

This project undertaken at the University of Bolton, and described in detail elsewhere,[41] combines the need to retain the highest level of aesthetics in a custom-designed product for the customer of an executive jet and for these fabrics to have fire performance requirements of the highest standard.

Recent changes in US Federal (and other national) Aviation Authority (FAA) regulations, which now demand high levels of fire resistance, have created a paradoxical situation where formerly acceptable fabrics fail the new specification, and fabrics which do pass fail to have the desired level of aesthetic quality. Thus, textiles used as décor inside commercial, and now executive, aircraft have to pass the stringent flammability requirements defined by the FAA specification FAR 25.853 Part IV Appendix F discussed above. At the present time only a few fabrics having the desired aesthetic quality, and often comprising exotic animal hair fibres such as mohair, alpaca and cashmere which when treated with conventional flame retardants, often suffer changes in appearance and handle. Therefore, a main aim of the project was to develop a flame retardant system, that could be applied to these fabrics without any loss of aesthetics while enabling them to pass the '65/65' heat release requirement defined above. A secondary aim was to undertake experiments using the more recently developed oxygen depletion or 'cone' calorimeter',[42] considered to provide a superior means of measuring heat release properties and relate these to OSU results.

Six 180 gsm fabrics were selected, comprising a variety of exotic animal hair fibres (mohair, alpaca and cashmere) and Sea Island cotton as weft yarns and silk or polyester warps:

- mohair weft/ silk warp (MS)
- mohair weft/polyester warp (MP)
- cashmere weft/silk warp (CS)
- alpaca weft/silk warp (AS)
- alpaca weft/polyester warp (AP)
- Sea Island weft/silk warp (SS).

Using factorial analysis, samples of the mohair/silk (MS) fabric were variously treated with a selection of flame retardants, back-coating formulations and adhesive, mounted on a typical aramid honeycomb board specimen, and each composite was tested using cone calorimetry at the preferred heat flux of 50 kW m^{-2} (shown to be equivalent to the 35 kW m^{-2} flux used in the OSU calorimeter).[41] An optimum combination of flame retardant, back-coating and adhesive at specific application levels was found to yield the lowest heat release values, and this system was applied to each of the above six fabrics. Testing in both the OSU at 35 kW m^{-2} heat flux and at 50 kW m^{-2} in the cone calorimeter gave the results for peak heat release in Table 4.5 below. From this it is seen that all fabrics have PHRR values < 65 kW m^{-2} and that OSU and cone calorimeter results are equivalent.

Table 4.5 PHRR values for fabric/board composites tested to FAR 25.853 Part IV Appendix F and cone calorimetry[41]

Fabric	PHRR (kW m^{-2})	
	OSU calorimeter	Cone calorimeter
Mohair/silk (MS)	58	48
Mohair/polyester (MP)	62	62
Cashmere/silk (CS)	61	56
Alpaca/silk (AS)	58	46
Alpaca/polyester (AP)	61	65
Sea Island/silk (SS)	41	46

Additional work examined whether fabric design in which significant differences in weft yarn face densities occurred might adversely influence PHRR values. In the event, the flame retardant system developed proved to be so tolerant that all designs were found to pass the OSU test and hence achieve the stringent FAA specification FAR 25.853 Part IV Appendix F. Thus, the project had fulfilled its aims and enabled fabrics having high aesthetic characteristics also to achieve very high levels of technical performance.

4.5.2 The development of fabric composite assemblies for use in acoustic and fire insulation aircraft fuselages

This project demonstrates how a flexible composite comprising technical textile structures has been designed to withstand the new fire break-through test for commercial aircraft that came into force in September 2005. In 1998 Swissair Flight 111 crashed in Nova Scotia, with the loss of 229 lives. The crash was caused by electricity arcing from faulty wiring in a newly-fitted entertainment system into the fuselage insulation, causing it to ignite. As a consequence, the US FAA have required a tightening of the specification for fire and acoustic insulation in the lower fuselage area. In July 2003, the FAA published an improved set of test standards. The proposed revision of FAA 25.856(a) Appendix F, Part VI required that the component materials of any insulation composite pass a radiant panel flame propagation test and, in FAA 25.856(b) Appendix F, Part VII, that the whole composite pass a burn-through test when subjected to a burner having a flame temperature of about 1050°C and output energy of about 160 kW m^{-2}. In addition, the FAA drafted a list of other specified features including thermal, acoustic, weight, water absorption, non-corrosive, non-conductive and durable requirements.

Prior to the new regulations, the barrier materials were based on composites comprising layers of glass fibre non-wovens which, although inert, melt at about 500°C and are relatively dense. The non-woven layers are contained in a sealed poly(vinyl fluoride) or PVF film bag. In addition to the fire, acoustic and related performance criteria, questions were also asked regarding the environmental acceptability of this insulating fibrous content.

The main aim of this project was to develop novel composite structures based on our previous successes in the previous Case Study 1 using flame-retarded mohair structures in order to achieve both the desired levels of fire performance and environmental sustainability. Of the two fire tests above, the burn-through requirement is deemed to be the more rigorous, and Fig. 4.2 schematically represents the composite–burner geometry. An assembly is considered to have passed if the burner flame can impinge upon the rear face of the assembly and not burn through both sides of the assembly and the calorimeter mounted to the face of the composite records an incident energy flux of not greater than 22 kW m^{-2} in less than 240 s. This is an extremely harsh test with a burner having a flame temperature of about 1050°C and a heat output equivalent to a forced air fuel fire. Present thermal/acoustic materials based on glass fibre fail the test when contained in currently acceptable bag assemblies. However, work in our laboratories has shown that we are able to achieve passes if flame-retarded

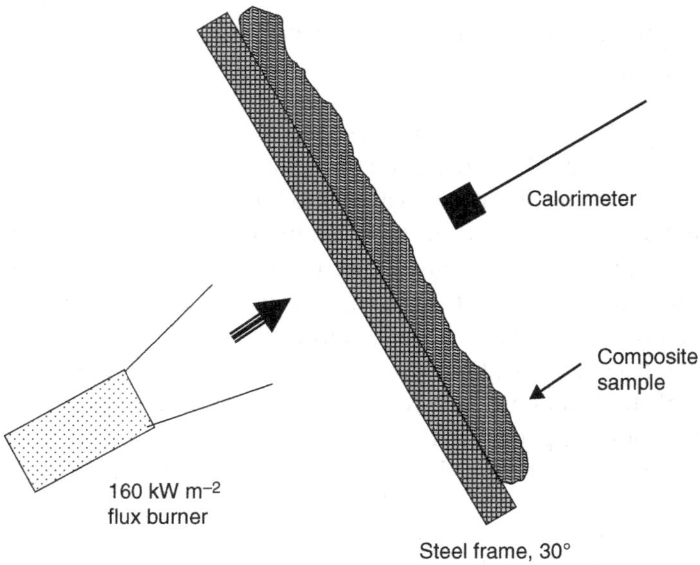

4.2 The new FAA thermal/acoustic barrier burn-through apparatus, FAA 25.856(b) Appendix F, Part VII.[43]

Table 4.6 Fire performance (burn-through: FAR 25.856(b) Appendix F Part VII) of experimental flame-retarded mohair-containing fabric assemblies[43]

Fibre/fabric	Film	Failure time (s)	Pass/fail
3 × 1 ply (30 mm) 75% mohair/25% polyester non-woven	PVF	9 (inner face penetrated)	Fail
3 × 1 ply (30 mm) 75% mohair/25% polyester non-woven	Polyimide	33 (integrity); 115 s (≤ 20kW/m^2)	Fail
3 × 1 ply (30 mm) 75% mohair/25% polyester non-woven plus 1 Nextel barrier paper opposite flame source	Polyimide	> 240	Pass
3 × 1 ply 25 mm ceramic/25 mm mohair/25 mm ceramic	PVF	> 1200	Pass

(a) (b)

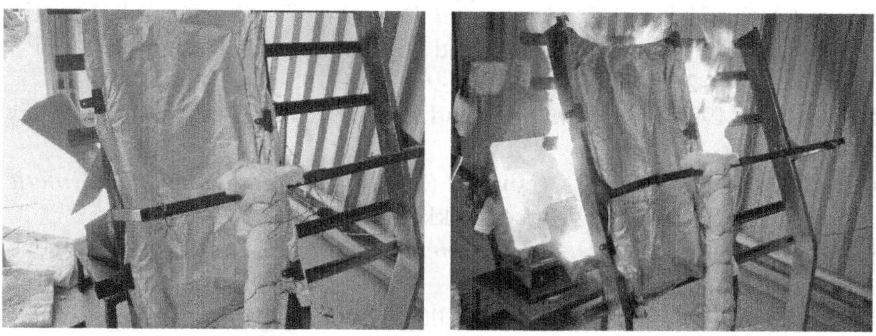

4.3 Experimental ceramic/mohair/ceramic composite contained in a PVF bag under test according to FAA 25.856(b) Appendix F, Part VII: (a) before testing commenced, (b) during the test (unpublished results).

mohair non-woven structures having particular characteristics are combined with a ceramic paper (Nextel, 3M) and contained in a polyimide (PI) bag. Table 4.6 describes these results in addition to our more recent ones, in which a composite comprising the non-woven layers ceramic/mohair/ceramic contained in the presently used PVF bag also demonstrates an excellent pass (see Figs 4.3a and 4.3b).

Clearly the need to develop heat and fire resistant textiles grows in challenge as the requirements of a modern and evolving world demand higher levels of fire safety on the one hand and higher performance as applications become more severe. The above examples in the commercial aerospace markets exemplify these challenges, which are no less so in other markets including the military and civil emergency fields as well as other transport areas.

4.6 Sources of further information and advice

The cited references in this chapter generally have been selected to be of a research review character, as opposed to reporting the outcomes of frontline research. The references within them will lead the reader to current research in specified areas. The list below also presents a complementary selection to enable the reader to widen understanding across a field that demands not only knowledge of fibre and textile technologies and science, but also of fire science and relating testing methodologies.

Alongi J, Horrocks A R, Carioso F and Malucelli G (2013), Update on Flame Retardant Textiles: State of the Art, Environmental Issues and Innovative Solutions, Shawbury, UK, Smithers RAPRA.

Barker R L and Coletta G C (1986), *Performance of Protective Clothing,* ASTM Special Technical Publication 900, Philadelphia, ASTM, Philadelphia.

Bischof Vukušić S (2012), *Functional Protective Textiles,* Zagreb, Craotia, University of Zagreb, (funded by European Commission FP7-REGPOT-2008-1-229801; ISBN -978-953-7105-45-7)

Drysdale D (1999), *Fire Dynamics,* 2nd Edition, Chichester, UK, John Wiley and Sons.

Grand A F and Wilkie C A (editors) (2000), *Fire Retardancy of Polymeric Materials,* New York, Marcel Dekker.

Hearle J W S (2001), *High Performance Fibres,* Cambridge, Woodhead Publishing Limited.

Horrocks A R and Anand S C (2000), *Handbook of Technical Textiles,* Cambridge, Woodhead Publishing Limited.

Horrocks A R and Price D (editors) (2001), *Fire Retardant Materials,* Cambridge, Woodhead Publishing Limited.

Horrocks A R and Price D (editors) (2008), *Advances in Fire Retardant Materials,* Woodhead Publishing Limited.

Johnson S and Mansdorf S Z (1996), *Performance of Protective Clothing, Fifth Volume,* ASTM Special Technical Publication 1237, West Conshohocken, USA, ASTM

Mansdorf S Z, Sager R and Neilsen A (1988), Performance of Protective Clothing, Volume **2**, ASTM Special Technical Publication 989, Philadelphia, ASTM.

Scott R(editor) (2005), *Textiles for Protection*, Cambridge, Woodhead Publishing Limited.

Selcen Kilinc F (2013), *Handbook of Fire Resistant Textiles,* Cambridge, Woodhead Publishing Limited.

Stull J O and Schwope A D (1997), *Performance of Protective Clothing, 6th Volume,* ASTM Special Technical Publication 1273, Philadelphia, ASTM

Troitzsch J (2004), Plastics Flammability Handbook, 3rd Edition, Munich, Hanser.

4.7 References

1. Klitzke R A (1959), 'Roman building ordinances relating to fire protection', *Am. J. Legal History*, **3**(2), 173–187.
2. Browne C (2003), 'Salamander's wool: The historical evidence for textiles woven with asbestos', *Text. History*, **34**(1), 64–73.
3. Plot R (1685), Letter to Mr. Arthur Bayly Merchant, and Fellow of the R. Society, *Philosophical Transactions (1683–1775)*, **15**, 1051–1062.
4. Wyld O, British patent BP551 (1735).
5. Gay-Lussac J L (1821), *Ann.Chim.Phys.*, **18**(2), 211.
6. *Consumer Protection Act (1987), the Furniture and Furnishings (Fire) (Safety) Regulations, 1988*, SI1324 1988, London, HMSO.
7. Horrocks A R (2001), 'Textiles' in *Flame Retardant Materials*, Horrocks A R and Price D (editors), Cambridge, Woodhead Publishing Limited, 128–181.
8. *Fire Statistics, United Kingdom, 2004*, (2006), London, The Home Office, The Government Statistical Office, UK, ISBN 13-978 185 112 8433.
9. Babrauskas V (2000), 'Fire test methods for evaluation of FR efficiency', in *Fire Retardancy of Polymeric Materials,* Grand A F and Wilkie C A (editors), New York, Marcel Dekker, 81–114.
10. Troitzsch J (2004), *Plastics Flammability Handbook,* 3rd Edition, Munich, Hanser.
11. Horrocks A R (2005), 'Thermal (heat and fire) protection', in *Textiles for Protection*, Scott R (editor), Cambridge, Woodhead Publishing Limited, 398–440.
12. Horrocks A R (2003), 'Flame retardant finishes and finishing', in *Textile Finishing,* Heywood D (editor), Bradford, Society of Dyers and Colourists, 214–250.
13. Drysdale D (1999), *Fire Dynamics,* 2nd Edition, Chichester, UK, John Wiley and Sons, 193–232.
14. Shields T J, Silcock G W and Murray J J (1994), 'Evaluating ignition data using the flux time method', *Fire Mater.*, **18**, 243–253.
15. Horrocks A R (1996), 'Developments in flame retardants for heat and fire resistant textiles – the role of char formation and intumescence', *Poly. Degrad. Stab.*, **54**, 143–154.
16. Horrocks A R, Eichhorn H, Schwaenke H, Saville N and Thomas C (2001), 'Thermally resistant fibres', in Hearle J W S (editor), *High Performance Fibres,* Cambridge, Woodhead Publishing Limited, 289–324.
17. Horrocks A R, Price D and Tunc M (1987), 'The burning behaviour of textiles and its assessment by oxygen index measurements', *Text. Prog.,* **18**(1–3), 1–205.
18. Bourbigot S and Flambard X (2002), 'Heat resistance and flammability of high performance fibres: a review', *Fire Mater.,* **26**(4–5), 155–168.
19. Lavin J G (2001), 'Carbon fibres', in Hearle J W S (editor), *High Performance Fibres,* Cambridge, Woodhead Publishing Limited, 156–190.

20. Bourbigot S, Flambard X, Ferreira M and Poutch F (2002), 'Blends of wool with high performance fibers as heat and fire resistant fabrics', *J. Fire Sci.*, **20**(1), 3–22.
21. Bourbigot S, Duquesne S, Bellayer S, Flambard X, Rochery M and Devaux E (2005), 'Novel developments in flame retardancy of textiles', in *Advances of Flame Retardant Polymeric Materials-Current Perspectives*, Schartel B (editor) Germany, Herstellung and Verlag, Norderstedt, 2007, 159–180.
22. Jones F R (2001), 'Glass fibres', in Hearle J W S (editor), *High Performance Fibres,* Cambridge, Woodhead Publishing Limited, 191–238.
23. Koch K (2004), 'New roof of the Olympic Stadium in Berlin'. *TUT Textiles a Usages Techniques,* **3**(53), 10–13.
24. Bunsell A R and Berger M-H (2001), 'Ceramic fibres' in *High Performance Fibres,* Hearle J W S (editor), Cambridge, Woodhead Publishing Limited, 239–258.
25. Le Bras M, Camino G, Bourbigot S and Delobel R (editors) (1998), *Fire Retardancy of Polymers: The Use of Intumescence*, London, Royal Society of Chemistry.
26. Tolbert T W, Dugan J S, Jaco P and Hendrix J E (1989), Springs Industries Inc., US Patent, 333174, 4 April 1989.
27. Damant G (2006), 'New flammability standards for residential furnishings are imminent in the USA', in *Flame Retardants 2006*, London, Interscience Communications, 215–224.
28. Horrocks A R, Anand S C and Hill B J (1997), UK Patent, G B 2279084 B, 20 June 1995; US Patent 5645926, 8 July 1997.
29. Horrocks A R and Kandola B K (1998), 'Flame retardant cellulosic textiles', in *Fire Retardancy of Polymers: The Use of Intumescence*, Le Bras M, Camino G, Bourbigot S and Delobel R (editors), London, Royal Society of Chemistry, 343–362.
30. Kandola B K. and Horrocks A R (1999), 'Complex char formation in flame-retarded fibre-intumescent combinations: III Physical and chemical nature of char', *Text. Res. J.*, **69**, 374–381.
31. Cazé C, Devaux E, Testard G and Reix T (1998), 'New intumescent systems: An answer to the flame retardant challenges in the textile industry', in *Fire Retardancy of Polymers: The Use of Intumescence*, Le Bras M, Camino G, Bourbigot S and Delobel R (editors), London, Royal Society of Chemistry, 363–375.
32. Horrocks A R, Kandola B K, Davies P J, Zhang S and Padbury S A (2005), 'Developments in flame retardant textiles – A review', *Polym Deg Stab.*, **88**(1), 3–12.
33. Horrocks A R and Zhang S (2000), UK Patent, GB 2352447 B, 20 July 2000.
34. Horrocks A R, Kandola B K and Padbury S A (2003), 'The effect of functional nanoclays in enhancing the fire performance of fibre-forming polymers', *J Text Inst.*, **94**, 46–66 (Published 2005).
35. Horrocks A R (2008), 'Nanocomposites II: Potential applications for nanocomposite-based flame retardant systems' in *Advances in Fire Retardant Materials,* Horrocks A R and Price D (editors), Cambridge, Woodhead Publishing Limited, 124–158.
36. Bourbigot S, Devaux E, Rochery M and Flambard X (2000), 'Nanocomposite textiles: New routes for flame retardancy, in *47th International SAMPE Symposium*, 12–16 May 2000, **47**, 1108–1118.

37. Scott R (editor) (2005), *Textiles for Protection*, Cambridge, Woodhead Publishing Limited.
38. Horrocks A R and Price D (editors) (2008), *Advanced Fire Retardant Materials*, Cambridge, Woodhead Publishing Limited.
39. Wilusz E (editor) (2008), *Military Textiles,* Cambridge, Woodhead Publishing Limited.
40. Troitzsch J (2004), *Plastics Flammability Handbook,* 3rd Edition, Hanser, Munich, 457–471.
41. Kandola B K, Horrocks A R, Padmore K, Dalton J and Owen T (2006), 'Comparison of cone and OSU calorimetric techniques to assess the flammability behaviour of fabrics used for aircraft interiors', *Fire Mater.,* **30**(4), 241–256.
42. Babrauskas V and Grayson S (editors) (1992), *Heat Release in Fires*, New York, NY, Elsevier Applied Science.
43. Davies P J, Horrocks A R, Kandola B K, Dalton J and Owen T (2005), 'Fire resistant acoustic insulation media for commercial aircraft: Recent changes in regulations and barrier developments', in conference proceedings, *Fire and Materials*, Interscience Communications, London, 439–448.

5
High performance medical textiles: an overview

M. MIRAFTAB, University of Bolton, UK

DOI: 10.1533/9780857099075.176

Abstract: This chapter gives an overview of progress made in understanding fibre fundamentals and their subsequent development into targeted application areas. It describes the key roles that textiles play in maintaining health and hygiene in everyday life. In *in vivo* application, it differentiates between temporary and permanent implantable materials and their underlying functional requirements. It also elucidates the role grafts, sutures and tissue engineering play in repairing malfunctioning organs and damaged surface/subsurface layers. It concludes by highlighting the roles that new technologies, including e-textiles, will increasingly play in recording health statistics, wellbeing standards and in actively prompting action long before an alignment is due.

Key words: disposables, reusable, infection, implantable, monitoring, e-textiles.

5.1 Introduction

Use of textiles, within the context of healthcare, goes back a long way to centuries before Christ. In fact, reports concerning the use of cotton, flax, silk and the like in the form of wound dressings, and even as sutures, dates as far back as 5000 BC. Since then, reference to use of textiles in medical applications in preceding civilisations, including the Egyptian, Persian, Roman, Chinese and the Indian, are plentiful in various literatures (Hiller, 1927, Behrouz *et al.*, 1993, p115; King, 2001, p34). Much of textile use in medical applications, however, remained unchanged, and by and large primitive, up till the late nineteenth century, when in-depth understanding of chemistry and physics of materials in general led to greater interest in building blocks and internal arrangements of natural fibres and what could possibly be done to alter their physical, mechanical and morphological characteristics. Mercerisation, for example, is a method by which cotton is made to feel softer, swell and become stronger, whilst also becoming lustrous in appearance, was achieved by treating standard cotton with sodium hydroxide. The treatment alters the chemical structure of cotton by converting the alpha-cellulose to a more thermodynamically favourable beta-cellulose (Sekar, 1999, p31). Morphological insights into wool fibres led to the notion of removing scales from this fibre to prevent material shrinkage and provide better comfort to the wearer (Hearle, 2000, p123). In the 1890s and beyond, artificial silk, or what became known as 'Rayon' or 'viscose', was developed (Woodings, 2001). This method, essentially, relied on greater understanding of cellulose, the main constituent of cotton, and its behaviour. This in turn led to the use of wood pulp as the main source of pure cellulose, and its manual manipulation to produce first ever artificial fibres, thus revolutionising the textile industry for ever. With the discovery of oil and the advent of synthetic fibres a few years later, a whole raft of textile fibres appeared on the scene where tailor-made properties could be engraved in the fibre structure or developed by secondary processes (Porczynski, 1960). From this point onwards, textile materials began their true journey of becoming 'specialised' where for the first time, application criteria dictated the type and form of the fibres to be used. Since then, many other artificial fibres have been developed, with greatly refined characteristics (Yang, 2003, p199; Hamouda, 2005, p261). However, lately the trend, particularly in some areas of the healthcare industry, has changed somewhat, towards natural or natural-based fibres where environmental friendliness, disposability, biodegradability and biocompatibility, as well as inherent antimicrobial properties and natural healing, have become the forefront issues.

5.2 Textiles for hygiene and infection control

Such words as cleanliness and hygiene have implied connotations of a dirt- and bacteria-free environment, which strike a chord with all living beings where survival is the common desire. Hence, textiles have always played a crucial role in alienating and preventing the spread of infection or, indeed, in causing infections where hygiene and sanitation have been dismal. Textiles used for general hygiene are wide and varied, ranging from standard clothing, uniforms, beddings, gloves and surgical gowns to incontinence nappies/wipes and general hygiene products. They are often disposable and made from pure natural fibre, or in blends with one or more synthetic fibres. There are also reusable hygiene products, where costs and convenience as well as performance are the determining factors. Traditionally, flax or linen, cotton and the like have been used in hospital bedding, mattresses and curtains, where durability and resistance to repetitive washing and cleaning are paramount. Conventional fibres and polymeric materials have little or no resistance to microorganisms; in fact, they are prone to accumulation, multiplication and proliferation of microorganisms into their surrounding environment. In situations where right temperature and humidity, presence of dust particles, stains from food and drinks, dead body tissues from the skin, natural body oils such as secretion from glands as well as inherent textile additives are present, rapid multiplication of microorganisms would occur.

At a time when resistance to classic antibiotics and antiseptics is becoming more persistent, due to the evolving nature of bacteria, strict hygiene rules in hospitals and other public places have become an unavoidable norm; and textiles, as an indispensable part of this scenario, have become even more important. To be hygienically clean or sterilised are no longer adequate criteria of textiles that are intended for clinical settings. Today, textiles are expected to be much more proactive and smarter in prevention and proliferation of foreign elements or pathogens. Ideally, such textiles need to be:

- effective against a whole spectrum of microbes, fungi and viral infections
- resistant to nourishment of microorganisms
- skin-friendly and non-allergic
- harmless to the human body.

Depending on where the textile material is used, a combination of one or more criteria maybe necessary. Traditionally, iodine, chlorhexidine and hydrogen peroxide have been used (Khan, 2006, p6) as general antiseptics or for wound management including irrigation and overall treatments, with a number of pros and cons regarding their efficacy and suitability. Honey as an ancient remedy for treatment of wounds and promotion of wound healing has recently been rediscovered and much praised for its antimicrobial ability, mainly governed by localised generation of low-dose hydrogen peroxide and its characteristic suffocation of pathogens by starving bacteria from oxygen (Adewumi and Ogunjinmi, 2011). Its potency against chronic and infected wounds in particular has had a large impact within the healthcare community. Many wound-care manufacturers today routinely produce dressings containing different forms of medicated honey, and a large number of published literatures support their antibacterial effectiveness. Honey dressings received the approval of the NHS in 2004, and have since been available on prescription in UK. Incorporation of inorganic materials, due to their active agents, into textiles is another area that has received much attention. Silver salts/nano-particles, copper, titanium oxide, gallium, gold nano-particles, carbon nano-tubes and nano-clays are some examples of what is now included within textiles (Silver and Phung, 2009, p220; Carter *et al.*, 2010, p668). Silver, as a potent antimicrobial, in particular, has received a huge upsurge in demand due to emergence of methicillin-resistant *Staphylococcus aureus* (MRSA) in the last decade or so. Its antimicrobial potency is due to the silver ions' ability to inhibit bacterial enzymes and to bind with the DNA of microorganisms, thus making them ineffective. Silver, however, is expensive and its excessive use can cause skin discolouration, irritation and even penetrate into the urinary tract, but on the whole, use of silver in moderate quantities and with sensible dispensation, is regarded as safe (Atiyeh *et al.*, 2007, p139).

Biodegradable fibres with inherent antimicrobial abilities are another class of materials that are increasingly explored and utilised in different applications. Bamboo fibre, for instance, reportedly displays excellent resistance to a number of Gram-positive and Gram-negative bacteria as well as common fungi (Owokotomo and Owoeye, 2011, p5030). It is very soft and highly absorbent, making it suitable for most kinds of hospital bedding/sanitation and underwear including socks, for combating infection and body odour. Research studies suggest that the antimicrobial elements within bamboo fibres are highly stable and can withstand numerous washing/cleaning cycles.

Chitin, a polysaccharide-based material, is present in the shells of such sea creatures as shrimps and crabs. Chitin is biodegradable by nature, and when acetyl groups i.e. CH_3-CO, are removed from the chitin molecules, chitosan is produced with exposed amine groups. The antimicrobial capabilities of chitin/chitosan are based on its cationic nature, which is capable of binding with anionic pathogens and making them ineffective (Aranaz *et al.*, 2009, p203).

Chitosan has been made into a number of commodity products, including fibre for wound dressing applications. Chitosan is also believed to promote healing through its susceptibility to disintegration by lysozyme, an enzyme present in the exudates that breaks down chitosan into glucosamine which is believed to play a major role in assisting cellular and tissue regeneration at the point of injury (Rhoades and Roller, 2000). Another desirable attribute of chitin and chitosan is their ability to decrease treatment time with minimum scar formation (Mezzana, 2008, p81).

To enhance the biomedical capabilities of chitosan, alginate, another natural polysaccharide extracted from brown sea weed, which has been used as wound dressing for a number of years, has recently been successfully combined with chitosan to generate hybrid fibres with superior properties of high liquid absorption, antimicrobial, anticoagulant and wound healing. Research at University of Bolton has developed this concept and acquired the universal patent for its imminent commercial exploitation (Miraftab *et al.*, 2008; Miraftab *et al.*, 2011, p345)

5.3 High performance implantable textiles

High performance implantable textiles include implants, grafts, sutures, prosthetics and medical devices. The human body is an extremely well-designed and constructed structure, surpassing any machine that man has thus far been able to make. Its complex system and organisation is governed by an array of well-coordinated components that, whilst acting independently, are managed by a pivotal command centre that keeps everything in check. When any malfunction is registered, warning alarms are sounded, symptoms are displayed and attention is demanded. If the system cannot readjust itself and address the problem, external interventions become inevitable. One such malfunction could be as a result of a foreign element entering the system i.e. bacteria or virus, leading to a disease, or as a consequence of an accidental or intentional disruption in the working operation of an otherwise healthy part or organ i.e. the body is either involved in an accident or is subjected to a surgical operation. Under these circumstances, inclusion of an intervening material or materials may become necessary. However, since the body is sensitive to foreign elements and intruders, the challenge of identifying and using the 'right' type of alien material in the body has been the subject of much research and investigation (Novikova, 2011, p372). So-called implantable materials have today become the norm in helping and supplementing the needs of a malfunctioning body.

5.3.1 Implantable materials

Ideally, natural healing by the body is the best and most effective way of dealing with malfunctioning or damaged parts of the body. The second best option, in certain cases, is repair or replacement of the damaged organs by transplantation, using the patient's own parts or utilising parts from appropriate donors. However, for various reasons that are beyond the scope of this chapter, this may not always be possible and artificial or synthetic body parts becomes an inevitable need. Materials that may be used in the body must generally be biocompatible, non-toxic and, where applicable, biodegradable. Some of the most suitable materials used in the body are categorised in Table 5.1.

5.3.2 Biodegradable implants

Non-biodegradable implants, such as silicones, ceramics, titanium, steel, carbons, polyesters and the like, are meant to stay in the body for life. Their role is often to support or enhance endurance under high static or cyclic loading/unloading or other repetitive expansive/contractive motions. However, some implants need to be removed after they have rectified a malfunction or disorder in bodily functions. Historically, there exists a lot of evidence supporting the use of implantable materials in the body; however, their systematic use really took off from the late 1800s, when aseptic 'without microorganisms' techniques were adopted as standard practice (Encyclopaedia, 2007). In the late nineteenth to early twentieth centuries, the use of metals

Table 5.1 Examples of biodegradable and non-biodegradable implantable materials

Biodegradables	Non-biodegradables
Cotton	Polyester
Viscose	Polypropylene
Polysaccharides (e.g. alginate, chitosan etc.)	Polyethylene
Collagens	Polytetrafluoroethylene
Silk	Nylon
Polylactic acid (PLLA)	Carbon
Polyglycolic acid (PGA)	Steel
Poly-DL-lactic acid (PDLLA)	Titanium
Poly-p-dioxanone (PDS)	Nitinol
Poly-beta-hydroxybutyric acid (PBHBA)	UHMWPE
Trimethylene carbonate (TMC)	PEEK
	Polyurethanes
	Ceramics
	Silicone

for fixing bone fractures and for hip replacements became a routine practice, even though only a limited number of metallic, ceramic and polymeric materials existed (Geertsma *et al.*, 2007). These materials have since been supplemented by an array of newly found suitable metallic and synthetic absorbable and non-absorbable materials. Today, implantable medical devices have surpassed their original application areas and have expanded into tissue engineering and all forms of regenerative medicine.

To keep within the context of this chapter, non-biodegradable materials such as polyolefin (polyethylene, polypropylene), nylon, rayon, polyester, polyvinyl chloride (PVC), polytetrafluoroethylene (PTFE), polyurethane (PU), polyvinylidene fluoride (PVDF) and polyacrylonitrile (PAN) are the main synthetic fibres used in implants. Depending on the functional requirement of the implants, there are, generally, four main methods by which these materials maybe prepared for use i.e. woven, non-woven, knitted and braided. For instance, in thoracic aortic repairs, where considerable resistance to stress/pressure and relatively low porosities are essential, structurally resilient grafts in the form of woven structures would be used, and in implants where access for fibroblasts is needed for migration and laying down collagens, warp-knitted velour structures would be used. Braided implants in the form of tapes or cords could be used in joints as ligaments, tendons and the like. Where stress and load-bearing abilities are not the prime criteria, relatively weak and open structures, such as those possible through non-woven techniques would be used. These could include performs /platforms for tissue engineering and scaffold structures where their main aim is facilitation of cell growth and structural support of the affected organ or system/area, either temporarily or permanently. Biodegradation is probably the desired outcome when a foreign implant, having served its purpose, is expected to eventually wither away and give way to natural bodily functions. Hence, a secondary operation for removal is not needed, and further complications maybe avoided. Furthermore, biodegradable implants, unlike metallic implants, do not release metallic ions into surrounding tissues and are not affected by magnetic resonance imaging (MRI), a technique routinely used to visualise detailed internal structures. The decomposition or biodegradation of an implant is normally initiated by material hydrolysis, causing molecular chain scissions or shortening of molecules, ultimately leading to material weakness and breakdown. The fragmented pieces are gradually absorbed by the body, and metabolic eliminations occur via respiration, urinary tract and the large intestine. If the degradation rate is excessive, or out of tune with what the tissues can tolerate, inflammatory reactions due to the presence of foreign bodies could take place and, if not dealt with appropriately, may lead to further complications (Tormala, 1998, p101). Most natural and some synthetic products are biodegradable, but might not necessarily be biocompatible.

In the recent past there has been a growing tendency to use biodegradable implants, where possible, based on the positive advantages of these materials. Orthopaedic surgery is one such area where low to medium loading in elbows, ankles, joints and the like are most suitable for biodegradable implants. Other than natural materials, synthetic biodegradables have also appeared on the scene since the late 1960s in the form of sutures made from polylactic acid and polydioxanones that are mainly used in bone pins and the like (Suuronen and Lindqvist, 2002). Polyanhydrides have also been introduced for implantable drug delivery systems. Today, the list of biodegradable implants is growing by the day, and includes most suture materials, staples, tacks, anchors, interference screws, scaffolds, rods and various other devices for body part repairs (Dhillon *et al.*, 2008, p319). Despite their acceptability and wide application area, there are still relatively few *in vivo* test results confirming their all-round suitability compared to those carried out *in vitro*. This is because *in vitro* results do not necessarily replicate what actually happens in the human body and therefore some controversial assessments have emerged (Hofmann *et al.*, 1995, p123).

5.3.3 Grafts

Grafting is a universal process by which damaged or diseased organs, tissues, muscles and skin are repaired by placement or implantation of natural or artificial materials. These implants/materials can come from the patient's own body parts (i.e. autologous), or from a related individual (i.e. isogenic), or they may come from a different species, such as an animal (i.e. xenogeneic). In the absence of such possibilities and/or for a range of good medical reasons, artificial or so-called prosthetic materials may be the only suitable option. Prosthetics could include biodegradable or non-biodegradable implants. This is an area where textile materials play a crucial part, provided that a number of medical and clinical prerequisites are met. The textile materials to be used must be biocompatible, non-toxic, inert and sufficiently adaptable to the physical and mechanical stresses that could be endured. Vascular grafts are one such example where damaged or diseased arteries maybe replaced by a prosthetic graft. The need for such medical devices arises when the heart suffers from thrombosis, a condition caused by the formation or presence of blood clot inside a blood vessel or heart cavity, or when aneurysm is detected, i.e. a ballooning out of the wall of an artery caused by weakening of the artery due to disease or injury. If left untreated, aneurysm can lead to rupture of the vessels and serious blood haemorrhage could occur. Arteriosclerosis, or deposition of fatty substances such as cholesterol, calcium or clotting materials in the inner lining of arteries, is another type of vascular malfunction that would require external intervention (Fishbein and Fishbein, 2009, p1309).

Development of prosthetic arteries to be used in these conditions started off as flexible plastics, mainly nylon tubes in the 1940s; however, these early attempts were largely unsuccessful because of their inherent lack of permeability and stress concentrations caused at suture joints, leading to thrombosis (Spronk *et al.*, 2004). A few years later, woven and knitted PTFE or Teflon, and polyethylene terephthalate (PET) or Dacron, were introduced and Dacron prosthetics gained the upper hand due to their softer handle, better porosity and better blood cell adhesion characteristics. A new type of ePTFE was introduced in 1970s with much improved properties. Considerable further developments in the past 10–20 years have taken place in manufacturing, processing and application criteria of prosthetic grafts, mainly to better address issues of porosity (depending on fabric structure), burst strength, elasticity and *in situ* adaptability e.g. inclusion of stents, and so forth. Currently, other manufacturing methods, such as 3D braiding, and non-wovens such as electrospun, are possibilities that are under investigation. What continues to be a technical challenge is the manufacturing and processing of narrow grafts i.e. less than 6 mm, where obstructions and malfunctions within these systems remain largely unresolved (How *et al.*, 1992).

5.3.4 Sutures

One of the earliest reports depicting stitches, or a kind of suture intervention, refers to ancient South Africans who would use giant black ants to bite wound edges together with their forceful jaws, literally clipping the two sides together. To ensure sustained grip, the bodies of the ants would then be snapped off leaving the head *in situ* to ensure natural healing take its course

(Gudger, 1925). Although the concept of joining two surfaces together is still the same, much has changed in the technology of suture materials and their application. Silk threads from silkworm, and catgut strings retrieved from the intestines of animals, were later developments of suture materials and are still used in some parts of the world (Malik, 2002). However, modern sutures are mainly synthetics, with the choice of bio-absorbability or non-bio-absorbability i.e. those that do not need to be removed because of their ability to disintegrate and are hence naturally removed from the body, and those which must be taken out. In addition to the basic ability to close wounds, sutures today are required to do much more. Treatment with antimicrobials or such inherent abilities in sutures prevents bacterial infection, inflammation and medical complications. Knotting and casting off sutures could also be difficult to achieve, or be inadequate, due to material slippage and suture failure caused by localised stresses. Later generations of suture designers have proposed many interesting theories and solutions for these problems (Muffly *et al.*, 2011). Other than expectation of rapid healing, exterior wounds, in particular, must today be cosmetically acceptable and leave no or minimal scars. This is a fast increasing growth area as plastic surgery and its wider availability to all becomes prevalent.

5.4 Textiles for tissue engineering applications

Tissue engineering is the marriage between biological matter and engineered materials in such a way as to assist re-growth and restructure of biological process or even, in some instances, to replace what is biologically impossible to reproduce. Compared to other techno-biological advances, tissue engineering is a relatively recent phenomenon, with a modest history of no more than 50 years at most. The need for tissue engineering intervention arises for a number of pressing reasons including: shortage of suitable organs for transplant or regeneration; inadequacy of some prosthetic or biological substitutes to perform well; and the versatility in design and shape formation now offered by tissue engineering (Griffith and Naughton, 2002).

Textile materials, for a number of reasons, are ideal candidates for tissue engineering applications. Their fineness provides flexibility and large surface areas which are ideal for cell growth and multiplication. Their construction in whatever form, e.g. woven, knitted or non-woven, provides porosity, accessibility for cell attachment, and gaseous exchanges, as well as biodegradability, where required, besides being highly efficient and minimally disruptive.

Tissue engineering allows surgeons the possibility of cultivating living cells/tissues that would ultimately amalgamate fully with the patients' own tissues. Textile materials made into suitable scaffolds are the most important part of the structure which supports the tissue forming cells. When functionally correct types of scaffold material are used, the cells attach themselves to the underlying structure and begin to divide, developing into extracellular matrices. Depending on the type of polymer used, the encapsulated fibres could form the overall permanent structure or, if biodegradable polymers are used, they would gradually break down and leave a self-supporting biological structure (Liu and Czernuszka, 2007).

Ideally, any material being considered for tissue engineering must meet the following requirements:

- The surface characteristics of the scaffold must allow cell adhesion, allow cell growth and permit the retention of cell functions.
- The biocompatible material must be non-toxic and not produce cytotoxic degradation products.
- If bioresorable materials are used, they must degrade with time and be removed through natural body functions without causing any side effects.
- The porosity within the scaffold structure must be such a way as to allow space for cell adhesion, and extracellular matrix generation, and minimise diffusional restriction during tissue growth.
- The material should have adequate physical strength and mechanical integrity to enable the scaffold to be handled and put into position.
- The material should be non-allergic and not sensitive to sterilisation.

Apart from woven, knitted, or non-woven, scaffolds could be made by embroidery, braiding and the like; but by and large, non-wovens are often preferred, allowing better penetration of cells into the structure and their subsequent proliferation. Finer spread of individual filaments within non-wovens also resembles more closely the collagen matrices found within humans. Lately, other techniques of producing scaffolds are being considered, where random or regulated nanostructures in 2D or 3D configurations could be formed via electrospinning. These methods of scaffold production are potentially better, because they create much greater surface area, porosity and versatility for cells to grow, and have better efficiency in disintegration, if required, or absorption by the body (Cui *et al.*, 2010).

5.5 Textiles for healthcare monitoring

As living standards improve and people live longer, their medical and healthcare requirements become greater, causing unsustainable strains on the healthcare providers which will inevitably affect the quality of the care being offered. To address these eventualities, patients in the twenty-first century increasingly need to be cared for at their homes and, where possible, to be monitored remotely, thus leaving the much needed hospital beds for critical cases and emergency use. Moreover, since labour and medication costs are unlikely to decline, much greater emphasis needs to be put on preventative and prognostic health measures rather than diagnostic and reactive solutions, if everyone is to receive a decent level of care and dignified existence.

Electronic textiles, or smart clothing, are one such example where inclusion of sensors and smart monitoring devices within clothing can sense and record basic bodily functions such as body temperature variation, breathing anomalies, heart activities, etc., and subsequently transmit the data via available accessories such as land or mobile phones, internet, and so on, to hospitals/health centres or even alert emergency units (Lymberis, 2003). When fully adopted, this approach would considerably reduce costs and would not only ensure patients' welfares are not compromised but in fact be substantially enhanced, due to continuous and uninterrupted patient surveillance. More refined and targeted extension of this approach to healthcare has already led to clothing and garments that predict muscular overloads and stress concentrations, helping to prevent strain injuries, monitor babies' heart beats whilst in the womb, and biochemical signal analysis of perspiration generated in normal daily activities or when engaged in strenuous physical exercises (Milenković *et al.*, 2006; Solomonow, 2006).

Combating bacteria via smart means has also attracted much attention as the number of drug resistant pathogens multiply. For instance, use of oscillating magnetic field generators, plasma and electrical zapping are some of the methods currently being assessed for full deployment in appropriate circumstances. Use of treated or loaded nanofibres as a coating medium or standalone air/water filters, or drug dispensation units, are some of the other textile-based advances in this area. The fashion industry has also capitalised on electronic textiles, creating all kinds of visual and entertainment effects, possibly influencing the psychology and social interaction of the wearers. Wireless keyboard fabrics with exciting prospects and even greater influence on daily lives of people are the latest addition of this multidisciplinary approach to use of textiles (Nakad *et al.*, 2010).

5.6 Future trends

Textile materials today are addressing a far greater number of fundamental problems than they were ever intended or expected to do in the past. Where fibre fineness and general flimsiness was once an inconvenience and a headache to manage in traditional textile manufacturing, today these very attributes have given textile materials a raised profile and rebirth in a broad range of multidisciplinary areas. Textile materials are today considered to be imperative ingredients of most engineering and biological assemblies, where their presence contributes to structural reinforcement, absorbency, degradation efficiency, and controlled and targeted drug delivery and dispensation, as well as serving as a base on which living cells grow and nourish. With the advent of micro/nanofibres and inclusion of nano-tubes, nano-particles and the like, future prospects for textiles in general, and medical and hygiene in particular, looks increasingly bright. Technologists, engineers and biologists are no longer working in isolation, and through their combined skills and interdisciplinary collaborations, they are pushing the boundaries of science and technology to new heights. The challenge is increasingly to mimic and reproduce Mother Nature in all her forms and shapes. Intricacy in design, complexity in function and efficiency in performance are the key features of these new outlooks, and will keep many researchers busy for now and the foreseeable future.

5.7 References

Aranaz I., Mengíbar M., Harris R., Paños I., Miralles B., Acosta N,Galed G. and Heras A. (2009), Functional characterization of chitin and chitosan, *Current Chemical Biology*, **3**, 203–230.

Adewumi A.A. and Ogunjinmi A.A. (2011), The healing potential of honey and propolis lotion on septic wounds, *Asian Pacific Journal of Tropical Biomedicine*, **1** (1), September 2011, S55–S57.

Atiyeh B.S. Costagliola M., Hayek S.N. and Dibo S.A. (2007), Effect of silver on burn wound infection control and healing: Review of the literature, *Burns*, **33** (2), March 2007, 139–148.

Behrouz R., Ourmazdi M. and Reza'I.P. (1993), *Iran – The Cradle of Science*. 21st edn., Iran Almanac, 115–118.

Carter M.J., Tingley-Kelley K. and Warriner R.A. (2010), Silver treatments and silver-impregnated dressings for the healing of leg wounds and ulcers: A systematic review and meta-analysis, *Journal of the American Academy of Dermatology*, **63**(4), October 2010, 668–679.

Cui W., Zhou Y. and Chang J. (2010), Electrospun nanofibrous materials for tissue engineering and drug delivery, *Science and Technology of Advanced Materials*, **11**, 014108 (11 pp).

Dhillon M.S., Prabhakar S. and Prasanna C. (2008), Preliminary experience with biodegradable implants for fracture fixation, *Indian Journal of Orthopaedics*, July–September, **42**(3), 319–322.

Encyclopaedia of Surgery (2007), A Guide for Patients and Caregivers: A-Ce. Aseptic Technique forum. Aseptic Technique. www.surgeryencyclopedia.com/A-Ce/Aseptic-Technique.html.

Fishbein G.A. and Fishbein M.C. (2009), Arteriosclerosis, rethinking the current classification, *Archives of Pathology & Laboratory Medicine,* **133,** 8 August 2009, 1309–1316.

Geertsma R.E., de Bruijn A.C.P., Hilbers-Modderman E.S.M., Hollestelle M.L., Bakker G. and Roszek B. (2007), New and Emerging Medical Technologies, A horizon scan of opportunities and risks, Report 65/07 BMT/RB/RG/cvr.

Griffith L.G. and Naughton G. (2002), Tissue engineering – current challenges and expanding opportunities, *Science,* 8 February 2002, **295** (5557), 1009–1014.

Gudger E.W (1925), 'Stitching wounds with the mandibles of ants and beetles', *Journal of the American Medical Association,* **84,** 1861–1864.

Hamouda H. (2005), *Handbook of Environmental Degradation of Materials.*

Hearle J.W.S. (2000), A critical review of the structural mechanics of wool and hair fibres, *International Journal of Biological Macromolecules,* **27** (2), April 2000, 123–138.

Hiller, L. (1927–1938), *Sutures in Ancient Surgery,* Brooklyn, NY: Davis & Geck.

Hofmann G.O. (1995), Biodegradable implants in traumatology: a review on the state-of-the-art, *Archives of Orthopaedic and Trauma Surgery,* **114** (3), 123–132, DOI: 10.1007/BF00443385.

How T.V., Guidoin R. and Young S.K. (1992), Engineering design of vascular prostheses, *Journal of Engineering in Medicine,* **206,** 61–71.

Khan M.N. and Naqvi A.H. (2006), Antiseptics, iodine, povidone iodine and traumatic wound cleansing, *Tissue Viability Society,* **16** (4), November 2006, 6–10.

King M.W. (2001), Overview of opportunities within medical textiles, *Canadian Textile Journal,* July/August, **118** (4), 34–36.

Liu, C.Z and Czernuszka, J.T. (2007), Development of biodegradable scaffolds for tissue engineering: a perspective on emerging technology, *Materials Science and Technology,* **23** (4), P379–P391.

Lymberis A. (2003), Intelligent biomedical clothing for personal health and disease management: state of the art and future vision, *Telemedicine Journal and e-Health* **9** (4), © Mary Ann Liebert, Inc.

Malik N.A. (2002), *Textbook of Oral And Maxillofacial Surgery,* Published by: Jaypee Brothers Medical Publishers (p) Ltd., New Delhi. First published in 2002.

Mezzana P. (2008), Clinical efficacy of a new chitin nanofibrils-based gel in wound healing, *Acta Chirurgiae Plasticae,* **50** (3), 81–84.

Miraftab M., Barnabas J., Kennedy J.F. and Masood R. (2011), Antimicrobial properties of alginate-chitosan (alchite) fibers developed for wound care applications, *Journal of Industrial Textiles,* **40** (4), April 2011, 345–360.

Miraftab M. and Smart G.M. (2007), Composite fibre of alginate and chitosan, Patent: international:A61L15/28;C08J5/18;D01F/18, (WO2007GB00537)20070215; GB20060003122 20060216.

Milenković M., Otto C. and Jovanov E. (2006), Wireless sensor networks for personal health monitoring Issues and an implementation, *Computer Communications,* **29** (13–14), 21 August 2006, 2521–2533.

Muffly T.M., Tizzano A.P. and Walters M.D. (2011), The history and evolution of sutures in pelvic surgery, *Journal of The Royal Society of Medicine,* **104**(3): 107–112.

Nakad Z., Jones M., Martin T. and Fawaz W. (2010), Networking in E-textiles, *Computer Communications,* **33** (6), 15 April 2010, 655–666.

Novikova G.E. (2011), Introduction to corrosion of bioimplants, *Protection of Metals and Physical Chemistry of Surfaces,* **47** (3), May 2011, 372–380.

Owokotomo I.A. and Owoeye G. (2011), Proximate analysis and antimicrobial activities of Bambusa Vulgaris L. leaves' beverage, *African Journal of Agricultural Research,* **6** (21), 5 October 2011, 5030–5032.

Porczynski C.Z.C. (1960), *Manual of Man-Made Fibres,* Guildford, Astex Pub. Co.

Rhoades J. and Roller S. (2000), Antimicrobial actions of degraded and native chitosan against spoilage organisms in laboratory media and foods, *Applied*

Environmental Microbiology, January, **66**, 80–86; doi:10.1128/AEM.66.1.80-86.2000.
Sekar, N. (1999), Single-stage preparatory processes for cotton – an update, *Colourage,* **46** (5), 31–32.
Silver S. and Phung L.T. (2009), Heavy Metals, Bacterial Resistance, Encyclopedia of Microbiology (Third Edition), 220–227.
Solomonow M. (2006), Sensory – motor control of ligaments and associated neuromuscular disorders, *Journal of Electromyography and Kinesiology,* **16** (6), December 2006, 549–567.
Spronk H.M.H., der Voort D.V. and Cate H.T. (2004), Blood coagulation and the risk of atherothrombosis a complex relationship, *Thrombosis Journal*, **2**, 12.
Suuronen R. and Lindqvist C. (2002), Bioresorbable materials for bone fixation: review of biological concepts and mechanical aspects, *Craniomaxillofacial Reconstructive and Corrective Bone Surgery*, Section II, 113–123.
Tormala P., Pohjonen T. and Rokkanen P. (1998), Bioabsorbable polymers: Materials technology and surgical applications, *Journal of Engineering in Medicine*, 1 February 1998, **212** (2), 101–111.
Woodings C. (2001), *Regenerated Cellulose Fibres*, Woodhead Publishing Limited, ISBN: 978-1-85573-459-3.
Yang H.M., Comprehensive Composite Materials, Volume 1, 199–229.

6

High performance textiles for wound care

S. RAJENDRAN and S. C. ANAND, University of Bolton, UK

DOI: 10.1533/9780857099075.190

Abstract: This chapter discusses the physiology of wounds and their management. The classification of wounds and appropriate dressing selection for successful wound management are outlined. High-tech wound dressings and the current state-of-the-art novel dressings, such as odour adsorbent and antimicrobial dressings, are critically reviewed. The role of compression therapy in the treatment and prevention of venous leg ulceration is discussed. The merits and limitations of the current compression therapy regime, and research into the development of novel orthopaedic padding and compression bandages, are highlighted.

Key words: wound, debridement, dressing, antibacterial, antiodour, venous leg ulcer, padding bandage, compression bandage, spacer fabric bandage.

6.1 Introduction

The market potential for healthcare and medical textile devices is rapidly increasing. In the EU alone, sales of medical textiles are worth US$ 7 billion, and already account for 10% of the market of technical textiles. The EU sector consumes 100 000 tonnes of fibre per annum, and this is growing in volume by 3%–4% a year. The global medical device market was valued at over US$100 billion, of which US$43 billion was generated from the US market. Western Europe is the second largest market, and accounts for nearly 25% of the global medical device industry. The UK has one of the largest medical device markets in the world. The market is dominated by the National Health Service (NHS), accounting for approximately 80% of healthcare expenditure, even though there are fewer private sectors. It is forecast that the share of hygiene and medical textiles would be 12% of the global technical textiles market and would account for US$4.1 billion.

The healthcare and medical devices market is driven by:

- population growth;
- the ageing population;
- rising standards of living and higher expectations of quality of life;

- changing attitudes to health; and
- the emergence of innovations and the availability of increasingly high technology.

It has been predicted that there is a substantial market potential for advanced wound dressings. The woundcare industry generated between US$3.4 and 3.6 billion for the period from 2010 to 2012, mostly from the US and Europe. The forecast for annual growth would be between 7% and 10% in 2019. In the US alone there are over 100 000 surgeries performed daily involving surgical wounds. Ageing population creates increased demand for all types of surgical intervention, particularly cardiovascular, orthopaedic, urological, and dermatological procedures. The annual cost of treating pressure ulcers accounts for 4% of the total NHS expenditure in the UK, and the annual cost of treating diabetic foot ulceration accounts for 5%. The treatment of venous leg ulcers creates considerable demands upon healthcare professionals throughout the world. In the UK alone, the treatment of this condition costs the NHS £600 million per year. The cost of treating an individual venous leg ulcer has been estimated to vary between £557 and £1366 over 1 year period. In the US, venous leg ulcer affects 3.5% of people over the age of 65, and has been estimated to cost between $2.5 and $3.5 billion per year.

6.2 Wounds and wound management

Wounds can be classified into acute wounds and chronic wounds. While acute wounds take only a few weeks to heal, chronic wounds require several months to heal completely. Chronic wounds include venous leg ulcers and pressure sores. The classification of wounds and healing process is depicted in Figs 6.1 and 6.2. Generally, the wound healing process involves three phases:

- the inflammatory phase, which occurs immediately after injury to tissue, and during which swelling takes place;
- the proliferation period, in which new tissues and blood vessels are formed; and
- the maturation phase, in which tissues laid down during the proliferation stage are remodelled.

The care and treatment of wounds are matters of clinical judgement. The wounds should first be examined, to determine the amount of exudate and infection that exists. In addition, the surrounding tissue should be assessed to establish the extent of ulceration. The first stage of treatment (debridement) often requires the removal of foreign material and dead tissue from the wound, in order to prevent infection and promote healing. Subsequent

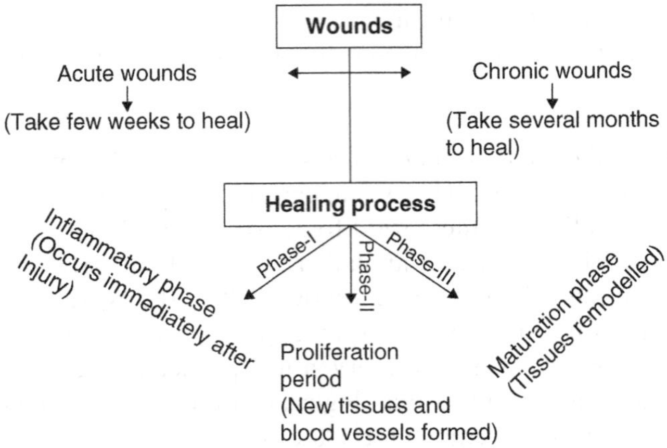

6.1 Classification of wounds and healing process.

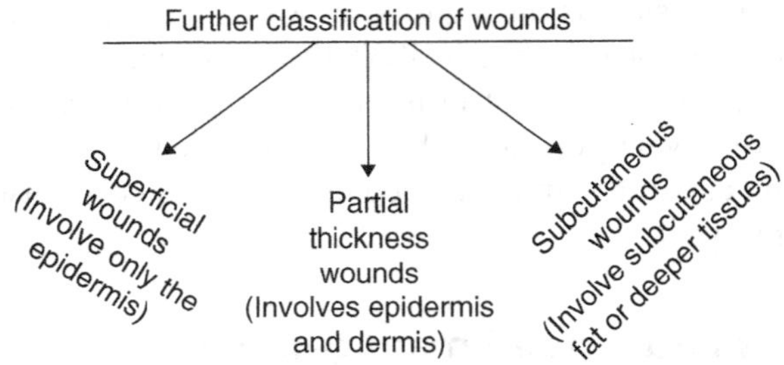

6.2 Classification based on depth of wounds.

stages in wound management include control of exudate, stimulation of wound healing, and wound protection.

6.2.1 Autolytic debridement

There are a variety of methods that can be used to debride a wound, including autolytic, chemical, mechanical, and surgical.[1] Autolytic debridement utilises the body's own enzymes and moisture to dissolve and clean the wound of necrotic tissue. This method has the advantage of being virtually painless for the patient, but can take many weeks to achieve. A moist wound environment is the key factor to autolytic debridement, and is obtained by using occlusive or semi-occlusive wound dressings.

6.2.2 Chemical debridement

Chemical debridement involves the use of topical enzymatic gels and solutions to dissolve and remove necrotic tissue from the wound. Different enzymes target specific components of necrotic tissue (i.e. fibrin, collagen) and are categorised as proteolytics, fibrinolytics, and collagenases.[2] This method is regarded as safe and effective, but should be limited to the area of necrosis or slough. The proteolytic enzyme fibrinolysin, which targets fibrin, is also said to release growth factors into the wound bed.[3] Fibrinolysin is normally prepared in combination with deoxyribonuclease (DNase) to form a topical ointment. A randomised study of 84 patients with chronic venous leg ulcers concluded that no long-term clinical benefit was gained by using proteolytic ointments when compared to a placebo treatment.[4]

6.2.3 Mechanical debridement

The various methods available for mechanical debridement include the use of wet-to-dry dressings, whirlpool baths, and wound irrigation (pulse lavage). The wet-to-dry technique utilises saline-moistened gauze, which is applied over the wound and allowed to dry.[5] As the gauze dressing dries, superficial wound debris and necrotic tissue adhere to it. Wound debridement occurs as the dressing is removed from the patient. However, this particular technique can be very painful, since the dressing also adheres to living as well as necrotic tissue. Ulcer wounds may also be cleansed and debrided by using a pressurised saline solution (pulse lavage). In this technique, the saline solution is passed through a catheter via a syringe in order to generate a fluid pressure of around 15 psi. The pressurised solution irrigates the wound and removes loose debris and necrotic tissue without causing pain to the patient.[5] The use of whirlpool baths has declined, mainly due to the noted increase in cross-contamination and wound infection.[5]

6.2.4 Surgical debridement

The most thorough and accurate method to debride an ulcer wound is to surgically remove any necrotic tissue.[6] Surgical instruments (scalpel, forceps, scissors, or laser) are used to cut away the dead tissue. Surgical debridement may be associated with bleeding, which can be controlled by direct pressure, electrocautery, or by cauterising the wound with silver nitrate.[5]

6.2.5 Biodebridement therapy

An emerging technique, maggot debridement therapy (MDT), that uses live maggots for wound debridement, is now gaining popularity in managing chronic wounds. The treatment is more effective and efficient at debriding many types of infected and gangrenous wounds, and ultimately enhances wound healing. The maggots are branded as 'World's Smallest Surgeons' and MDT is described as 'Biotherapy'. The therapy has been around for centuries, but much attention is now focused on the resurgence of this treatment. A surgical materials testing laboratory (SMTL), Cardiff, UK is doing a pioneering work and is the sole UK breeder of maggots. The treatment involves the placement, using restrictive dressings, of live medical-grade maggots into non-healing wounds to provide for cleansing of necrotic tissues and initiation of the healing process. Maggots clean wounds by eating dead tissues. They first spit out enzymes that liquefy the tissue and then suck it up like soup. Maggots clean wounds by dissolving dead and infected tissues, disinfect the wound by killing bacteria, and stimulate wound healing. The mechanism of maggot therapy is discussed elsewhere.[7] A study highlighted the pain level in diabetic and non-diabetic patients treated with maggot therapy.[8] An *in vivo* study conducted on 64 patients to determine the effective methods of maggot therapy revealed that containment method reduces the effectiveness of MDT.[9] It is reported[10] that MDT enhances the growth of granulation tissue and promotes greater wound healing rates in non-healing foot and leg ulcers. MDT was used repeatedly for the first time on a patient in ICU to avoid amputation of the leg.[11] There are challenges ahead in full use of MDT at hospitals: sourcing medical-grade maggots is a logistical challenge that remains a problem in many parts of the world; high cost; patients' perception and acceptability; storage and handling; training and education for tissue viability nurses; and lack of evidence-based research.

6.2.6 Antibiotic treatment

If the ulcer wound is infected with bacteria and the patient develops associated cellulitis or lymphangitis (erythema, swelling, tenderness, pain, and fever) treatment with systemic antibiotics is required. Various microorganisms are commonly found in wounds, and wound healing can be severely impaired if a sufficient number of pathogenic species of bacteria multiply in the healthy tissue. Bacteriological examination of the ulcer wound base using swabs is required in order to determine the type of pathogens present. Culture swabs should be taken directly after wound cleansing and debriding, making sure the bacteria sample is from the wound base and not from the wound edge. Information concerning the type of bacteria present allows the clinician to select an appropriate oral antibiotic therapy.

6.3 Wound dressings

The healing of wounds depends not only upon medication but also upon the use of proper dressing techniques and suitable dressing materials. Dressings should be easy to apply and painless on removal. They should be able to create the optimal environment for wound healing, and should be designed to reduce nursing time by requiring fewer dressing changes.

Table 6.1 Commercial wound dressings

Commercial name	Major application
Low-adherent dressing Adaptic, release, Melolin, Telfa, Skintact, Mepore, Primapore, Tricotex, NA Ultra, Jelonet, Paranet, Paratulle, Unitulle, Vasaline gauze, Inadine, Sofra-Tulle, Silicone NA, Mepitel, Metalline	Reduces risk of adherence to wounds. Less trauma
Semipermeable film dressing Mefilm, Tegaderm, Bioclusive, Cutifilm, Epi View, Opsite Flexigrid	Lightly exuding wounds, superficial pressure sores, primary dressing and secondary dressing in combination with alginates and hydrogels
Odour adsorbing dressing Actisorb Plus, Carbonet, Kaltocarb, Metrotop, Cliniflex	Undesirable odour producing wounds
Hydrocolloid dressing Aquacel, Granuflex, Comfeel, Combiderm, Cutinova Foam, DuoDerm, Tegasorb, Hydrocoll	Light to medium exuding wounds. Not suitable for infected wounds
Hydrogel dressing AquaForm, Sterigel, Purilon Gel, Intrasite Gel, Granugel Hydrocolloid Gel	Dry and necrotic wounds. Lightly exuding wounds and granulating wounds. Not suitable for infected and heavily exuding wounds
Polyurethane foam dressing Allevyn, Tielle, Lyofoam, Sterigel, Nu-Gel, Purilon Gel, Intrasite Gel, Granugel, Flexipore, Spyrosorb	Light to medium exuding wounds. Not recommended for dry superficial wounds
Alginate dressing Sorbsan, Tegagel, Kaltostat, Algosteril, Algisite, Algoderm, Melgisorb, Kaltogel, Tegagel, Algosteril and Comfeelseasorb	Mainly primary dressing for medium to heavily exuding wounds and cavity wounds. Not suitable for dry wounds
Alginate- collagen dressing Fibracol	Suitable for dressing foot ulcers and heel pressure sore

There are numerous types of wound dressings available for the management of different kinds of wounds (Table 6.1). The ideal dressing should protect the wound, keep it moist and warm, remove exudate, promote healing, and reduce the risk of infection.[12] Primary wound dressing should have considerable capacity to absorb liquids. The absorbent dressing should be changed frequently, in order to avoid wound odour and also to prevent the development of dermatitis of the surrounding skin. Switching to an occlusive dressing after the initial treatment stage provides the optimal moist environment for promoting wound healing. Occlusive dressings (i.e. hydrocolloids, hydrogels, and alginates) retain wound fluids that contain growth factors, enzymes, and immune cells, which help to accelerate wound healing. In comparison to non-occlusive dry dressings, occlusive dressings also prevent bacteria from entering the wound, thereby reducing the likelihood of infection.[13] Alginate dressings exchange sodium ions with calcium ions that are exuded from the wound. The exchange in ions creates a fibrous gel, which helps to provide a moist and warm wound environment. Hydrocolloid dressings produce a similar warm and damp environment by liquefying and swelling on contact with the wound. In addition, hydrocolloid dressings provide a barrier to microorganisms and help to reduce pain by keeping surrounding nerve endings moist.[13]

Generally, the dressing is placed directly over the wound (primary dressing) and is covered with an absorbent pad (secondary dressing). The whole dressing is then retained with adhesive tape or a suitable bandage, depending on the location of the wound in the body. The primary dressing is expected to maintain the wound temperature and moisture level to permit respiration and to allow epithelial migration. The secondary dressing must not be too absorbent, as it may cause the primary dressing to dry out too quickly. Different shapes are available which are suitable for dressing wounds in difficult positions such as heels, joints, digits, and the perineal area.

Wound dressing materials are mainly classified as absorbent and non-absorbent, depending on the types of fibres used. Dressings vary with the type of wound and wound management, and no single dressing is universally applicable.

An ideal dressing is normally expected to:

- provide a barrier against microorganisms, dirt, and other foreign bodies;
- provide a humid environment at the wound surface;
- control exudates; and
- be capable of removal without trauma.

Dressings are also used to protect against further injury and abrasion by acting as a cushion, and also promote rather than interfere with the healing process. Traditionally, cotton gauze is used for dressings because of its good absorption properties and soft handle. However, it has been established that cotton gauze allows moisture to evaporate from the wound, which means that cotton gauze dressings do not maintain the moist environment to facilitate faster wound healing. Furthermore, cotton gauze adheres to the wound bed. This means that patients suffer pain when the gauze is removed. Also, cotton gauze requires frequent changes. A list of commercial wound dressings is published elsewhere.[14]

A modern wound dressing consists of absorbent layers held between a wound contact layer and a base material. The absorbent layers absorb blood,

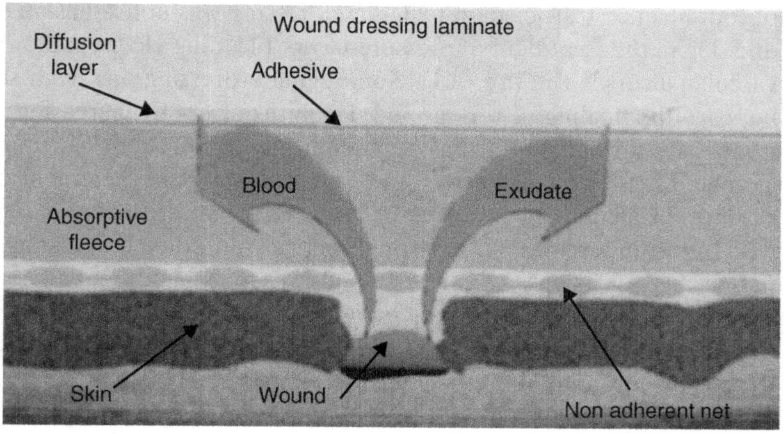

6.3 Wound dressing concept.

body fluids, and exudates. The wound contact layer is non-adherent and can easily be removed without disturbing new tissue growth. The general wound dressing model is represented in Fig. 6.3.

6.4 Hi-tech wound dressings: non-adherent, odour adsorbent, activated charcoal cloth (ACC), and antimicrobial dressings

There are several different shapes and sizes of dressings suitable for treating wounds in difficult or awkward positions such as heels, joints, digits, and perineal area. Dressing varies with the type of wound and wound management, and no single dressing is universally applicable. The ribbon gauze dressing soaked in antiseptic solution is still used for wound management because it is inexpensive. A study comparing the use of ribbon gauze soaked with antiseptic, proflavine, and a hydrofibre dressing showed that the ribbon gauze was painful for the patient, extended the length of stay in hospital, and necessitated more analgesia compared with the hydrofibre dressing.[15] Alginate dressing is increasingly used for the treatment of diabetic foot ulcer.[16] It forms a hard occlusive matt over the ulcer and thus prevents the continuous drainage. A novel alginate dressing (AGA-100) has been developed that does not have an inhibitory effect on proliferation of fibroblasts.[17] Seven to eight rectangular 22 mm × 22 mm wounds, 0.3 mm deep, were made in the paravertebral areas of pigs and the closure rate of the wounds was measured. The results showed that the closure rate of full thickness wounds treated with AGA-100 is significantly higher (98%) on day 15 compared with that of well-known Kaltostat and Sorbsan alginate dressings. Collagen-alginate wound dressing is also an effective dressing for the management of foot ulcers.[18] Calcic-sodium alginate dressing was utilised as an effective dressing in the treatment of pressure ulcers, bleeding ulcers, and/or infected vascular ulcers.[19] The use of calcium alginate dressing has been shown to improve the healing of venous and decubitus ulcers by increasing granulation tissue formation.[20] It has been clinically proved that calcium/sodium alginate dressings are effective and easy to use for the treatment of cavity wounds.[21] Calcium alginate dressing reduces the severity of pain to patients with burns undergoing skin grafting, and is favoured by the nursing personnel because of its ease of care.[22] It is also used as an effective haemostatic dressing to reduce post-operative pain after haemorrhoid surgery,[23] as the standard dressing for split skin donor sites[24] and as the drug delivery system for the treatment of various surgical infections.[25]

A study has shown[26] that a lyocell fibre dressing can replace the alginate dressing for the treatment of chronic wounds. Lyocell is chemically converted to produce a new fibre, Hydrocel, using carboxymethyl cellulose. The dressing made from Hydrocel gels on contact with wound fluid, like alginate, provides a moist environment and is non-adherent. A wound dressing comprising a chitin wound facing layer and an expandable polytetrafluoroethylene (PTFE) non-woven has been developed.[27] The chitin layer is absorbed by the body while the PTFE serves as a barrier to bacteria and water. Stone *et al.*[28] studied the use of chitosan dressings for healing at skin graft donor sites. In this study, half of the split skin graft donor sites were dressed with chitosan and the other half with conventional dressings, and the rate of healing was evaluated. It was observed that chitosan dressing facilitated rapid wound re-epithelialisation and the regeneration of nerves within a vascular dermis. Biagini *et al.*[29] developed a chitosan derivative, N-carboxybutyl chitosan, dressing for treating plastic surgery donor sites. The solution of N-carboxybutyl chitosan was dialysed and freeze-dried to produce a soft and flexible pad, which was sterilised and applied to the wound. The dressing promoted better histoarchitectural order and better vascularisation. Muzzarelli[30] introduced a chitosan derivative, 5-methylpyrrolidinone, which has proved promising in medical applications. He claimed that this polymer is compatible with other polymer solutions including gelatin, poly(vinylalcohol), poly(vinylpyrrolidone), and hyaluronic acid. The advantages claimed by the inventor include healing of wounded meniscal tissues, healing of decubitus ulcers, depression of capsule formation around prostheses, limitation of scar formation, and retraction during healing.

Mepitel is a non-adherent silicone dressing made of a medical-grade silicone gel bound to a soft and pliable polyamide net.[31] It can be left in place for up to 2 weeks, and during this time the secondary dressing can be changed without risk of damaging delicate epithelial tissue. Similarly, Tegapore consists of a fine polyamide net containing pores of 90 microns that permit the free passage of exudate into the secondary absorbent layer, but the pores are too small to allow the ingress of granulation tissue.[32] The perforated net prevents the dressing adhering to the surface of the wound, and the holes allow the passage of exudate to the absorbent layer. It may be mentioned that the ability of perforated net dressings to absorb the body fluids is limited and, therefore, a secondary absorbent dressing/pad is normally used over the primary net dressing. In this situation, the absorbency of the dressing is less important than its low-adherent properties. Non-adherent dressing can also be produced by coating with a thin layer of aluminium by vacuum deposition, for example Metalline.[33] It should be noted that a recent survey involving nurses to study the effect of various commercial dressings such as alginates, film dressings, foam dressings, hydrocolloids, hydrogels, non-adherent dressings, paraffin tulle dressings, silicon products, knitted viscose dressings, and gauze on pain, trauma, and tissue damage at dressing changes showed that hydrogel dressings performed well. Seventy per cent of hydrogel dressings did not cause pain, trauma or tissue damage at dressing changes, followed by 50% of hydrofibre products and 40% of silicone products. Contrarily, only 10% of non-adherent dressings did not cause pain, trauma and tissue damage at dressing changes. However, 80% of non-adherent dressings investigated rarely caused pain, trauma, and tissue damage at dressing changes.[34,35]

6.4.1 Non-adherent wound dressings

The limitations of the currently available commercial non-adherent dressings are highlighted below:

- the majority are non-absorbent and/or hydrophobic;
- a secondary absorbent dressing/pad is normally used over the primary net dressing;
- the above dressings are often covered or retained in place by a tap, bandage or stocking, etc.; and
- the majority of high absorbent dressings based on alginate, hydrogel, hydrocolloid, etc., materials do not retain their original shape or integrity once they have been used and thus wounds have to be cleaned or washed to remove the remnants of such dressings left in the wound.

Novel two-in-one high-absorption and non-adherent dressings have been designed and developed to overcome the above drawbacks.[36] A number of novel knitted and crochet structures with enhanced absorbency have been designed by making use of weft-knitting and crochet technologies, and this has proved that a careful examination of all possible alternatives available for the manufacture of a product could lead to the design and development of these products. A novel non-adherent finish for the developed structures has also been formulated with superior properties. The salient features of the developed products for use as a primary wound dressing are listed below:

- A single-layer non-adherent and high-absorption dressing.
- The dressing retains its original shape and size even after use as a dressing material.
- Produces gels on the dressing surface that facilitate non-traumatic dressing removal.
- Maintains moist environment at wound site that promotes healing.
- Enhances the absorption of body fluids.
- The dressing is non-toxic, non-irritant, and non-allergic to humans.
- High absorbent capacity and absorption rate, which promote free flow of body fluids and prevent fluid pooling at the wound site thus reducing the risk of bacterial infection and wound maceration.
- Gelling facilitates non-adherence to the wound, which permits non-traumatic dressing removal. Gelling also maintains moist environment at the wound site that promotes wound healing.
- New products are soft and easy to handle, viz. no gelling in the dry state. They can also be used on dry wounds, unlike alginate dressing.
- A range of novel structures and finishing components have been developed and fully characterised by using commercial and relatively inexpensive medical-grade fibres such as bleached cotton, bleached viscose, bleached lyocell, and other natural, biodegradable and non-toxic fibres and finishing components by using conventional materials, equipment, and chemicals.

6.4.2 Odour adsorbent wound dressings

Wound malodour is increasingly becoming a problem. It mostly affects patients suffering with chronic wound types. It can affect the patients in numerous social and psychological ways. Patients often find the odour too embarrassing to socialise. These patients often withdraw socially from their friends, family and loved ones, which often leads to depression. The embarrassment is due to the foul odour being often misinterpreted as unhygienic and dirty. For some patients, the malodorous wound is an incurable chronic condition and is surrounded by a constant repulsive smell. Wound malodour generally affects chronic wound types, such as venous leg ulcers, fungating (malignant cancerous) lesions, diabetes foot ulcers, and pressure ulcers.

It is known that wound malodour is due to the presence of devitalised, necrotic (dead) tissue or is the result of severe colonisation/infection of bacterial microorganisms.[37] If the wound malodour is due to the presence of necrotic tissues, debridement can drastically reduce the malodour. If the wound malodour is due to the severe colonisation/infection, then it could be due to aerobic and anaerobic bacteria. It is now believed that anaerobic bacteria (*bacteroides* and *clostridium* species) are instrumental in the production of volatile odorous molecules. Often malodour is a mixture of different volatile agents, such as organic acids (n-butyric, n-valeric, and n-caproic) produced by anaerobic bacteria, and amines and diamines (cadaverine and putrescine) produced during metabolic processes.[38]

Methods of malodour treatment involve a combination of: (a) debridement to remove the necrotic dead tissue; (b) a short dose of an antibiotic treatment; and (c) a specialist odour adsorbent dressing. Currently, all the commercial odour adsorbent dressings contain a layer of highly adsorbent ACC. It is very efficient in the adsorption of the volatile odorous molecules bacteria; however, there are some issues regarding the use of ACC. The characteristics of these specialist odour adsorbent dressings can differ from relatively simple constructions to complex multilayer composites.

6.4.3 ACC wound dressings

Since 1976 the number of commercial odour adsorbent wound dressings has increased, but it was not until the late 1990s that they were available on prescription in the UK. There are a wide variety of odour adsorbent wound dressings available, but these can differ in construction, composition and, therefore, physical characteristics including odour adsorption. These dressings are composed of many different materials, ranging from the traditional naturally absorbent fibre types such as cotton and viscose, specialist materials such as polyurethane film, and fibres such as alginate and carboxymethylcellulose fibres (CMC).

At present all the commercial odour adsorbent dressings contain a layer of ACC (Fig. 6.4). These dressings can be loosely split into two categories, relating to their fluid up-take/absorption, either being absorbent or non-absorbent (Table 6.2).

The (fluid) absorbent activated charcoal wound dressings are generally composed of multilayer composites, with layers of highly adsorbent fibrous fleece or foam along with the ACC. Some of these dressings contain alginates and CMC fibres as the wound contact layer. They possess the characteristic of forming a hydrogel, thereby creating the moist microenvironment which promotes wound healing. The sophistication of some of these dressings can make them relatively expensive when compared to other dressings, and this often limits them being considered in deciding on the most appropriate course of treatment.

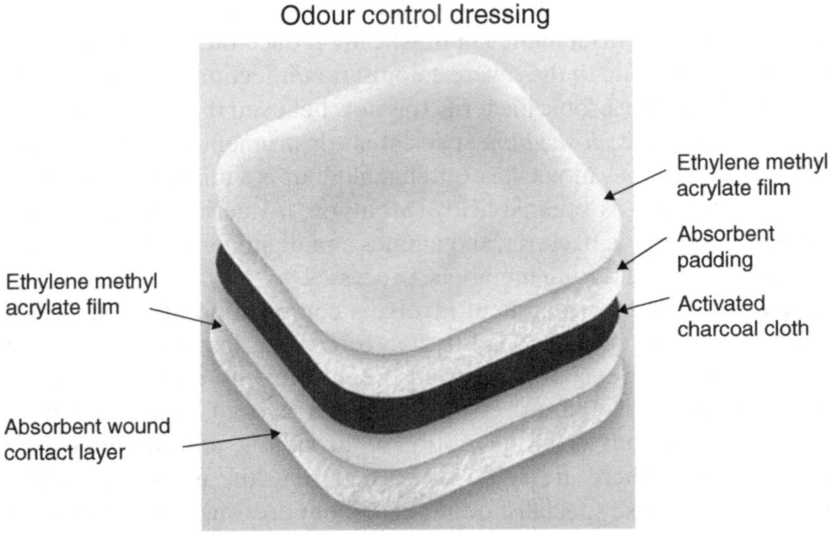

6.4 Odour adsorbent dressing.

The odour adsorbent ACC is sandwiched/encapsulated between two layers of a polyamide non-woven fabric. A popular example is Actisorb silver 220 (formally known as Actisorb plus) by Johnson & Johnson, which is the first to incorporate a layer of ACC to make an odour adsorbent dressing. This is also, currently, the only odour adsorbent dressing that possesses antimicrobial properties. The antibacterial silver is incorporated into the ACC layer. The ACC with the silver has to be in close contact with the wound's surface in order for the silver ions to react with the bacteria that are causing the infection. Therefore, there is a risk of ACC prematurely becoming saturated with the wound fluids, which can highly compromise the odour adsorption efficiency. Currently, a range of naturally occurring polymers is being investigated at the University of Bolton as alternative odour adsorbent materials to activated charcoal dressings.

6.4.4 Antimicrobial wound dressings

Bioactive fibres and polymers are high molecular weight natural and synthetic macromolecules and their complexes. The biomaterial research is focused in wound healing and antimicrobial materials where the design and mechanism of the biologically active molecule plays a key role in the textile fibre function. By achieving more insight into the actual activity of the molecule on the textile surface, it is possible to develop novel wound healing and antimicrobial materials. The mechanism of fibre activity is directly related to the complex biological environment surrounding the fibre. Hence,

Table 6.2 Commercial odour adsorbent dressings

Dressing	Manufacturer	Main components (ACC – activated charcoal cloth)
Carboflex	ConvaTec	ACC, absorbent multilayer with an alginate and CMC fibrous wound contact layer
Lyofoam C	Medlock	ACC, polyurethane foam
Carbonet	Smith & Nephew	ACC, absorbent multilayer laminate with a 100% viscose warp-knitted wound contact layer with claimed low adhesion
Sorbsan plus carbon	Pharma-plast Ltd	ACC absorbent multilayer laminate with an alginate fibrous wound contact layer
Actisorb silver 220	Johnson & Johnson	ACC with silver, non-woven polyamide
Carbopad VC	Vernon Carus	ACC, non-woven polyamide
Clinisorb	CliniMed Ltd	ACC, non-woven polyamide laminate
(Std) Melolin	Smith & Nephew	Absorbent cotton and polyester fibrous fleece with a perforated polyester film contact layer

this interdisciplinary subject area has brought together physical science disciplines from textile synthetic, and analytical and polymer chemistry, with the life science disciplines of medicine, biochemistry, biophysics, and microbiology.

Scientists have been working on the issues that have underpinned the making of more efficient wound dressings and antiseptic textiles for more than a century. The molecular bases of disease processes are better understood now, and our basic understanding of the structure and function of biologically active molecules enables the creation of bioactive fibres that can selectively interact with their biological environment. Some scientists have coined the term 'smart fabrics' to depict the targeted function these types of textiles have and their ability to perform a specific function in wound healing, arterial implants, or antimicrobial activity.

Many types of wound dressings have been developed, both non-medicated and medicated. Commercially available synthetic wound dressings consisting of a polyurethane membrane are capable of minimising evaporative water loss from the wound and preventing bacterial invasion, and thus are useful in the management of superficial second-degree burns. The ideal structure of an illuminate dressing consists of an outer membrane and an inner three-dimensional matrix of fabric or sponge. The outer membrane prevents body fluid loss, controls water evaporation, and protects the wound from bacterial invasion. On the other hand, the inner matrix encourages wound adherence by tissue growth into the matrix. Silver/nanosilver is mostly incorporated in the wound dressing which provides an antimicrobial shield against a wide range of bacteria. The antibacterial effect of silver was already known in ancient times. Silver tools and containers (from around 4000 BC) were used for storing and transporting water, to prevent the formation of germs, and to ensure high water quality. A number of wound dressings containing silver have recently been developed. Thomas[39] critically discussed the role of silver in wound dressing. These dressings function through the sustained release of low concentrations of silver ions over time, and generally appear to stimulate healing as well as inhibiting microorganisms. The evaluation of silver-impregnated dressings, as with other topical therapies, includes *in vitro* antibacterial studies, animal models, and clinical testing.

It has been argued that antimicrobial efficacy alone is of insufficient benefit in modern wound dressings, and that additional properties promoting wound healing are required. Based on this, the ability to remove any undesirable bacterial products in the wound environment that impinge on healing would be a bonus; for example, binding bacterial endotoxin (toxins released on cell death) to a silver dressing would be of benefit. Materials incorporated into modern silver-based dressings such as hydrocolloids, charcoal, and polymers are included as an aid to wound management, but also modulate the release of silver ions. Silver exhibits a selective toxicity in bacterial cells and yeasts through its action on cell membranes, respiratory enzymes, and DNA. Silver-impregnated polyamide cloths are effective antibiotics, and are designed to deliver silver ions to wound sites without potential side effects; the silver is rendered harmless, as it is lost naturally as the wound heals. The systemic toxicity of silver is not well documented, but silver sulphadiazine used in the treatment of burn wounds is implicated as a cause of leucopenia and renal damage. In addition to silver, natural products such as honey, aloe vera, and neem are potential antibacterial agents for modern wound dressing. A systematic review of the use of honey in wound dressing has been published elsewhere.[40]

6.5 Treatment of venous leg ulcers

About 1% of the adult population suffer from venous leg ulcers in the UK.[41] The condition mainly affects the elderly, especially women, and is a result of a previous venous thrombosis or incompetent venous valve within the superficial or communicating veins.[42] Approximately 400 000 patients have initial symptoms of leg ulcers, and 100 000 have open leg ulcers that require treatment.[43] The prevalence of leg ulcers increases with age, affecting 1.69% of patients aged between 65 and 95 years. The incidence rate for patients in this age group is estimated at 0.76% for men and 1.42% for women.[44] Venous ulcers are the most common type of leg ulceration. Approximately 80% of patients who have leg ulceration suffer with a venous ulcer. Arterial disease, which may coexist with venous disease, accounts for a further 10–25% of all leg ulcer cases.[11] Chronic leg ulcers are defined as those lasting 6 weeks or more.[45] Some patients may have more than one episode of venous ulceration, with estimated recurrence rates ranging from 6% to 15%.[46]

Venous ulcers appear in the gaiter area of the lower limb between the ankle and mid-calf. They can vary in size, ranging from very small, to large ulcers that extend beyond the gaiter area. The wound is characteristically shallow, irregular in shape, and has sloping well-defined borders. Typically, the skin surrounding the wound is thickened and hyper-pigmented, indicating lipodermatosclerosis.[47] With chronic ulcers a yellow-white exudate is observable, signifying the presence of slough. A shiny appearance indicates a fibrinous base, which inhibits new tissue formation and wound healing. Varicose veins and ankle oedema often accompany a venous ulcer.[48] Approximately 80% of patients who have a venous leg ulcer suffer from some form of discomfort, while 20% experience severe or unremitting pain.[49]

It is important that the arterial and venous systems should work properly without causing problems to blood circulation around the body. Pure blood flows from the heart to the legs through arteries taking oxygen and food to the muscles, skin and other tissues. Blood then flows back to the heart carrying away waste products through veins. The valves in the veins are unidirectional, which means that they allow the venous blood to flow in upward direction only. If the valves do not work properly or there is not enough pressure in the veins to push back the venous blood towards the heart, the pooling of blood in the veins takes place and this leads to higher pressure to the skin. Because of high pressure and lack of availability of oxygen and food, the skin deteriorates and eventually the ulcer occurs. The initial indications of venous leg ulcers are swollen veins (varicose veins) and blood clots in the veins – deep vein thrombosis (DVT), a growing problem for long-haul flights.

The diagnosis of lower limb ulceration must start by determining the patient's full clinical history, together with a physical examination of the condition. It is essential to identify possible risk factors that could cause ulceration or impact on the treatment of the ulcer. These risk factors could include arterial insufficiency, trauma, diabetes, sickle cell disease, infection, malignancy, and inflammatory disorders.[50]

A number of non-invasive test methods are available to the clinician for investigating the cause of leg ulceration and venous insufficiency. These test methods help to assess the arterial and venous circulation of the patient, and can provide information on the location of blood reflux or an obstruction within the veins.

6.5.1 Doppler ultrasonography

Doppler ultrasonography is used to measure the ankle-to-brachial blood pressure index (ABPI) of the patient. This ultrasound technique produces a signal that identifies the presence of blood flow within the arteries. The ABPI is obtained by measuring the systolic blood pressure within the dorsalis pedis or posterior tibial artery of the lower limb and the ipsilateral brachial artery of the arm.[51] The ratio between the ankle systolic pressure and the brachial systolic pressure provides the ABPI value. Measurement of the ABPI is important in order to exclude arterial disease as the cause of ulceration or as a possible risk factor that might inhibit treatment. An ABPI of > 0.9 is normal, whereas patients with moderate to severe arterial disease have an ABPI that is between 0.8 and 0.5.[50] These patients should be excluded from high-compression bandage therapy since its use could lead to further ulcer complications or even limb amputation.[52] The Doppler measurement technique produces elevated readings when diagnosing patients that may have diabetes and other conditions with calcified arteries.[53]

6.5.2 Ultrasound scanning

Colour duplex ultrasound scanning is currently the technique of choice in order to assess the venous system of the lower limb. This technique combines ultrasound imaging with pulsated Doppler ultrasound, and provides detailed anatomic information of the superficial, deep, and perforating venous systems. It can identify specific veins in which blood reflux occurs, or obstructions which may be contributing to venous hypertension.

6.5.3 Photoplethysmography and air plethysmography

Photoplethysmography and air plethysmography are simple tests designed to evaluate calf muscle dysfunction and degree of venous reflux. These techniques are used to observe the change in blood volume within the lower limb before and after exercise. Application of a tourniquet to restrict blood flow within the superficial system allows the deep venous system to be assessed for a potential obstruction.

Invasive venous tests, such as ascending and descending phlebography, are also used to assess venous insufficiency. Phlebography combines electromagnetic radiation (X-rays) and fluorescent materials to provide a technique that allows the veins to be clearly visualised. These immunofluorescence methods can detect venous outflow obstructions, provide information of valvular incompetence, and also highlight the presence of pericapillary fibrin.[54] Phlebography is usually used before a patient undergoes valvular surgery.

6.6 Venous leg ulcer treatment using compression bandages

It should be stated that venous leg ulcers are chronic and there is no medication or surgery to cure the disease other than the compression therapy. A sustained graduated compression mainly enhances the flow of blood back to the heart, improves the functioning of valves and calf muscle pumps, reduces oedema, and prevents the swelling of veins. Mostly elderly people are prone to develop DVT, varicose veins, and venous leg ulcers. Venous leg ulcers are the most frequently occurring type of chronic wound, accounting for 80–90% of all lower extremity ulceration.

It has been established that compression therapy making use of compression bandages is an efficient treatment for healing various leg ulcers, despite surgical strategies, electromagnetic therapy, and intermittent pneumatic compression. Venous leg ulcers are the most common type of ulcers and their prevalence increases with age. They are chronic and are caused due to poor venous return from the calf to the heart.

Compression bandages are mainly classified as elastic and non-elastic. Elastic compression bandages (Table 6.3) are categorised according to the level of pressure generated on the angle of an average leg. Class 3a bandages provide light compression of 14–17 mmHg, moderate compression (18–24 mmHg) is imparted by class 3b bandages, and 3c type bandages impart high compression between 25 and 35 mmHg.[55] The 3d type extra high-compression bandages (up to 60 mmHg) are not often used because the very high pressure generated will reduce the blood supply to the skin. It must be stated that approximately 30–40 mmHg at the ankle, which reduces to 15–20 mmHg at the calf, is generally adequate for healing most types of venous leg ulcers.[56] Compression stockings provide support to treat DVT and varicose veins, and to prevent venous leg ulcers. They are classified as light support (Class 1), medium support (Class 2), and strong support (Class 3).[57]

It should be noted that compression bandages may be harmful if not applied properly. They provide high tension as well as high pressure. A thorough assessment involving several criteria is therefore essential before applying a compression bandage on a limb. For example, it is important to consider the magnitude of the pressure, the distribution of the pressure, the

Table 6.3 Elastic bandage classification

Class	Bandage type	Bandage function
1	Light weight conforming	Apply very low levels of sub-bandage pressure and is used to hold dressings in place.
2	Light support	Apply moderate sub-bandage pressure and is used to prevent oedema or for the treatment of mixed aetiology ulcers.
3a	Light compression	Exert a pressure range of 14–17 mmHg at the ankle.
3b	Moderate compression	Exert a pressure range of 18–24 mmHg at the ankle.
3c	High compression	Exert a pressure range of 25–35 mmHg at the ankle.
3d	Extra high compression	Exert a pressure of up to 60 mmHg at the ankle.

duration of the pressure, the radius of the limb, and the number of bandage layers. The ability of a bandage to provide compression is determined by its construction and the tensile force generated in the elastomeric fibres when extended. Compression can be calculated by Laplace's Law, which states that the pressure is directly proportional to the bandage tension during application and the number of layers applied but inversely proportional to limb radius.[58] The structure of a compression bandage is, therefore, regarded as an important factor in producing a uniform pressure distribution. An ideal compression bandage should:

- provide compression appropriate for the individual;
- provide pressure evenly distributed over the anatomical contours;
- provide a gradient pressure diminishing from the angle to the upper calf;
- maintain pressure and remain in position until the next change of dressing;
- extend from the base of the toes to the tibial tuberosity without gap;
- function in a complementary way with the dressing; and
- possess non-irritant and non-allergenic properties.

6.6.1 Compression and pressure in bandaging

Compression can be exerted to the leg either by a single-layer bandage or multilayer bandages. In the UK the four-layer bandaging system is widely used, while in Europe and Australia the non-elastic two-layer short-stretch bandage regime is the standard treatment. A typical four-layer compression bandage system comprises padding bandage, crepe bandage, high-compression bandage, and cohesive bandage. Both the two-layer and four-layer systems require padding bandage (wadding or orthopaedic wool) that is applied next to the skin and underneath the short-stretch or compression bandages.

A plaster type non-elastic bandage, Unna's Boot, is favoured in the USA. However, compression would be achieved by three-layer dressing that consists of Unna's Boot, continuous gauze dressing followed by an outer layer of elastic wrap. It should be realised that Unna's Boot, being rigid, is uncomfortable to wear and medical professionals are unable to monitor the ulcer after the boot is applied. Unna's Boot provides a high working pressure when the calf muscle contracts, but very little pressure while the patient is at rest.[59] The high working pressure serves to increase blood flow, while the low resting pressure facilitates deep venous filling. The Unna's Boot is only effective in ambulatory patients and requires constant re-application as leg volume decreases due to a reduction in oedema. Used widely in USA, the Unna's Boot system is uncomfortable to wear due to its rigidity, and is both expensive and difficult to apply. Short-stretch bandages function in a similar manner to the rigid/inelastic Unna's Boot. They consist of 100% high twisted cotton yarns and are applied onto the limb at full extension. Unlike elastic bandages, short-stretch bandages firmly hold the calf, thereby providing a high working pressure when the patient walks.[60]

A variety of padding bandages are used beneath compression bandages as padding layers in order to evenly distribute pressure and give protection. They absorb high pressure created at the tibia and fibula regions. It will be noticed that the structure of a padding bandage is regarded as an important factor in producing a uniform pressure distribution. Research has shown that the majority of commercially available bandages do not provide uniform pressure distribution.[61,62]

The main function of a compression bandage is to exert the required level of pressure onto the leg. The ability of the compression bandage to perform this task is determined by its elastic properties. Sub-bandage pressure is a function of the tension induced into the compression bandage during application, the number of layers used, the width of the bandage, and the circumference of the limb. Applying the bandage with a 50% overlap effectively produces two layers, which generates twice the pressure. The principle of compression bandaging is based on the Laplace principle. When a compression bandage is applied at a constant tension on a limb of increasing circumference it will produce a sub-bandage pressure gradient with the highest pressure exerted on the ankle. Therefore, sub-bandage pressure is directly proportional to bandage tension but inversely proportional to limb circumference. The ability of a bandage to maintain sub-bandage pressure is determined by the elastomeric properties of the yarns, the fabric structure, as well as the finishing treatments applied to the fabric.

Compression bandages are mostly used during the initial therapy phase where the aim of treatment is to reduce oedema and overcome venous insufficiency. A number of different types of compression bandage systems are commercially available; the bandages are classified as either rigid/inelastic, short-stretch, long-stretch, or multilayered. The type of fabric construction influences the degree of extensibility that the bandage will have. At some point the bandage will not be able to extend or stretch any further (lock-out) under a predetermined tension. Evidence suggests that a sub-bandage pressure of 35–40 mmHg at the ankle, which gradually reduces to 17–20 mmHg at the knee, is required to overcome venous hypertension and successfully treat venous leg ulcers.[63] The authors found that initial narrowing of the veins occurred at a pressure of 30–40 mmHg in both the sitting and standing positions. Complete occlusion of the superficial and deep leg veins occurred at 20–25 mmHg (supine position), 50–60 mmHg (sitting position), and at 70 mmHg (standing position).[64]

6.6.2 Application of bandages

The elastic properties of the bandages help to provide a high recoiling force, which serves to increase venous flow and reduce venous hypertension. In addition, they conform easily around the lower limb and allow for frequent dressing changes. Skill is required to apply compression bandages at the correct tension and to avoid excessive sub-bandage pressures. Application of high sub-bandage pressure on patients with any type of micro-vascular disease can lead to further occlusion and pressure necrosis of these vessels.[65] Some manufacturers supply compression bandages with a series of geometric markers printed onto the bandage surface. The markers assist in the application of a predetermined level of compression by visually distorting when the bandage is stretched to a specific tension. For example, printed rectangles become squares when the correct bandage tension is reached.

In a multilayer bandaging system, three or four layers of different types of bandage are used to provide external compression. A multilayer system may include a combination of non-woven padding bandage, inelastic creep bandage, elastic compression bandages, and cohesive (adhesive) bandage. The different properties of each bandage type contribute to the overall effectiveness of the bandage system. The elastic bandage component provides sustained compression while the cohesive bandage offers rigidity thereby enhancing calf muscle pump function. The four-layer high-compression system developed by a clinical group at Charing Cross Hospital (London) has gained wide acceptance for use in UK hospitals. The four-layer system was developed specifically to incorporate different bandage types and properties in order

Table 6.4 Bandages used in compression therapy

Bandage name	Function	Manufacturer
Tensopress	Type 3c long-stretch bandage	Smith & Nephew
Setopress	Type 3c long-stretch bandage	Medlock Medical
SurePress	Type 3c long-stretch bandage	ConvaTec
Adva-co	Type 3c long-stretch bandage	Advancis Medical
Dauerbinde K	Long-stretch bandage	Lohmann & Rauscher
Silkolan	Type 2 short-stretch bandage	Urgo Limited
Tensolan	Type 2 short-stretch bandage	Smith & Nephew
Comprilan	Type 2 short-stretch bandage	Smith & Nephew
Actiban	Type 2 short-stretch bandage	Activa Healthcare
Actico (Cohesive)	Type 2 short-stretch bandage	Activa Healthcare
Rosidal K	Type 2 short-stretch bandage	Lohmann & Rauscher
Co-Plus	Cohesive bandage	Smith & Nephew
Tensoplus	Cohesive bandage	Smith & Nephew
Coban	Cohesive bandage	3M
Surepress	Padding bandage	ConvaTec
Soffban	Padding bandage	Smith & Nephew
K-soft	Padding bandage	Urgo Limited
Softexe	Padding bandage	Medlock Medical
Advasoft	Padding bandage	Advancis Medical
Flexi-ban	Padding bandage	Activa Healthcare
Cellona	Padding bandage	Lohmann & Rauscher
Ultra-soft	Padding bandage	Robinsons Healthcare
Ortho-band	Padding bandage	Millpledge Healthcare
Formflex	Padding bandage	Lantor (UK) Limited
Profore	Multilayer compression system	Smith & Nephew
Proguide	Multilayer compression system	Smith & Nephew
Ultra Four	Multilayer compression system	Robinsons Healthcare
System 4	Multilayer compression system	Medlock Medical
K-four	Multilayer compression system	Urgo Limited

to overcome the clinical issues of exudate, protection of bony prominences, and the ability to sustain sub-bandage pressure over a period of time.[66] In addition, the system was designed to apply the required 40 mmHg of pressure at the ankle, overcome disproportionate limb size and shape, and to remain in position on the leg without slippage. Application of the four-layer system involves first applying a padding bandage layer from the base of the toes to just below the knee. A crepe bandage is applied next, followed by an elastic compression bandage. Finally, a cohesive layer is applied in order to add durability and to complete the overall pressure profile. Examples of different types of compression bandages, cohesive bandages, padding bandages, and multilayer compression systems are shown in Table 6.4. The multilayer high-compression bandage system has been shown to provide a safe and effective treatment option for uncomplicated venous leg ulcers. Ulcer healing rates of up to 70% at 12 weeks have been obtained.[67] The four-layer bandaging technique has been shown to heal chronic ulcers that have failed to respond with traditional adhesive plaster bandage systems.[68]

6.7 Improving compression bandages: non-woven vari-stretch compression bandages (NVCB) and three-dimensional compression bandages

During the past few years there have been increasing concerns relating to the performance of bandages, especially pressure distribution properties, for the treatment of venous leg ulcers. This is because the compression therapy is a complex system and requires two or multilayer bandages, and the performance properties of each layer differ from other layers. The widely accepted sustained graduated compression mainly depends on the uniform pressure distribution of different layers of bandages in which textile fibres and bandage structure play a major role. The commercially available padding bandages are non-wovens that are mainly used to distribute the pressure, exerted by the short-stretch or compression bandages, evenly around the leg; otherwise higher pressure at any one point not only damages the venous system but also promotes arterial disease. Therefore there is a need to distribute the pressure equally and uniformly at all points of the lower limb and this can be achieved by applying an effective padding layer around the leg beneath the compression bandage. In addition, the padding bandages should have the capability to absorb high pressure created at the tibia and fibula regions. Wadding also helps to protect the vulnerable areas of the leg from generating extremely high pressure levels as compared to those required along the rest of the leg. The research carried out at the University of Bolton involving ten most commonly used commercial padding bandages produced by major medical companies showed that there are significant variations in properties of commercial padding bandages,[69,70] more importantly the commercial bandages do not distribute the pressure evenly at the ankle as well as the calf region (Fig. 6.5). For instance, the measured pressure for the bandage PB3 is 30 mmHg as against the applied pressure of 35 mmHg. In other words, PB3 allowed around 85% of the applied pressure to pass through at one point of the bandage structure and this created a high pressure at the specific region of the limb. An ideal bandage is expected to absorb all the applied pressure and distribute it uniformly over the limb instead of passing through a high pressure at one point.

The integrity of the non-woven bandages is also of great concern. When pressure is applied using compression bandages, the structure of the non-woven bandages may collapse and the bandage would not impart cushioning effect to the limb. The comfort and cushioning effect are considered to

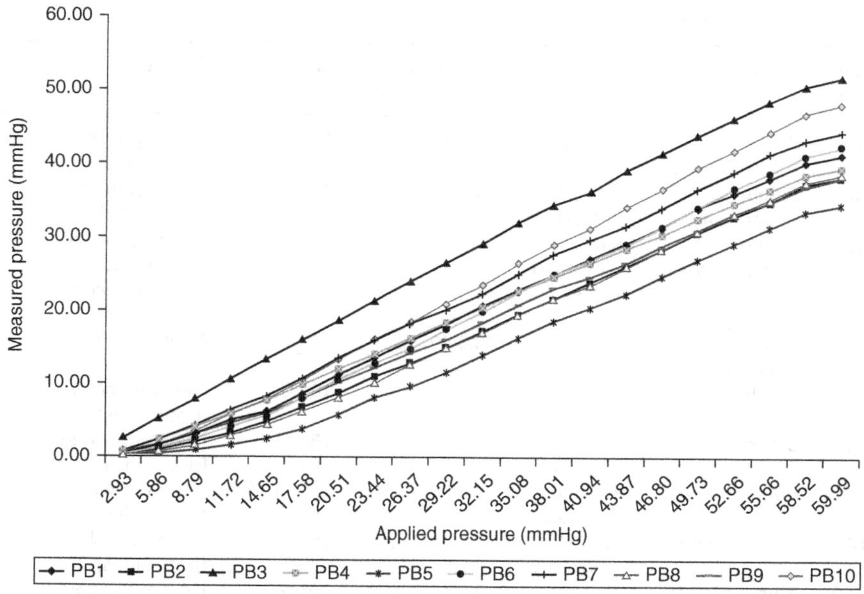

6.5 Pressure distribution of commercial padding bandages.

be essential properties for padding bandages because they stay on the limb for several days.

Twelve padding bandages, which consist of single component fibres, binary blends and tertiary blends incorporating polyester, bicomponent fibres, and natural fibres such as cotton and viscose, have been designed and developed at the University of Bolton (Table 6.5).

The salient properties of the developed bandages are:

- All the developed padding bandages possess suitable bulkiness.
- None of the bandages has lower tensile strength or breaking extension that hinders the performance characteristics of an ideal padding bandage.
- The tear resistance of bandages, except 100% hollow viscose (NPB5), is high and this means that the bandage cannot be easily torn by hand after wrapping around the leg. However, making perforations at regular intervals across the bandage facilitates easy tearing.
- The absorption of solution containing Na^+ and Ca^{++} ions (artificial blood) is significantly high, irrespective of fibre type and structure.
- The rate of absorption of all the developed bandages is also high.
- The pressure distribution of all the novel bandages is good up to 60 mmHg (Fig. 6.6).

In the UK, multilayer compression systems are recommended for the treatment of venous leg ulcers.[71] Although multilayer compression

Table 6.5 Novel padding bandages

Identification	Product	Fibre type	Fibre dtex; Length (mm)	Blend ratio	Structure
NPB1	Single component	Polyester	3.3;40	100%	Needlepunched (both sides)
NPB2	Single component	Polyester (bleached)	5.3;60	100%	Needlepunched (both sides)
NPB3	Single component	Hollow polyester	3.3;50	100%	Needlepunched (both sides)
NPB4	Single component	Viscose	3.3;40	100%	Needlepunched (both sides)
NPB5	Single component	Hollow viscose	3.3;40	100%	Needlepunched (both sides)
NPB6	Single component	Lyocell	3.3;38	100%	Needlepunched (both sides)
NPB7	Binary Blends	Polyester/ viscose	3.3;40/3.3;40	75%/25%	Needlepunched (both sides)
NPB8	Binary blends	Polyester/ viscose	3.3;40/3.3;40	50%/50%	Needlepunched (both sides)
NPB9	Binary blends	Polyester/ viscose	3.3;40/3.3;40	25%/75%	Needlepunched (both sides)
NPB10	Binary blends	Polyolefin/ viscose	2.2;40/3.3;40	20%/80%	Needlepunched (both sides) and Thermal bonded
NPB11	Tertiary blends	Polyester/ viscose/ Cotton (bleached)	3.3;40/3.3;40/ 1.8;22	33%/33%/ 33%	Needlepunched (both sides)
NPB12	Tertiary blends	Polyester/ viscose/ Polyolefin	3.3;40/3.3;40/ 2.2;40	60%/25%/ 15%	Needlepunched (both sides) and Thermal bonded

bandages are more effective than single-layer bandages in healing venous leg ulcers,[72] it is generally agreed by clinicians that multilayer bandages are too bulky for patients and the cost involved is high. A wide range of compression bandages is available for the treatment of leg ulcers, but each of them has different structure and properties and this influences the variation in performance properties of bandages. In addition, long-stretch compression bandages tend to expand when the calf muscle pump is exercised, and the beneficial effect of the calf muscle pump is dissipated. It is a well-established practice that elastic compression bandages that have the extension of up to 200% are applied at 50% extension and at 50% overlap to achieve the desired pressure on the limb. It has always been a problem for nurses to exactly stretch the bandages at 50% and apply without losing the stretch from ankle to calf, although there are indicators for the desired stretch (rectangles become squares) in the bandages. The elastic compression bandages are classified into four groups (Table 6.3) according to their ability to produce predetermined levels of compression and this

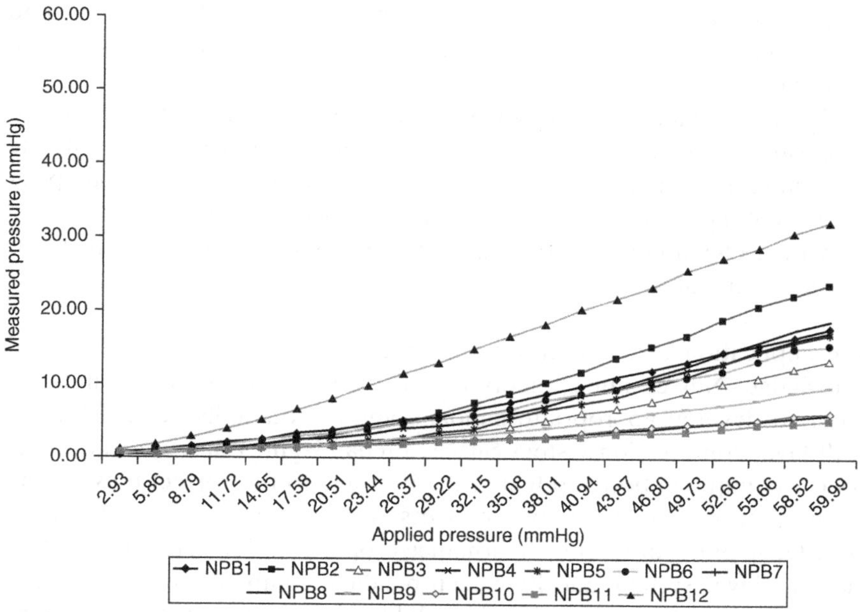

6.6 Pressure distribution of novel padding bandages.

has always been a problem to select the right compression bandage for the treatment. The inelastic short-stretch bandage (Type 2) system, which has started to appear in the UK market, has the advantage of applying at full stretch (up to 90% extension) around the limb. The short-stretch bandages do not expand when the calf muscle pump is exercised, and the force of the muscle is directed back into the leg, which promotes venous return. The limitations of short-stretch bandages are that a small increase in the volume of the leg will result in a large increase in compression, and this means the bandage provides high compression in the upright position and little or no compression in the recumbent position when it is not required. During walking and other exercises the sub-bandage pressure rises steeply, and while at rest the pressure comparatively drops. Therefore, patients must be mobile to achieve effective compression, and exercise is a vital part of this form of compression. Moreover, the compression is not intact with skin when reduction in limb swelling because the short-stretch bandage is inelastic, and it has already been stretched to its full.

In order to address the above problems, a novel NVCB has been designed and developed at the University of Bolton. The principal features of the NVCB are[70]:

- Novel non-woven technology was used to develop the variable compression bandages. It should be mentioned that no non-woven compression

bandages are listed in Drug Tariff. In the UK, the availability of wound dressings and bandages for use in patients' homes is dictated by the Drug Tariff.
- The performance and properties of the novel bandages are superior to existing multilayer commercial compression bandages. This fulfils the requirement of ideal variable pressure from ankle to below knee positions of the limb for the treatment of venous leg ulcers.
- Vari-stretch non-woven bandages also meet the standards and the tolerances stipulated by BS 7505.

The application of a multilayer bandage system requires expertise and knowledge. Nurses must undergo significant practice-based training in order to develop appropriate bandage application skills needed for the multilayer compression system. Successful bandaging relies upon adopting good technique in both stretching the bandage to the correct tension and ensuring proper overlap between layers. In addition, nurses need to have knowledge of the different performance properties of each bandage within the multilayer system, and how each bandage combination is to achieve safe and adequate compression. The ability of multilayer bandage systems to maintain adequate compression levels for up to 1 week has reduced the necessity for frequent dressing changes and has, therefore, decreased treatment costs. However, the cost of the multilayer compression system is still relatively high, due to the requirement for a specific bandage for each layer. Tolerance to a multilayer compression system is generally good, but non-compliance in some patients often results in prolonged or ineffective treatment. Some patients are unable to wear footwear due to the bulkiness of the multilayer compression regime. These patients often refuse treatment, since the requirement to remain house-bound is totally unacceptable. At night, patients find compression bandages too uncomfortable and often remove them in order to sleep. Since the application of multilayer compression systems is complex, most patients are unable to re-apply the bandages themselves.

Recently, spacer technology has been increasingly used to produce three-dimensional materials for technical textiles sectors such as automotive, medical, sports, and industrial markets. The spacer technology is flexible, versatile, cost-effective, and an ideal route to produce 3D materials for medical use. It is identified that the spacer is the right technology to produce novel compression bandages that meet the prerequisites of both ideal padding and compression bandages. The main reasons for the current interest in 3D spacer fabrics for producing novel compression bandages are several-fold. In 3D spacer fabrics, two separate fabric layers are combined with an inner spacer yarn or yarns using either warp-knitting or weft-knitting route (Fig. 6.7). The two layers can be produced from different fibre types

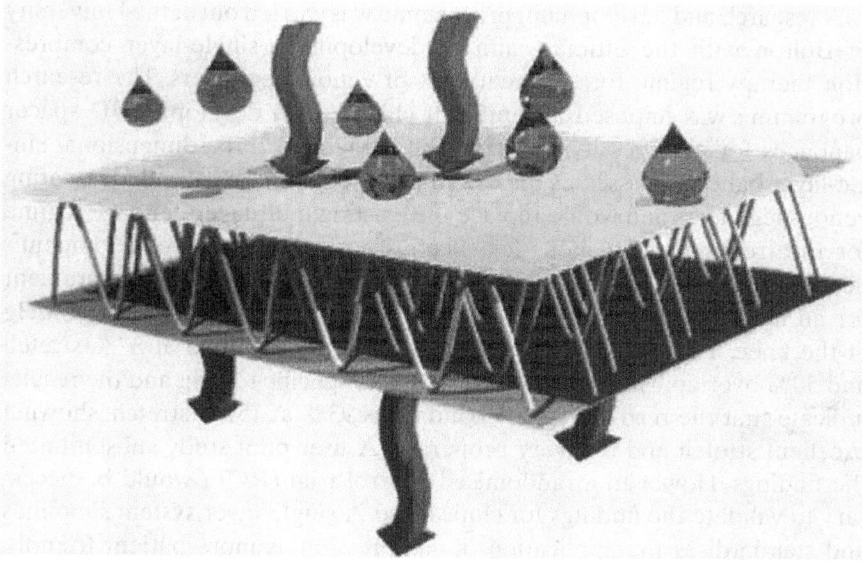

6.7 Illustration of a spacer structure.

such as polyester, polyamide, polypropylene, cotton, viscose, lyocell, wool, etc., and can have completely different structures.[73] It is also possible to produce low modulus spacer fabrics by making use of elastic yarns. Elastic compression could be achieved by altering the fabric structure. It should be mentioned that the 3D structure allows greater control over elasticity and these structures can be engineered to be uni-directional, bi-directional and multi-directional. Unidirectional elasticity is one of the desired properties for compression bandages. The three-dimensional nature of spacer fabrics makes for an ideal application next to the skin[74] because they have desirable properties that are ideal for the human body. 3D fabrics are soft, have good resilience that provides cushioning effect to the body, breathable, and ability to control heat and moisture transfer.[73] For venous leg ulcer applications, such attributes, together with improved elasticity and recovery, promote faster healing. It must be stated that 3D spacer fabrics can also be produced using double-jersey weft-knitting machines.[73] The main advantages of weft-knitted spacer fabrics over warp-knitted fabrics include cost effectiveness, because there is no need to prepare a number of warp beams, and spun yarns as well as coarser count hairy yarns can be used on weft-knitting machines.

Because of the problems associated with the currently available bandages for the treatment of venous leg ulcers as discussed under the heading 6.7, it is vital to research and develop an alternative bandaging regimen that meets all the requirements of an ideal compression system.

A research and development programme was carried out at the University of Bolton with the ultimate aim of developing a single-layer compression therapy regime for the treatment of venous leg ulcers. The research programme was imposed a significant challenge in developing 3D spacer bandages for compression therapy. The developed three-dimensional single-layer bandage possesses the essential bench marks required for treating venous leg ulcers and would replace the existing multilayer bandage regime for the treatment of venous leg ulcers. The research and development[75] demonstrated that the novel 3D spacer bandage fulfilled the requirement for an ideal compression of 30–40 mmHg at the ankle and 17–20 mmHg at the knee. The above sustained compression is obtained at 37% stretch and 50% overlap. The bandage is tested on a specific test rig and the results indicate that the recovery of 3D bandage is 93% at 151% stretch, showing excellent stretch and recovery properties. A user pilot study substantiated the findings. However, a randomised control trial (RCT) would be necessary to validate the findings for clinical use. A single-layer system simplifies and standardises the application of compression, is more patient friendly, reduces the nursing time, and significantly decreases the treatment cost.

6.8 Conclusions

This chapter provides an insight into the physiology of wounds, various dressings, and the management of different kinds of wounds. Successful wound management requires a careful selection of wound dressing, which is crucial for medical personnel as each and every type of wound needs appropriate dressing. Treatment of venous leg ulcers requires careful clinical consideration in order to determine the correct care regime. Often the first course of action is to debride the wound so that infection is minimised and wound healing is stimulated. The presence of bacteria in wounds can severely impair healing and hinder the formation of healthy tissue. Therefore, suitable antimicrobial dressing enhances the wound healing process. Wound dressings have several important functions to perform throughout the entire treatment phase. In the initial stage, dressings are required to remove and absorb exudate, to provide a moist and warm wound environment to promote healing, and to prevent bacterial infection. Once new granulated tissue has formed, different types of dressings may be used to simply protect the wound area from damage. Advances in bio-engineering and biochemistry have resulted in new treatments becoming available for various wounds. Bio-engineered skin contains various growth factors and proteins that help to encourage and increase wound healing. Cloned growth factors from specific proteins are now available, and research studies continue to determine their effectiveness in accelerating wound healing. Continued research and innovation provides evidence that these new technologies are effective for better wound management. Currently innovations mainly come from large textile medical devices companies which have their own research and development departments. However, many novel wound dressing products continue to be developed by small and medium size companies – albeit largely through collaboration with Universities and other allied research establishments.

Extensive literature has been cited in this chapter, constituting a 'ready reckoner' for the readers to understand the multidisciplinary subject areas of wounds as well as textile-based wound dressing medical devices.

6.9 References

1. Falabella, A.F. (1999) Debridement and management of exudative wounds. *Dermatol Ther*, **9**, 36–43.
2. Maklebust, J. (1996) Using wound care products to promote a healing environment. *Crit Care Nurs Clin North Am*, **8**, 2, 141–158.
3. Tallon, R.W. (1996) Wound care dressings. *Nurs Manage*, **27**, 10, 68–70.
4. Falabella, A.F., Carson, P., Eaglstein, W.H. and Falanga, V. (1998) The safety and efficacy of a proteolytic ointment in the treatment of chronic ulcers of the lower extremity. *J Am Acad Dermatol*, **39**, 5, pt 1, 737–740.
5. Steed, D.L. (2004) Debridement. *Am J Surg*, **187**, 5 (1), 71S–74S.
6. O'Brien, M. (2003) Methods of debridement and patient focused care. *JCN*, **17**, 11, 17–25.
7. Wang, J., Wang, S., Zhao, G., Wang, Z., Lineaweaver, W.C. and Zhang, F. (2006) Treatment of infected wounds with maggot therapy after replantation. *J Reconstr Microsurg*, **22**, 4, 277–279.
8. Steenvoorde, P., Budding, T. and Oskam, J. (2005) Determining pain levels in patients treated with maggot debridement therapy. *J Wound Care*, **14**, 10, 485–488.
9. Steenvoorde, P., Jacobi, C.E. and Oskam, J. (2005) Maggot debridement therapy: free-range or contained? An *in-vivo* study. *Adv Skin Wound Care*, **18**, 8, 430–435.
10. Sherman, R.A. (2003) Maggot therapy for treating diabetic foot ulcers unresponsive to conventional therapy. *Diabetes Care*, **26**, 2, 446–451.
11. Reschreiter, H., Wong, P. and Kapila, A. (2004) An understimated tool in wound care? Biosurgery, the use of larval therapy. *Care of the Critically Ill*, **20**, 6, 189–192.
12. Anon. (1991) Local applications to wounds – II: dressings for wounds and ulcers. *Drug Ther Bull*, **29**, 25, 97–100.
13. Bolton, L.L., Johnson, C.L. and Rijswijk, L.V. (1992) Occlusive dressings: therapeutic agents and effects on drug delivery. *Clin Dermatol*, **9**, 4, 573–583.
14. Rajendran, S. and Anand, S.C. (2002) *Developments in Medical Textiles, Textile Progress*. The Textile Institute, Manchester., 32, 4.
15. Foster, L. and Moore, P. (1998) *Proceedings of the 7th European Conference on Advances in Wound Management*, Madrid.
16. Lawrence, I.G., Lear, J.T. and Burden, A.C. (1997) 'Alginate dressings and the diabetic foot ulcer', *Pract Diabetes Int*, **14**, 61.

17. Suzuki, Y., Tanihara, M., Nishimura, Y., Suzuki, K., Yamawaki, Y., Kude, H., Kakimara, Y. and Shimizu. Y. (1999) '*In vivo* evaluation of a novel alginate dressing'. *J Biomed Mater Res*, **48**, 522.
18. Donaghue, V.M., Chrzan, J.S., Rosenblum, B.I., Giurini, J.M., Habershaw, G.M and Veves, A.J (1998) 'Evaluation of a collagen-alginate wound dressing in the management of diabetic foot ulcers'. *Adv Wound Care*, **11**(3), 114.
19. Torres-de-Castro, O.G., Carlos, A.G and Torra-i-Bou. (1997) 'Pure calcium-sodium alginate dressing. Multicenter evaluation of chronic cutaneous lesions'. *Rev de Enfermeria*, **20**, 23.
20. McMullen, D. (1991) 'Clinical experience with a calcium alginate dressing'. *Dermatology Nursing/Dermatology Nurses' Association*, **3**, 219.
21. Berry, D.P., Bale S. and Harding K.G. (1996) 'Dressings for treating cavity wounds'. *J Wound Care*, **5**, 10.
22. Bettinger, D., Gore D. and Humphries, Y. (1995) 'Evaluation of calcium alginate for skin graft donor sites'. *J Burn Care Rehabi*, **16**, 59.
23. Ingram M., Wright, T.A. and Ingoldby, C.J. (1998) 'A prospective randomized study of calcium alginate (Sorbsan) versus standard gauze packing following haemorrhoidectomy'. *J Roy Coll Surg Edin*, **43**, 308.
24. Davey, R.B., Sparnon, A.L. and Byard, R.W. (2000) 'Unusual donor site reactions to calcium alginate dressings'. *Burns*, **26**, 393.
25. Lin, S.S., Ueng, S.W., Lee, S.S., Chan, E.C., Chen, K.T., Yang, C.Y., Chen, C.Y. and Chan, Y.S. (1999) '*In vitro* elution of antibiotic from antibiotic-impregnated biodegradable calcium alginate wound dressing'. *J Trauma – Injury, Infect Crit Care*, **47**, 136.
26. Anon. (1997) 'Lyocell-based fibre to replace alginate in wound dressing'. *Med Text*, **4**, 2.
27. Anon. (1989) 'Protective dressing designed for minor wounds'. *Med Text*, **7**, 7.
28. Stone, C.A., Wright, H., Devaraj, V.S., Clarke, T. and Powell, R. (2000) 'Healing at skin graft donor sites dressed with chitosan'. *Brit J Plastic Surg*, **53**, 601.
29. Biagini, G., Bertani, A., Muzzarelli, R., Damadei, A., DiBenedetto, G., Riccotti, A.G., Zucchini, C. and Rizzoli, C. (1991) 'Wound management with *N*-carboxybutyl chitosan'. *Biomaterials*, **12**, 281.
30. Muzzarelli, R. (1995) 'Methylpyrrolidinone chitosan, production process and uses thereof'. US Patent 5 378 472.
31. Williams, C. (1995) 'Mepitel'. *Br J Nurs*, **4**, 51.
32. Thomas, S. (1994) 'Low-adherence dressings'. *J Wound Care*, **27**.
33. Thomas, S. (1990) *Wound Management and Dressing Pharmaceutical Press*, London.
34. Hollinworth, H. (2000) *Developing a Nursing Strategy to Change Woundcare Practice*, Poster presented at the 5th clinical Nurse Specialism Conference, Kensington, London.
35. Hollinworth, H. (2000) Overcoming Confusion; Selecting Dressings that Minimise Wound Pain and Tissue Trauma, Independent Survey leaflet, Ipswich.
36. Rajendran, S. and Anand, S.C. (2002) 'Insight into the development of nonadherent absorbent dressings'. *J Wound Care*, **11**, 191.
37. Bowler, P.G. (1999) 'Microbial involvement in chronic wound malodour'. *J Wound Care*, **18**, 216.
38. Williams, K. and Griffiths E. (1999) 'Malodorous wounds: causes and treatment'. *Nurs Residential Care*, **1**, 5, 276.

39. Thomas, S. and McCubbin, P. (2003) 'An *in vitro* analysis of the antimicrobial properties of 10 silver-containing dressings'. *J Wound Care*, **12**, 420.
40. Moore, O.A., Moore, R.A, Smith, L.A., Campbell, F., Seers, K. and McQuay, H.J. (2001) Systematic review of the use of honey as a wound dressing, *BMC Complementary Alternative Med*, **1**, 2.
41. Martson, W.A., Carlin, R.E., Passman, M.A., Farber, M.A., Keagy, B.A. and Parent III, F.N. (1999) Healing rates and cost efficacy of outpatient compression treatment for leg ulcers associated with venous insufficiency. *J Vasc Surg*, **30**, 3, 491–498.
42. Sarkar, P.K. and Ballentyne, S. (2000) Management of leg ulcers. *Postgrad Med J*, **76**, 674–682.
43. Margolis, D.J., Bilker, W., Santanna, J. and Baumgarten, M. (2002) Venous leg ulcer: incidence and prevalence in the elderly. *J Am Acad Dermatol*, **46**, 3, 381–386.
44. Valencia, I.C., Falabella, A., Kirsner, R.S. and Eaglstein, W.H. (2001) Chronic venous insufficiency and venous leg ulceration. *J Am Acad Dermatol*, **44**, 3, 401–424.
45. Moffatt, C.J., Franks, P.J., Doherty, D.C. Martin, R., Blewett, R. and Ross, F. (2004) Prevalence of leg ulceration in a London population. *QJM*, **97**, 7, 431–437.
46. Nicholaides, A.N. (1997) Investigation of chronic venous insufficiency: a consensus statement (France, 5–9 March). *Circulation* 2000, **102**, 20, E126–E163.
47. Reichenburg, J. and Davis, M. (2005) Venous ulcers. *Semin Cutan Med Surg*, **24**, 4, 216–226.
48. Zimmet, S.E. (1999) Venous leg ulcers: modern evaluation and management. *Dermatol Surg*, **25**, 3, 236–241.
49. Franks, P.J., Moffatt, C.J., Connolly, M., Bosanquet, N., Oldroyd, M., Greenhalgh, R.M. and McCollum, C.N (1994) Community leg ulcer clinics: effect on quality of life. *Phlebology*, **9**, 2, 83–86.
50. Reichenburg, J. and Davis, M. (2005) Venous ulcers. *Semin Cutan Med Surg*, **24**, 4, 216–226.
51. Sibbald, R.G. (1998) Venous leg ulcers. *Ostomy Wound Manage*, **44**, 9, 52–64.
52. Callam, M.J., Ruckley, C.V., Harper. D.R. and Dale, J.J. (1985) Chronic ulceration of the leg: extent of the problem and provision of care. *Br Med J (Clin Res Ed)*, **290**, 6485, 1855–1856.
53. McGuckin, M., Stineman, M., Goin, J. and Williams, S. (1996) Draft guideline: diagnosis and treatment of venous leg ulcers. *Ostomy Wound Manage*, **42**, 4, 48–78.
54. Vowden, P. and Vowden, K. (2001) Investigations in the management of lower limb ulceration. *Br J Nurs*, The Profore Supplement, **4**, 627.
55. Cullum, N. (2002) *Compression for Venous Leg Ulcers Cochrane Review*, The Cochrane Library, Oxford.
56. Simon, D. (1996) Ostomy Wound Management, **42**, 34.
57. Veraat, J.C., Pronk, G. and Neuman, H.A. (1992) Pressure differences of elastic compression stockings at the ankle region. *Dermatol Surg*, **23**, 10, 935–939.
58. Weiss, R.A. and Ramelet, A.A. (2002) 'Removal of blue periocular lower eyelid veins by ambulatory phlebectomy'. *Dermato Surg*, **28**, 1, 43-45.
59. Partsch, H. (1991) Compression therapy of the legs: a review. *J Dermatol Surg Oncol*, **17**, 10, 799–805.

60. Hampton, L. (1997) Venous leg ulcers: short stretch bandages for compression therapy. *Br J Nurs*, **6**, 17, 990–998.
61. Anand, S.C. and Rajendran, S. (2006) Effect of fibre type and structure in designing orthopaedic wadding for the treatment of venous leg ulcres. In Anand, S.C., Kennedy, J.F., Miraftab, M., Rajendran, S., Editors. *Medical Textiles and Biomaterials for Healthcare*, Woodhead Publishing Limited, Cambridge, 243.
62. Rajendran, S. and Anand, S.C. (2003) 'Design and development of novel bandages for compression therapy'. *British J Nurs*, Tissue Viability Supplement, **12**, 6; S20–S29.
63. Stemmer, R. (1969) Ambulatory-elasto-compressive treatment of the lower extremities particularly with elastic stockings. *Derm Kassenarzt*, **9**, 1–8.
64. Partsch, B. and Partsch, H. (2005) Calf compression pressure required to achieve venous closure from supine to standing positions. *J Vasc Surg*, **42**, 4, 734–738.
65. Simon, D.A., Freak, L., Williams, I.M. and McCollum, C.N. (1994) Progression of arterial disease in patients with healed venous ulcers. *J Wound Care*, **3**, 4, 179–180.
66. Moffatt, C.J. and Dickson, D. (1993) The charing cross high compression four-layer bandaging system. *J Wound Care*, **2**, 2, 91–94.
67. Nelzen, O., Bergqvist, D. and Lindhagen, A. (1991) Leg ulcer etiology – a cross sectional population study. *J Vasc Surg*, **14**, 4, 557–564.
68. Buchbinder, D., McCullough, G.M. and Melick, C.F. (1993) Patients evaluated for venous disease may have other pathological considerations contributing to symptomatology. *Am J Surg*, **166**, 211–215.
69. Rajendran, S. and Anand, S.C. (2002) *Development of Novel Bandages for Compression Therapy*, Wounds UK, Harrogate, 19–20 November 2002.
70. Rajendran, S. and Anand, S.C. (2003) *Evaluation of Novel Bandages for Compression Therapy*, Wounds UK, Harrogate, 11–12 November 2003.
71. Cullum, N., Fletcher, A., Semlyen, A. and Sheldon, T.A. (1997) Compression therapy for venous leg ulcers, *Qual Health Care* December, **6**, 4, 226–231.
72. Nelson, E.A., Prescott, R.J., Harper, D.R, Gibson, B., Brown, D. and Ruckley, C.V. (2007) A factorial, randomized trial of pentoxifylline or placebo, four-layer or single-layer compression, and knitted viscose or hydrocolloid dressings for venous ulcers. *J Vasc Surg*, **45**,134–41.
73. Anand, S.C. (2003) Spacers – at the technical frontier. *Knit International*, **10**, 38–41.
74. Anon (2002) Spacer fabric focus. *Knit International*, **109**, 20–22.
75. Lee, G., Rajendran, S. and Anand, S.C. (2009) 'New single-layer compression bandage system for chronic venous leg ulcers, *British J Nurs*, Tissue Viability Supplement, **18**, 15, S4–S18.

7
High performance textiles for industrial filtration

E. HARDMAN, formerly of Madison Filter UK (now Clear Edge Filtration UK), UK

DOI: 10.1533/9780857099075.223

Abstract: Industrial filtration operations involving textile filter media are found in the production of countless items that we encounter in our everyday lives, and yet more operations are engaged in protecting the environment. The chapter attempts to provide a brief overview of the filtration mechanisms that are involved both in industrial dust collection and solid–liquid separation processes, and the equipment types that are engaged in each. The chapter also describes the types of textile filter media that are employed in the various operations, their component strengths and weaknesses, and some of the finishing treatments that are applied to enhance their efficiency.

Key words: industrial filtration, dust collection, solid–liquid filtration, cake filtration, thermal and chemical conditions.

7.1 Introduction

The separation of solids from liquids, or solids from gases, is an essential part of countless industrial processes, many of which are inextricably linked to our everyday lives. The aim of this chapter is to provide the reader with a brief insight into the use of textile filter media in industrial filtration and the 'thought processes' that govern their design and selection. In so doing, it will hopefully demonstrate to the student of textile technology the value of his/her schooling – from a detailed understanding of fibre and yarn properties, through fabric production and finishing techniques, to QC test procedures and even cutting and sewing technology.

In carrying out their tasks, the filter media may in some cases be expected to operate for quite lengthy periods, for example 3–5 years, quite often in extremely challenging conditions. These may involve prolonged exposure to high temperatures or highly corrosive chemical agents, whilst at the same time subjected to attrition from abrasive particles and continuous stress from flexing and tensile forces. The penalty of failure during use could be very costly, especially if the final product is of high monetary value, as in

Table 7.1 Particle capture ranges for various filtration systems

Description	Reverse osmosis	Nanofiltration	Ultrafiltration	Microfiltration	Macrofiltration
Size range (μm)	< 0.001	0.0001–0.001	0.001–0.1	0.1–10	10–1000
Filter media type			Various membrane systems		Textile filter media

the case of precious metals or pharmaceuticals, or if the environment were to become heavily polluted and require extensive clean-up. In addition, the cost of lost production and subsequent remedial work could also represent an unacceptable drain on financial resources.

The field of filtration in its widest sense covers the capture of particles ranging in size from several millimetres down to the molecular scale. Table 7.1 provides an indication of the various particle size ranges, and where textile filter media fit into this picture. The industrial filtration processes on which the chapter will concentrate are separated into two broad sections, namely, the separation of solids from gases, more commonly referred to as dust collection, and the separation of solids from liquids, usually referred to as liquid filtration.

7.2 Dust collection

Dust collection is carried out whenever a process generates sufficient airborne dust as to require some form of control, or where an operation requires the removal of airborne pollutants to ensure satisfactory performance. Some examples where control is needed include milling or grinding processes, mining operations, bag or hopper filling, and thermal combustion processes. Alternatively, for equipment with air-intake systems, as found for example in gas turbine engines, the removal of air pollutants will be necessary to avoid damage to internal parts, in this example to turbine blades. The dust particles in question may range in size from < 0.1 μm, for materials such as carbon black, to > 50 μm for flour. They may also be of some monetary value e.g. cement, flour or sugar, or simply be of nuisance value, as in the case of emissions from coal-fired power stations.

Some of the dust collection techniques employed do not employ textile filter media e.g. settling chambers, electrostatic precipitators, cyclones and wet scrubbing devices, and whilst these are mentioned here for completeness, for the remainder of this section it is intended to concentrate solely on dust collectors that engage textile filter media comprising woven or (mainly) needle felt constructions.

7.2.1 Dust particle capture and cleaning mechanisms

The mode of operation of an industrial dust collector may in very simplistic terms be compared with that of a traditional domestic vacuum cleaner i.e. dust-laden air is drawn by vacuum on to the surface of the 'dust bag', the filter medium retains the dust whilst the dust-free air is expelled into the atmosphere. In an industrial unit, the filter media comprise either woven or non-woven materials, and are normally sewn in the form of tubes (usually referred to as filter sleeves), flat enveloped shaped bags, or cartridges of various design e.g. flat panel or pleated.

It is widely accepted that the fibres in the filter fabric capture the dust particles by one of several mechanisms, these being identified as: inertial impaction, interception, diffusion (Brownian motion) and electrostatic forces (see Fig. 7.1). Inertial impaction occurs when particles are of sufficient mass and bulk as to leave the gas stream and make a direct impact on a fibre in the filter.

Interception occurs when the particles, although following the gas stream, are of such a size as to be unable to negotiate a path around the fibres, and so become 'intercepted'. Electrostatic attraction affects relatively small particles, and occurs when the dust particles carry a different electrical charge from that of the fibre. Diffusion is essentially due to the erratic flow pattern often found in gas streams, and is frequently referred to as 'Brownian motion'. This effect also normally involves relatively small particles, which are drawn into this erratic pattern until they too eventually become attached to a fibre through attractive surface forces.

These mechanisms notwithstanding, if the particles were significantly larger in size than the pores in the fabric, a sieving mechanism would be a more appropriate description. It may also be added that some particles fail to be captured by the fabric due to gravitational force.

It is almost inevitable that on start-up of filtration, due to the powerful suction of the exhaust fan, a few particles may pass through a clean fabric into the atmosphere; however, after a short period of time, a porous 'dust cake' begins to form on the fabric's surface, which effectively restricts further emission. As this dust cake accumulates, the pressure drop across the

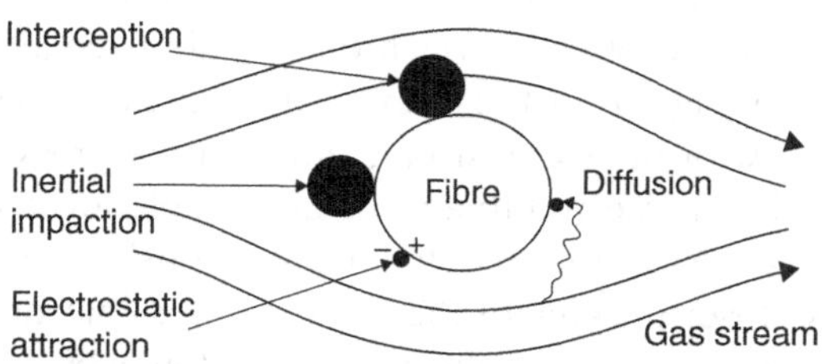

7.1 Particle capture mechanisms.

fabric also increases until a point is reached where the pressure drop is of such magnitude that the exhaust fan becomes largely ineffective. At this point it becomes necessary to clean the material, which in an industrial filter is carried out *in situ*, by (a) a shaking mechanism, or (b) by replacing the vacuum with a flow of air in the opposite direction, or (c) by a more dynamic pulse of highly compressed air. Whilst from the foregoing it will be appreciated that the filter will be so designed as to retain as much dust as possible on the filter fabric's surface, it should also be mentioned that some cartridge filters are designed to capture the dust within the body of the filter medium. This is referred to as depth filtration, the filter elements in these cases being replaced when fully loaded.

In traditional 'shake' collectors, the 'closed' ends of the filter sleeves are suspended from an upper beam (complete with shaking mechanism), whilst the open ends are anchored to a cell plate, the dust being drawn through the plate to be collected onto the inside of the sleeves (see Hardman, 2000a). After a period of time, the exhaust fan is switched off and the sleeves are subjected to a rigorous shaking or vibrating action. This causes the sleeves to buckle and the dust to be shaken off, where it falls by gravity into a hopper.

Reverse air collectors also collect dust on the inside of the sleeves, though on this occasion cleaning is achieved by simply reversing the direction of the airflow, in some cases possibly with the additional assistance of a vibration mechanism for greater effectiveness. In this type of collector the sleeves, which, as before, are inflated through the suction of the exhaust fan, will deflate when the opposing airflow is introduced, complete collapse being prevented by the sleeve tensioning system or through the attachment of metal rings, which are sewn into the sleeve at strategic points along its length.

The final dust collector style incorporates a so-called 'pulsejet' cleaning mechanism (see Hardman, 2000b), in which the sleeves are mounted on wire cages (see Fig. 7.2) and the dust is collected on the outside. At the appropriate time, a short pulse (typically 0.1 s.), of approximately 12 L of dry compressed air at c.600 kPa is injected into the top of each cage. A Venturi tube located at the entrance intensifies the impact of the pulse, which causes a brief expansion of the sleeve. This is sufficient to overcome the draw of the exhaust fan and dislodge the dust cake, which once again is collected in a hopper. Following the cleaning action, the sleeve is immediately drawn back on to the cage by the draw of the exhaust fan, and with the pressure drop across the fabric having been restored to a more acceptable level, the dust collection process resumes. As filtration velocities are considerably higher with pulsejet collectors, closer packing of the sleeves is possible, and hence the construction of more compact dust collection units. It may be noted that the same pulse-cleaning mechanism is also applied, where appropriate, to

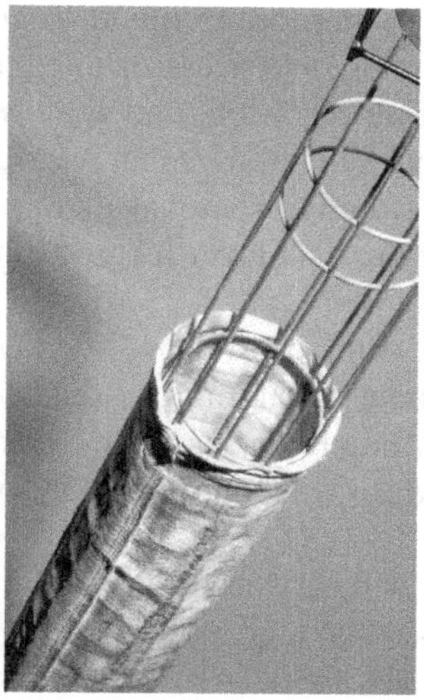

7.2 Filter sleeve on wire cage.

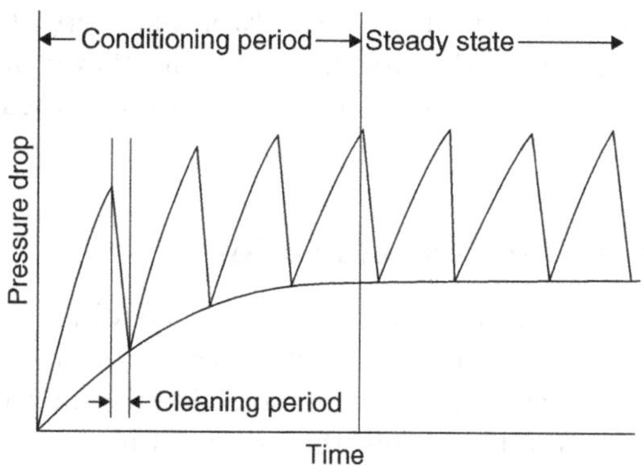

7.3 Pressure drop *vs* time.

cartridge filters, typically made up of pleated non-woven material usually (though not exclusively) supported on a wire frame.

The filtration and cleaning stages can be demonstrated graphically as shown in Fig. 7.3. The graph shows that whilst the pressure drop increases during filtration, it does not return to its original level after cleaning – indeed, for a period it continues to rise. This is because the fabric retains a thin layer of dust, even after cleaning. This dust layer is critical to the operation, as it effectively takes over the filtration duty from the filter fabric, the latter reverting to the more subordinate role of support medium. Were this layer to be destroyed e.g. due to excessive cleaning, it is likely that the incoming dust would penetrate too deeply into the body of the structure, which in turn would probably necessitate even more frequent cleaning. As a consequence of the more frequent cleaning, an excess of finer particles will be emitted into the atmosphere during filtration, and may well also lead to premature sleeve failure due to mechanical damage. A careful balance must therefore be found between effective removal of the bulk of the dust cake and avoiding destruction of the dust layer formed during the conditioning period.

After the conditioning period, the pressure drop curve ideally begins to stabilise to a 'steady state', at which point the amount of dust collected becomes virtually equal to the amount that is discharged. Unless some unforeseen incident should occur, this pattern will continue, quite often for several years of service.

The actual size of a fabric dust collector will vary according to the volume of dust that has to be removed from a process in a given period of time. The velocity at which the dust is carried into the filter (sometimes referred to as the air-to-cloth ratio) is also a major factor in determining the total fabric surface area requirement. To give the reader some perspective as to the diversity of sizes, a filter may contain just a few sleeves in order to capture a small volume of nuisance dust, from say a woodworking shop, whereas in one of the largest fabric filters in the world, the Eraring Power Station in Australia, some 190 000 sleeves are used, with a total filtration area of over 500 000 m^2.

7.3 Fibre selection for dust collection

Because of the wide diversity of thermal and chemical conditions that prevail in dust collectors, the correct choice of fibre type is the most fundamental of all considerations when designing or selecting the filter fabric. A selection of the more common fibre types that are used in dust collection and their properties is shown in Table 7.2.

From this table if it is known that the dust-laden gas stream comprises dry air at a temperature less than 150°C, the most likely choice of fibre would be of polyester origin, this being the most widely used material in dust collection. Although it has been shown (ICI Research, 1975) that continuous operation at 150°C with this fibre will cause a steady reduction in tensile strength, this reduction is widely considered to be tolerable for most purposes. By comparison, Polyamide 66 would very quickly lose virtually all its strength at this temperature. Conversely, as polyester is notoriously susceptible to hydrolytic attack, if moisture were present in the gas stream at say 70°C or higher, degradation of the fibre will take place, resulting in a progressive deterioration of the filter sleeve. In the presence of saturated steam at temperatures above 100°C fibre failure would be very rapid. As a consequence, in situations where both heat and moisture are anticipated, the use of hydrolysis resistant 'homopolymer acrylic' fibres would be the

Table 7.2 Fibres and their properties (dust collection)

Fibre type	Trade names	Maximum operating temp (°C)	Resistance to:		
			Abrasion	Acid	Alkali
Polyester	Dacron Trevira	150	VG	G	P
Polyacrylonitrile (Homopolymer)	Dolanit, Ricem Zefran	140	G	G	F
Polyacrylonitrile (Copolymer)	Orlon Dralon	120	G	G	F
Polyaramid	Nomex Conex	200	VG	P	VG
Polyimide	P84	260	VG	P	VG
PPS	Ryton Procon	190	G	VG	E
PTFE	Teflon Rastex Profilen	260	F	E	E
Glass	Fibreglass	260	P	F	E
Polypropylene	Moplefan Trol	90 125	F–G	E	E
Cellulose	Cotton	85	F	P	G

E = Excellent, VG = Very Good, G = Good, F = Fair, P = Poor.

preferred choice. The power plant at Eraring in Australia referred to previously is understood to fall into this category.

Continuing the selection process, it will be appreciated that in all combustion operations, before filtration can commence, the gas stream must be cooled to a level that the fibre can handle, which almost invariably involves the use of expensive heat exchangers, fans and associated trunking, etc. DuPont offered a partial solution to this problem with the development of a polyaramid fibre, trade named Nomex®, which can operate continuously at temperatures up to 200–220°C. Furthermore, in most high temperature situations, this fibre does not support combustion, does not melt and exhibits extremely good dimensional stability. A subsequent development by Upjohn of a polyimide fibre, trade named P84™ (manufacturing technology currently held by Evonik Fibres GmbH), enabled a still higher continuous operating temperature of 260°C. A comparison of these two polymers is shown in Fig. 7.4.

On the other hand the combustion of fossil fuels in general creates yet another complication, due to the formation of combustion products such as oxides of sulphur (SOx), and nitrogen (NOx), which in the presence of moisture can lead to highly acidic conditions. In certain circumstances, SO_3 may be

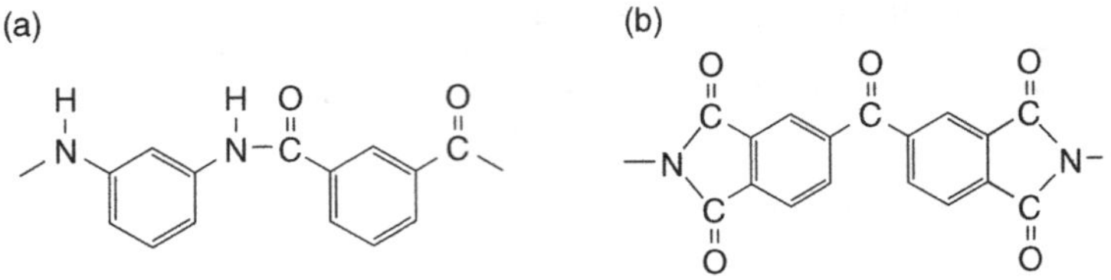

7.4 The structure of (a) polyaramid and (b) polyimide.

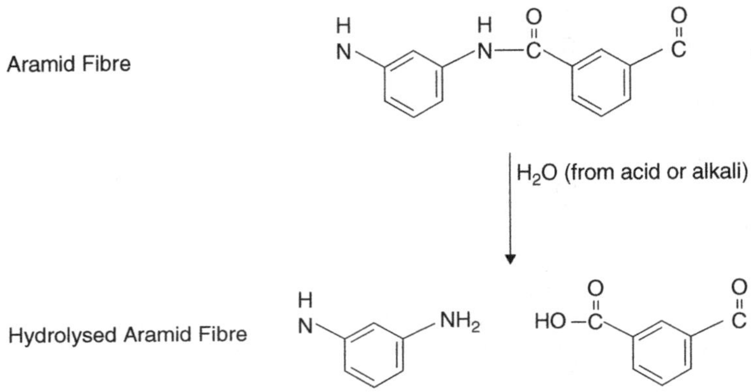

7.5 Condensation of aramid fibre.

produced, which is particularly hazardous as, depending on the SO_3 concentration in the gas stream, the 'acid dew point' may be quite high e.g. 160°C. If the temperature inside the collector were to fall below this dew point, as with all condensation polymers, rapid degradation of both polyaramid and polyimide fibres would ensue, as shown in the aramid example given in Fig. 7.5.

In a development by Philips USA, a polyphenylene sulphide (PPS) or 'Sulfar' fibre, originally trade named Ryton®, went some way to overcoming this deficiency, and this is widely used where high temperatures and hydrolysis conditions are prevalent. The structure of this fibre is shown in Fig. 7.6. It is today produced by a number of companies e.g. in Japan, by Toyobo and Toray, trade named Procon™ and Torcon™ respectively, in Europe by EMS-Griltech, trade named Nexylene® and numerous companies in China, the first producer of which is said to be Jiangsu Ruitai Science and Technology Co. Ltd.

Unfortunately, the maximum temperature (190°C) at which this fibre can continuously operate is somewhat lower than the previous fibres, and if the oxygen content of the gas stream were also high, e.g. 15% or more, degradation due to thermal oxidation could be expected, thus reducing the maximum operating level still further. The presence of bromine as found in certain coals is also known to have a deleterious effect on this fibre.

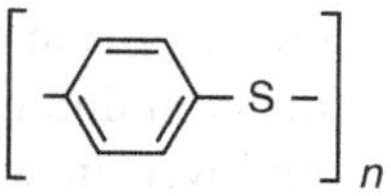

7.6 The structure of polyphenylenesulphide.

In the event that the thermal and chemical nature of the gas stream is too aggressive for the preceding polymers, it may be necessary to employ filter media produced from polytetrafluoroethylene (PTFE), which can operate continuously at temperatures up to 260°C and will be unaffected by the chemical nature of most gases. It is, however, significantly more expensive and, as such, only used where local conditions override the cost factor.

Filter media produced from glass are also employed where high temperatures are involved, although the physical characteristics of this material tend to restrict its use to applications involving less abrasive dusts and less rigorous cleaning regimes. These limitations notwithstanding, glass fabrics are widely used in power plants in North America.

Where the filter medium is in the form of a needle felt (as found in pulse-jet collectors), some manufacturers have offset the deficiencies found in some of the above fibres by producing composite structures, e.g. polyaramid, polyimide or PPS fibres on the surface of the felt with a PTFE scrim or base-cloth providing a framework of support. Although under certain physicochemical conditions the surface fibres become degraded, it has been found that the skeleton of support provided by the PTFE base-cloth is often sufficient to enable satisfactory performance to continue.

From this broad-brush approach it will be appreciated that in selecting the correct fibre, it is vital that the technologist is (a) in possession of a complete dossier on the gas being filtered, and (b) is fully aware of long-term fibre performance under the said conditions.

7.3.1 Fibre fineness and cross-sectional shape

The linear densities of the fibres that are used in needle felt constructions are usually in the range 1.66–3.33 decitex. In many cases, such fibres will be blended, typically in ratios of 50/50 or 66/34; the greater the proportion of finer fibres, the greater the fibre surface area and thus the more efficient the final structure. For example, it can be shown that the specific surface area of 1 kg of 3.33 decitex polyester fibre is 166 m^2, whereas for the same weight of 1.66 decitex fibre, the surface area is 234 m^2. For this reason, in recent times the use of 'micro-denier' fibres (typically < 1 decitex or ≤ 10 μm) has found favour where fine dusts are collected, ensuring a much greater surface filtration area and therefore less dust penetration into the body of the felt. However, as productivity rates are much slower with these fibres (they require a much higher degree of carding power), they are normally only applied as a surface layer, the main body of the structure containing coarser fibres as indicated above. NB, even finer fibres are employed in some cartridge filters, as will be described in Section 7.4.3..

Similar filtration efficiency benefits may be obtained by applying a layer of fibres with irregular or multi-lobed cross-sectional shape as, for a given count, these also provide a greater surface filtration area. P84™ is such a fibre, and is often used where superior particle capture is required.

7.3.2 Electrostatic effects

In the transfer of dust to the collector, particularly at low levels of humidity, the particles may acquire a static charge, an accumulation of which in the sleeve compartment can pose a potentially hazardous risk of explosion through arcing. Morden (1994) has examined this in relation to dust collection in the sugar industry. By incorporating antistatic fibres into the fibre blend, and ensuring that the filter sleeves are properly earthed, such risks are significantly reduced.

Antistatic fibres may be in the form of stainless steel fibres (see Bekaert), or polyester fibres coated with a layer of conducting material such as carbon (see 'Epitropic' fibres) or silver (see 'SilveR.STAT®'). Depending on the efficiency of the blending system, the addition of 2–5% of these fibres is normally recommended to ensure satisfactory dissipation of any static charge. The effectiveness of such additions may be assessed by measurement of the fabric's surface resistivity (see BS 6524 1995 part 4), a value of $10^{-10}\ \Omega\ \text{square}^{-2}$ being considered adequate for most applications (NB, the size of the square is irrelevant).

7.4 Fabric types for dust collection

A variety of fabrics styles are employed in dust collection, these being chosen according to the physical forces that will be imposed by the filter and the quantity and abrasive nature of the dust being collected. As a consequence, for reverse air and pulsejet filters needlefelts are preferred whereas for shake filters both woven and needlefelt structures are used. By comparison for cartridge filters spunbonded, meltblown and electrospun fabrics are more likely to be found.

7.4.1 Woven structures

Primarily because of their good resistance to frequent and aggressive flexing during the cleaning phase, filter sleeves made from woven fabrics are

Table 7.3 Typical (polyester) woven fabric constructions for shake and reverse air dust collection

	Shake collector	Reverse air collector
Finished fabric area density	500 gm^{-2}	500 gm^{-2}
Air permeability	200–250 $Ldm^{-2}min^{-1}$ at 200 Pa	200–250 $Ldm^{-2}min^{-1}$ at 200 Pa
Weave	2/1 twill	5/1 warp satin
Finish	Heat set and raised on one side	Heat set and raised on one side
Warp ends (cm^{-1})	15	25
Weft picks (cm^{-1})	11	15
Warp count	175 tex (woollen spun)	550 d'tex (multifilament)
Weft count	175 tex (woollen spun)	120 tex (cotton spun)

traditionally used in shake collectors. Such fabrics are usually woven in simple 2/1 or 2/2 twill designs, often with bulky staple-spun yarns as produced by the woollen spinning system, which also ensures excellent particle collection efficiency. Although there are many specifications to suit individual needs, the construction shown in Table 7.3 provides an example of what may be found in a shake dust collector.

As some reverse air filters also employ shake or vibration mechanisms, it is not uncommon to find the same fabric in both types of filter. This notwithstanding, conventional reverse air filter sleeves are most likely to feature multifilament yarns in the warp combined with staple fibre yarns in the weft. On this occasion the twill weave will be replaced by a 5- or 6-shaft warp satin construction, which is designed to facilitate superior dust release from the smooth multifilament warp yarns during the less aggressive cleaning phase. Particle capture will be further enhanced if the fabric is subjected to a raising or napping process on the reverse side. A typical polyester construction, as may be found in a reverse air filter, is also shown in Table 7.3 for comparison with the previous construction.

7.4.2 Needle felts

Needle felts in this instance are made by needle punching a 'batt' of carded and cross-lapped fibres to both sides of a woven scrim, followed by additional needle punching to further consolidate the structure to provide it with the necessary density and also obviate the tendency for (batt) delamination during use. The needles that are used on this occasion are triangular in cross-section and comprise a series of barbs on each edge, these being so designed as to engage the fibres on the down-stroke but release them on the up-stroke of the needle punching action. From this simplistic description, the two key elements in the construction of a needle felt are the batt of fibres and the scrim, the former providing the necessary filtration efficiency and resistance to abrasive forces and the latter providing strength to withstand the frequent pulsing forces (NB, in a limited number of low tensile applications some needle felts are actually produced without any scrim at all).

By their very nature, needle felts contain many more pores per unit area than woven fabrics. This is particularly advantageous in pulsejet filters, facilitating higher filtration velocities than would be possible with woven fabrics. Furthermore, as a result of the success of needle felts in pulsejet collectors, they have also found use in some reverse air and shake applications, though in some cases measures have to be taken to alleviate potential damage due to their relatively poor resistance to flexing e.g. by incorporating more flex-resistant woven sections into the sleeves where the flexing is most pronounced.

Whilst needle felts are produced to countless specifications in order to address the nature, volume and velocity of the dust being collected, the type of filter employed, to meet a standard test requirement or an equipment manufacturer's specification, their area densities will generally fall between 340 and 640 gm^{-2}; the greater the dust load and intensity of cleaning, the greater is the area density.

The area density of the woven scrim or base-cloth normally falls between 50 and 150 gm^{-2}, and may comprise 100% continuous filament yarns, or 100% staple-spun yarns, or a combination of both i.e. continuous filament warp and staple weft, depending on the demands of the application. As a result of the needle punching action, it is inevitable that some damage will be inflicted on the base-cloth, especially if it is woven from continuous filament yarns. The appropriate design and fineness of needle, the needle board pattern and the subsequent needling programme will be critical to minimising this damage. If the specification demands high tensile strength in a particular direction, the selection of needles with barbs on two rather than three edges may be appropriate, the latter being orientated in the needle board so as to ensure that minimum damage is inflicted on the yarns in the critical direction.

Of particular importance is the stretch propensity at relatively low stress levels, particularly in the 'weft' direction where the pulse-cleaning mechanism has its greatest impact. If the needle felt were to stretch excessively from repeated pulsing forces, this could lead to a slack fit on the cage and the risk of abrasion damage against the (possibly corroded) cage wires. It is not uncommon, therefore, to find that, in addition to the final breaking force, manufacturers also include extension at low stress levels in their specifications e.g. percentage extension at 20, 50 and 100 N 5 cm^{-1}.

7.4.3 Other non-woven fabrics

Spunbonded fabrics

As in synthetic yarn production, spunbonded fabrics are also produced by extruding molten polymer chip through a spinneret, the difference being that on this occasion the spinning head or dye contains a much larger number of holes. The filaments that emerge from the extruder initially pass through a quenching zone, where they are solidified by jets of cool air. This is followed by high velocity air, which attenuates the filaments to orientate the polymer chains and increase the filaments' strength. The filaments are then laid in random fashion on to a vacuum-assisted wire mesh belt to form a batt, which is carried to a bonding unit. This may involve chemical bonding, thermal bonding, hydroentanglement, or mechanical (needle punch) bonding. For further reading, a review of the spunbonding process is provided by Lim (2010).

Meltblown fabrics

In similar vein to spunbonded fabrics, meltblown fabrics are also produced by extruding molten polymer chip through a dye, the difference being that in this case the emerging filaments are fibrillated and fractured by a jet of high velocity hot air to produce much finer fibres, typically 0.5–5 μm. The airflow directs the fibres to a vacuum-assisted mandrel where they form a fibrous batt, being cooled and solidified en route by secondary air that is drawn into the process. However, being still quite hot, a degree of fibre bonding will have taken place during batt formation and, whilst the batt may be further consolidated by calendering, this process will inevitably reduce its air permeability, which may be detrimental to filtration performance. A more detailed overview of the meltblown process is provided by Dutton (2008).

Electrospun fabrics

Whilst the fibres in meltblown fabrics are considerably finer than those in spunbonded fabrics, electrospun fibres are still finer, typically in the range 50–500 nm. In this case, the polymer chip is usually dissolved in a suitable solvent (polymer melts are also possible) and the solution metered through a fine needle-like syringe. The latter is attached to a generator, which delivers a high voltage DC charge, typically 5–30 kV, and is positioned at a suitable distance from a screen or mandrel collector, which contains an opposing charge. The high intensity electric field that is formed causes the emerging droplet to become deformed into a cone shape and as the electric field intensifies, the electrostatic force pulls and stretches this droplet towards the collector in the form of a liquid jet. As the stretching intensifies the solvent evaporates, resulting in the deposition of a batt of solid nanofibres at the collector. In a subsequent development, one manufacturer (see Elmarco) produces electrospun fabrics, referred to as 'Nanospider™', without the use of the aforementioned needles. In this case the electrode is in the form of a rotating roller which is partially immersed in a bath of liquid polymer. As the roller rotates, the liquid is picked up by the charged roller and at the critical point thousands of cones are formed, which once again are pulled towards the collector. However, in this case the nanofibres are intercepted to form on a substrate or backing cloth that moves from roll to roll between the two electrodes. A comprehensive introduction to electrospinning is provided by Ramakrishna *et al.* (2005), whilst details of 'needleless' electrospinning are further described by Jirsak (2009).

Non-woven fabrics as described above are widely used in cartridge filters, some of which operate by surface filtration and pulsejet cleaning, especially where the dust load is expected to be high, or by depth filtration, where the dust load is not as high; the filter element in the latter case is being removed when fully loaded with dust. The use of composite structures made from a combination of webs with different fibre/filament diameters is also widely used. Such constructions may be designed for operation either by depth filtration or surface filtration. In the former the larger particles in the gas stream are retained by more open fabrics made from thicker filaments on the surface, whilst finer particles are captured within the body of the felt due to the smaller pores as found in, say, meltblown fibres. Conversely, in surface filtration the finer meltblown or nanofibre webs are applied to the surface of the medium where they restrict all dust particles, whilst being supported by more open fabrics made with tougher fibres or filaments (see Donaldson Torit, http://www2.donaldson.com/torit).

7.5 Fabric finishing techniques

Fabric finishing processes are designed essentially to ensure that the fabric will (a) remain dimensionally stable during use, (b) provide effective cake release during the cleaning cycle, (c) meet the fabric's designed air permeability (a control parameter widely used in the industry) and (d) be protected, as far as possible, from any chemical conditions that may be encountered in the filter.

7.5.1 Stabilisation processes

It is well known that most synthetic fibres and continuous filament yarns will to a greater or lesser extent shrink when exposed (in unrestrained state) to high temperatures. In addition, due to tensions imposed on fibres and yarns during processing, further shrinkage is to be expected due to relaxation effects. At the temperatures to which the textile components are likely to be exposed during use, fabric shrinkage can result in filter sleeves becoming too tight on filter cages, leading to inefficient cleaning, and in extreme cases can even generate such force as to lift the cages out of the cell plates into which they have been located. In order to reduce these effects, the fabrics are almost universally subjected to a thermal relaxation or heat-setting process. Here the normal practice is to raise the temperature of the material some 30–40°C degrees higher than what it may be expected to encounter during use, ideally in unrestrained form as on a heated cylinder, or 'overfed' on a stenter. For example, in the case of polyester sleeves, which have a maximum operating limit of nominally 150°C, with possible surges up to say 170°C, the heat-setting temperature is likely to be of the order 200°C. The effectiveness of this operation is normally checked by measuring the unrestrained shrinkage on a sample of fabric after an exposure period of say 2 h in a laboratory oven, a maximum shrinkage of 3% generally being considered acceptable for most applications. In addition, it is sometimes useful to be able to demonstrate that any residual shrinkage force in the heat-set material is unlikely to cause further problems. Tensile testing machines with environmental chambers can be used for this purpose, although other machines (see Testrite™) are equally suited to the task.

7.5.2 Surface treatments

Surface treatments in this context are aimed primarily at ensuring good cake release, either by removing any obstacles to the release process or by restricting dust to the surface of the material. For example, if the surface of the fabric were of a fibrous or hairy nature, effective cake discharge may be impeded through physical attachment to the protruding fibres. This problem is simply overcome by means of a singeing process in which the fabric is passed at speeds of 10–50 m min^{-1} over a gas flame. With thermoplastic fibres such as polyester, the heat of the flame causes the fibres to shrivel into molten beads, which quickly cool to leave a relatively harsh surface, but one that is devoid of protruding fibres. Fabrics produced from non-thermoplastic fibres e.g. polyaramids are also subjected to this process although, as they do not respond to the heat source in the same manner as thermoplastic fibres, more intensive singeing conditions are normally applied e.g. lower speed or increased gas pressure or a combination of both. NB, adequate ventilation must be provided to evacuate noxious fumes.

The relatively harsh surface created by the singeing process can be made smoother through the heat and pressure of a calendering process. The heated bowl in this operation is normally of chrome-plated steel composition, whereas the second bowl, which applies the pressure, usually comprises a somewhat softer material so as not to crush the fabric when the pressure is applied. Modern calender bowls are frequently fitted with a thick nylon sleeve for this reason.

The calendering process also increases the density of the fabric (especially needle felts); the pore size is reduced and dust penetration into the body of the needle felt thus restricted. In practice, the polymer type largely dictates the surface temperature of the heated bowl whilst the pressure, which may be up to 300 da N m^{-1}, and speed are adjusted to achieve the desired density and air permeability.

A surface treatment that is not engaged for improved cake release purposes is the raising or napping process. This is usually applied to satin weave fabrics made with staple-spun fibres in the weft, and is designed to enhance particle collection efficiency by teasing the fibres of the staple-spun yarn into a pile or nap. It will thus be appreciated that the treatment is normally applied to the reverse side of the fabric i.e. the side that does not collect the dust cake.

An alternative to singeing and calendering is the application of a microporous surface coating, typically in the form of wet latex foam, either acrylic or more abrasion resistant polyurethane, which is usually applied by knife-over-roller technique. The applied foam is dried, lightly crushed to produce an open cell structure, and then cured to secure the cellular structure as shown in Fig. 7.7. Whilst the application of a foam coating normally ensures good particle capture, retains the majority of particles on the surface and thus facilitates efficient cake release, in certain cases microscopic particles are able to plug the pores before the cake formation mechanism takes over. As a result, the steady-state condition previously mentioned (Fig. 7.3) is not reached, and the pressure differential across the sleeves continues to rise, leading to premature shut down of the filter.

The above problem can be avoided by attaching a membrane to the surface of the fabric, the term membrane on this occasion being viewed as a thin sheet of permeable PTFE capable of capturing particles < 0.1 μm. A number of such membranes are used in dust collection (see Goretex™ and Tetratex™), which not only ensure that extremely fine dust particles such as carbon black are restricted to the surface of the fabric, but also facilitate excellent cake release during the cleaning cycle (see Dubey *et al.*, 2005). Moreover, as it is unnecessary for the fabric to retain a portion of dust cake as previously described, steady-state filtration conditions are reached virtually instantaneously with very little risk of particle emission into the atmosphere. On the debit side, as PTFE membranes are of a somewhat delicate nature, in more arduous conditions mechanical damage, possibly due to flex fatigue, can ensue, leading to the formation of holes and hence the risk of particulate emission. NB, more robust membranes have recently been developed in an attempt to address this shortcoming. Special attention will also have to be given during fabrication to ensure that the sleeves are devoid of

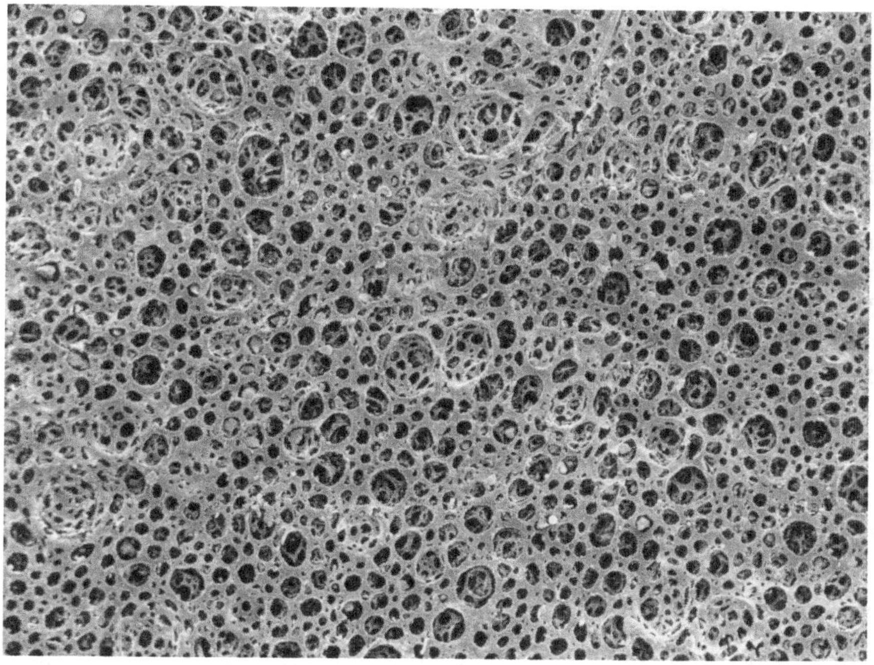

7.7 Microporous finish (mag 100:1).

stitching holes, which may otherwise also lead to dust emission. Although there has been a steady growth in the use of PTFE membranes since their introduction in the 1970s, their relatively high cost remains a deterrent to general use, and so the tendency is to use them where filtration conditions are particularly difficult e.g. where strict emission controls are in force or where very fine, sticky or toxic dusts are encountered.

7.5.3 Other treatments

Manufacturers routinely apply a variety of proprietary treatments, most likely fluorocarbon-based, to suit particular conditions e.g. hydrophobic treatments have been developed for applications where humid or moist conditions are likely, where oleophobic treatments to counter oily, sticky dusts, and anti-acid treatments are used as a protective coating in acidic environments.

Additional surface treatments, using graphite as a base, are also applied where the dust stream may contain incandescent particles leading to pin-holes (often found in the asphalt industry) or a mixture of silicone and graphite, as in the case of glass filter media, to reduce fibre–fibre friction. Still further treatments have been developed to counter specific problems arising from dusts such as those found in the ferrous or alumina industries.

7.6 Solid–liquid filtration

In the separation of solids from liquids, it is possible that the product of the operation may be (a) the liquid fraction, as in the beer industry, (b) the solids component as in pigment manufacture, or (c) both liquid and solids, as in metallurgical refining where the primary target is an electrolyte solution whilst the separated solids may contain a quantity of precious and/or semi-precious metals for separate recovery. There are still further cases where there is no product at all, the operation simply being used to clean up a liquid prior to it being used, re-used or discharged into a river or sea.

Although textile filter media are used in numerous types of filter, the mechanisms by which the particles are captured are commonly referred to as surface straining (alternatively screening), depth straining, depth filtration and cake filtration. These mechanisms are shown in Figs 7.8a–7.8d.

7.6.1 Surface straining

In surface straining/screening, the filter fabric retains particles of known size whilst smaller particles are permitted to pass through (see Fig. 7.8a). This mechanism, which frequently utilises monofilament fabrics with precise apertures or 'mesh' sizes, often features in food processing such as flour or corn milling, the separation of minerals or as a 'precursor' to more intensive filtration down line. To ensure that the apertures in the fabric do not become blocked, a vibration mechanism is frequently engaged to keep the particles moving.

7.6.2 Depth straining

By comparison, depth straining captures particles largely within the body of the structure, specifically where a particle is confronted by an aperture that is simply too small (or the path too tortuous) for it to pass through (see Fig. 7.8b). From this it will be appreciated that this mechanism is particularly applicable to needle felts. In some straining applications the fabric is sewn in the form of a bag or so-called 'strainer' bag, often used in the paint industry, which is fitted inside a pressure vessel. The suspension is pumped into the bag until a point is reached when the pressure and/or filtrate flow become unacceptable, at which point the bag is removed and usually discarded. Needle felts are frequently given (nominal) micron ratings for such purposes, which are widely assumed to be equivalent to their maximum pore size. Measurement of this property (strictly speaking 'equivalent pore size') is traditionally carried out by means of a 'bubble point' test (see BS 3321, 1986) in which the fabric is thoroughly wetted with an agent such as white spirit and clamped in a test chamber. Air is gradually introduced behind the specimen until sufficient pressure has built up in the chamber to force

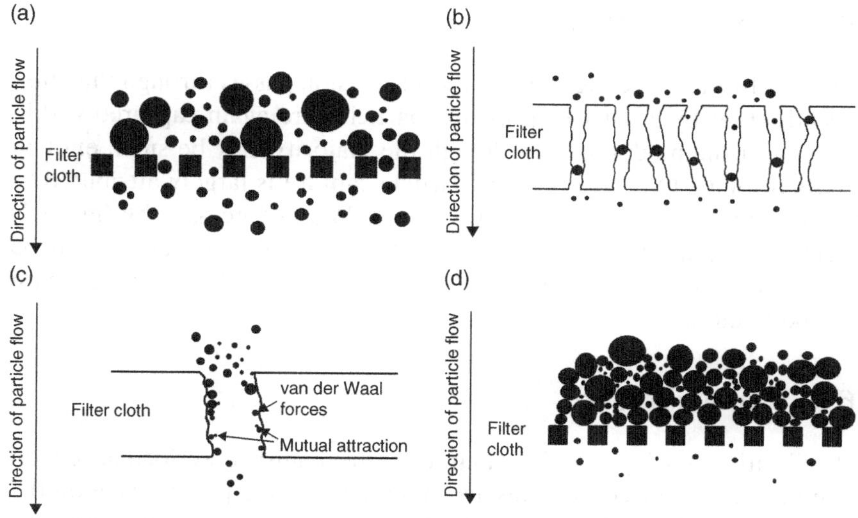

7.8 Filtration mechanisms: (a) surface straining (screening), (b) depth straining, (c) depth filtration and (d) cake filtration.

the liquid out of the fabric, appearing on the surface in the form of air bubbles. The pressure readings are noted (a) when the first bubble appears, and (b) when a steady stream appears; these readings are being used to calculate the largest and average pore size respectively using the equation:

$$D = \frac{4\sigma(\cos\theta)}{P}$$

where D = pore diameter (μm); σ = surface tension of the wetting fluid (Nm^{-1}); θ = fluid contact angle and P = pressure (Pa). Since wetting agents such as white spirit make it virtually impossible to measure any contact angle, this equation can be simplified to:

$$D = \frac{4\sigma}{P}$$

Although this technique is widely used to characterise pore size, other techniques are also used, not least amongst which are challenge tests (see Rideal, 2005) in which the filter medium is 'challenged' by a quantity of calibrated micro-spheres, which are fluidised, typically by sonic means, to assist in their passage through the filter. Comparison is then made of the particle size distribution of the micro-spheres before and after filtration.

7.6.3 Depth filtration

In depth filtration (see Fig. 7.8c), as the suspension passes through the filter, assisted either by gravity or pressure, particles are again captured within the body of the medium, even though the majority may be small enough to pass through. On this occasion, particle capture is largely attributed to attraction by electrical or attractive van der Waal's forces, or by (mutual) attraction to particles already deposited. Whilst this mechanism can be applied to fabrics such as needle felts, it is more normally associated with deep-bed sand filters.

7.6.4 Cake filtration

Cake filtration (see Fig. 7.8d) is similar to the mechanism described in Section 7.2, in that the captured particles accumulate to form a porous cake on the surface of the fabric, which thereafter takes over the role of filter medium. At the end of the filtration cycle, the cake is discharged and the filtration process resumes. This is the most widely used mechanism as far as filter fabrics are concerned and, as a consequence, the following section will concentrate on areas of industrial filtration where this mechanism is predominant.

7.7 Filtration equipment

The types and styles of filtration equipment are legion, but generally fall into one of the four categories viz, gravity, vacuum, centrifuge and pressure.

7.7.1 Gravity filters

Gravity filters are generally found in processes where the solids are free flowing e.g. sandy suspensions or where the cake needs to be handled with care i.e. not forced through the filter medium as in beer mash filtration. Open weave monofilament fabrics, possibly aided with a vibrating mechanism, may be found here and even non-woven fabrics are used in some gravity filters, as in the separation of swarf and debris from cooling oils. In a wider sense, gravity is also used to concentrate suspended solids by means of so-called thickeners or settling tanks, which operate without the need for filter media.

7.7.2 Centrifuge filters

In centrifuge filtration, the textile filter element is usually a tighter fabric and manufactured in the form of a suitably shaped bag, which fits inside a 'perforated basket' that rotates at high speed. The centrifugal force that is generated causes the liquor to pass through the fabric whilst the solids are retained in the form of a cake, which is subsequently collected and forwarded to the next process stage. Centrifuge filters are widely used where low cake moisture content is important and are frequently found in the pharmaceutical industry.

7.7.3 Vacuum filters

In vacuum filtration the filter fabric may take the form of a horizontal belt, or be cut into shapes and fitted to a variety of drainage surfaces. For continuous filtration, these drainage surfaces may comprise horizontal panels on the circumference of a rotating drum, segments of horizontal table/tipping pan filters, or an array of vertically mounted rotating discs. By comparison, for batch duties, the media will be attached to frames or 'filter leafs', an assembly of which are lowered into the suspension to be filtered. In both continuous and batch filtration, as the title suggests, a vacuum pump draws the liquid through the filter fabric where it exits via drainage channels to the next process stage, whilst the solids are retained as a cake on the fabric's surface. In continuous filtration systems, a series of sprays are often engaged to wash the filter cake whilst it is still under vacuum. The continuously moving filter medium, together with cake, eventually moves out of the vacuum zone, at which point the 'dewatered cake' is discharged from the fabric, usually with the aid of a doctor knife.

Vacuum filters consume a much larger volume of fabric than gravity and centrifuge filters, and in one form or other are widely used across the whole of the mining and chemical process industries. As the applied vacuum across the filter medium is < 100 kPa, the cakes that are formed in vacuum filtration are generally less dense than in centrifuge or in pressure systems. Whilst this is arguably better for processes where efficient cake washing is essential, the final cake moisture content will inevitably be higher, which means greater expense if the solids are to be subsequently dried.

7.7.4 Pressure filters

Compared with the previously mentioned systems, pressure filters consume by far the greatest volume of fabric, being employed in a host of industries including mining applications, food processing, chemical process industries and effluent treatment. The most extensively used type of pressure filter is the filter press (of which there are several variations), though other pressure filters are also widely found with titles such as 'candle filter', 'tower belt press', 'pressure leaf filter', 'tube press', 'multi-roll press' and 'pressure strainer'. In most cases the fabric is located within a pressure vessel into which the suspension is pumped. The solids are again captured through a cake forming mechanism whilst the liquid escapes via the filter's drainage system. Pumping continues until either the liquid flow through the filter falls to an unacceptable level or a predetermined pressure is reached. At this point, the pressure is switched off and the cake is removed, ideally with the minimum of manual assistance. For the majority of applications the operating pressure of filter presses is between 0.5 and 1 MPa, though for some applications this may be as high as 6 MPa and for tube presses, pressures in excess of 10 MPa are not uncommon. As a consequence of these pressures, pressure filters are able to force much more liquid out of the filter cake than vacuum filters, which is a major advantage if the cake has to be dried.

7.8 Fabric selection for solid–liquid filtration

The key parameters in fabric selection have been described (Hardman, 1994) as the thermal and chemical conditions, filtration requirements, equipment considerations and cost.

7.8.1 Thermal and chemical conditions

The thermal and chemical conditions that prevail in the application will effectively dictate what fibre or polymer type may be confidently used, the principal fibres that are used in liquid filtration, together with their general properties being shown in Table 7.4. The fibre properties shown in this table are somewhat different from those in Table 7.2 in the previous section, and reflect how the different polymers behave in liquid applications. From this table, the most widely used polymer in liquid filtration is unquestionably polypropylene. It is reasonably tough, resistant to strong acids and alkalis and, by virtue of its low density, provides a high yield for a given count of yarn. It is also commercially available in a number of forms i.e. fibres, multifilaments, monofilaments and fibrillated tape, which facilitates a wide range of fabric options. Compared with other polymers it is also 'competitively priced'. The only areas of concern are (a) its resistance to strong oxidising agents e.g. bleach, and (b) its dimensional stability as the filtration temperature approaches 100°C.

By comparison, as other polymer types are of higher density and generally more expensive, they tend to be used where their physical and/or chemical properties significantly out-perform those of polypropylene. For example, fibres and filaments in the polyamide family, whilst unable to compete with polypropylene in strongly acidic conditions, are notably tougher and more durable than polypropylene. As a result, they are often found in less demanding chemical environments where high resistance to harsh abrasive

Table 7.4 Fibres and their properties (liquid filtration)

Fibre type	Density (gcm^{-3})	Max operating (temp °C)	Resistance to: Acids	Alkalis	Oxidising agents
Polypropylene	0.91	95	E	E	P
Polyethylene	0.95	80	E	E	P
Polyester (PBT)	1.28	100	G	F	VG
Polyester (PET)	1.38	100	G	P	VG
Polyamide 66	1.14	110	P	VG	P
Polyamide 11	1.04	100	P	VG	P
Polyamide 12	1.02	100	P	VG	P
Polyvinylidene chloride (PVDC)	1.70	75	E	VG	VG
PVDF	1.78	100	E	E	G
PTFE	2.10	120+	E	E	VG
PPS	1.37	120+	VG	E	F
PVC	1.37	75	E	E	F
PEEK	1.30	120+	G	G	F
Cotton	1.5	90	P	G	F

E = Excellent, VG = Very Good, F = Fair, G = Good and P = Poor.

particles in the slurry is essential. Polyamide 11 and 12 are a particularly tough polymers with good flex resistance, although on the down side these are only available in monofilament form, the limitations of which will be explained later. Oxidation resistance is also superior with polyamides, as is their resistance to the relatively few applications where organic solvents are encountered.

Polyester fibres are also tougher than polypropylene, and are again used where oxidising conditions are present, but on this occasion acidic conditions preclude the use of nylon. Polybutylene terephthalate (PBT) is a particularly tough variation but, like Polyamide 11, is also only available in monofilament form. On the other hand, unlike polypropylene and polyamide, all polyester fibres/filaments are severely degraded by strong alkaline conditions, the ester link in the molecular chain being attacked as shown in Fig. 7.9.

The presence of steam, which may be used for example to express more product from the filter cake in oils and fats filtration, will also cause polymer chain scission through hydrolysis, again leading to rapid fibre degradation.

Although other polymers such as PTFE, PPS, polyetheretherketone (PEEK) and polyvinylidene fluoride (PVDF) are also included in the table, due to their high cost they are only used in applications where their unique characteristics outweigh their cost. PVDF, for example, being particularly resistant to available chlorine, may find limited use in brine filtration

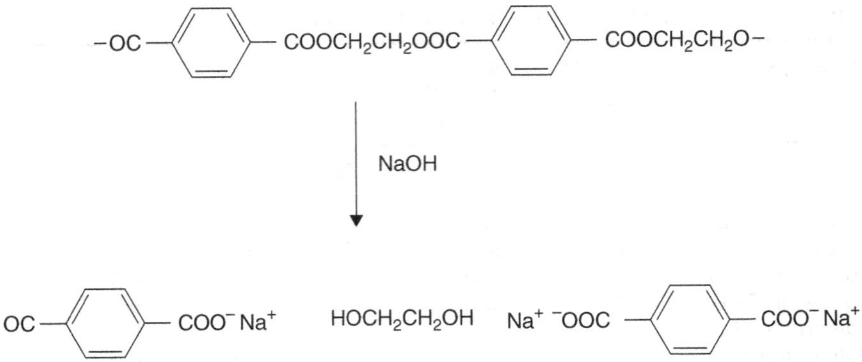

7.9 **Degradation of polyester by sodium hydroxide.**

applications, and PPS as a diaphragm fabric in the presence of hot potassium hydroxide in water electrolysis and PTFE in removing fines when filtering strong bleach solutions.

7.8.2 Filtration requirements

In terms of filtration requirements, in general terms the production engineer will be looking to see how much product or throughput he/she can achieve in the minimum of time and for the minimum of cost. If filtrate clarity is crucial to the operation, the production rate may be somewhat compromised, as a tightly woven fabric with a relatively low liquid flow rate will probably be required. On the other hand, if filtrate clarity is not too critical, then a more open structure, permitting a higher flow, will be used. Concerning this (flow rate) property, a general but still widely used parameter as to a filter fabric's 'fitness for purpose' is its air permeability value (see BS EN ISO 9237, 1995), i.e. the volume of air that passes through a sample of material in a given time and at a fixed pressure, often expressed as L dm^{-2} min^{-1} at 200 Pa. Whilst numerous research studies into the relationship between liquid flow, and fabric components have been undertaken (see Wakeman and Tarleton, 1999a), it is considered that the results of these studies are of limited practical value. As a consequence, filter media manufacturers prefer to rely heavily on previous experience when making their recommendations, supported where possible by 'bench tests', using a sample of the slurry to be filtered or challenge tests as previously mentioned.

In the event that the solids in the suspension are either too fine, even for the most efficient of filter fabrics, pre-coating the filter cloth with 'filter aids' such as diatomaceous earth or expanded perlite will significantly enhance particle capture. In similar vein, where filtration conditions are difficult due to slimy suspensions, such agents are often added as body feed

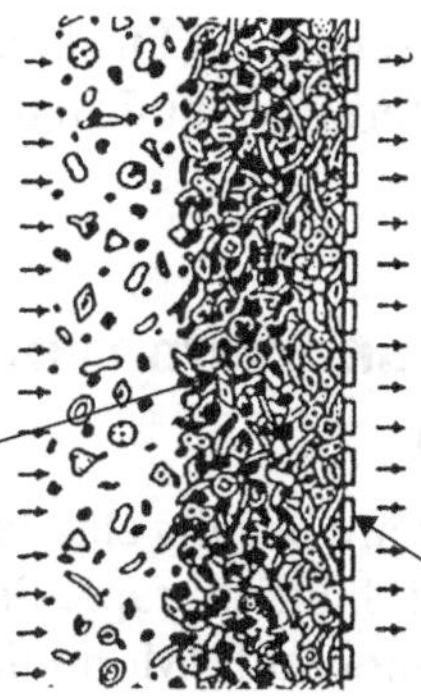

7.10 Addition of pre-coat and body feed.

to the suspension to assist in the formation of a more porous cake. Indeed, in special cases both body feed and pre-coating are employed in the same operation (see Fig. 7.10).

7.8.3 Equipment considerations

Without losing sight of filtration requirements, the media manufacturer must also take account of the physical forces that will be imposed on the fabric by the equipment itself; in particular, the effect that these forces will have on the fabric's resistance to stretch, its resistance to flexing, or its abrasion resistance, the last in this instance is through contact with rough, possibly corroded metal surfaces. In some cases, special reinforcement patches have to be stitched to the cloth during fabrication to avert potential abrasion damage inflicted by the filter.

7.8.4 Cost

As industrial process engineers strive to reduce their production costs, it is inevitable that they closely scrutinise the cost of all consumable items, including filter media, this despite the fact that in most applications the cost of this item is usually a fraction of the value of the product being filtered. Indeed, it could be argued that premature failure caused by a cheap, inferior filter fabric may have serious repercussions, not only in terms of fabric replacement costs but also lost production costs, contaminated product and remedial maintenance work. This notwithstanding, whilst top class filtration performance must be the designer's first consideration, he too must ensure that production costs are kept in check if a competitive edge is to be maintained. The cost of basic raw materials, consistent with efficient fabric production, is a crucial element in this equation, and the speed and efficiency by which the fabric can be cut and sewn into finished articles also plays a critical part in the final cost. It is therefore vital that all these aspects are kept under constant review.

7.9 Fabric constructions and yarn types for solid–liquid filtration

In the filtration industry, fabric types are usually classified according to their composition i.e. monofilament, multifilament, staple fibre, fibrillated tape or combination e.g. multi/staple or mono/multi. Monofilament yarns are normally chosen where blinding (a term that relates to a progressive reduction in flow through a filter as its pores become plugged) may be anticipated, or where high throughput and good drainage are the principal requirements. Being quite smooth, monofilaments will also facilitate efficient cake release at the end of the filtration cycle, a feature that is enhanced if woven in warp satin design. On the down side, they are unsuitable for use where very fine particles such as pigments are involved and are not particularly renowned for their good resistance to abrasion.

The diameters of the monofilaments engaged in filter media usually fall in the range 0.1–1.0 mm, the lower diameters being used in relatively lightweight constructions e.g. 200–300 gm^{-2} for the collection of finer fractions, typically < 50 μm, whereas larger diameters are usually found in heavy-duty filter belts with area densities up to 1700 gm^{-2}. The lighter fabrics are usually woven in warp satin designs for more efficient cake release, whereas the heavier styles are in duplex or herringbone designs for greater durability and stability. Although heavy-duty belts are normally associated with the capture of coarser fractions e.g. >1000 μm, by interweaving finer, tightly 'sett' monofilaments on the surface, in duplex style, with coarser monofilaments on the back, the filtration efficiency of a heavy-duty filter belt can be taken to a much higher level (see Sefar Tetex®).

So-called 'link' structures (Hardman, 2000c) have also found favour in some heavy-duty filter belt applications, largely due to their durability, particularly where coarse abrasive slurries are being processed. Unlike traditionally woven filter belts, there is also no requirement for a mechanical seam with link structures, which is frequently the first point of failure in a conventional filter belt.

Multifilament yarns are characterised by greater strength and flexibility than monofilaments, which enables them to be woven into much tighter constructions and therefore better suited to the capture of finer particles. Resistance to stretch is a further feature, which is used as an advantage in numerous applications e.g. avoiding excessive stretch caused by filter belt tensioning systems. On the debit side, multifilament fabrics are not renowned for their resistance to blinding and, where this is an issue, e.g. from crystal growth arising from the process itself, monofilament fabrics would, if practicable, be preferred.

Fibrillated tape or 'split film' yarns are made by splitting narrow polypropylene films into thin strips, then binding the strips together by twist to form a kind of crude multifilament yarn. Although they are too coarse for general filtration duties, tape yarns do find a use in the production of backing cloths, providing support, where needed, for lightweight filter cloths. Fibrillated tape yarns are usually woven in plain or, for increased drainage, 'mock-leno' designs.

Staple fibre yarns are normally produced via the woollen, semi-worsted or cotton spinning systems, the last mentioned involving either open-end or ring spinning. Normally used in quite tight, bulky fabrics, the essential characteristics of fabrics woven from staple fibre yarns are (a) good particle capture, and (b) the ability to withstand abrasive forces, especially from rough surfaces. Provided that the particles in the suspension are of a 'particulate nature', and thus readily able to form a porous cake, blinding need not be a problem with such yarns. Conversely if the filter cake were of a slimy nature or, due to process conditions, crystal growth may be expected, fabrics incorporating staple-spun yarns would very quickly become blind. Relatively poor cake release and resistance to stretch from high tensile forces may also be limiting factors with staple fibre yarns.

To some extent, the limitations of one yarn may be countered by producing composite fabrics. The use of multifilament warp yarns for resistance to stretch in the vertical direction and staple-spun weft yarns for increased bulk, particle capture and abrasion resistance in the horizontal direction is a common practice.

Needlefelts are not as widely used in liquid filtration as they are in dust collection. Unless required for depth straining, as in the viscose industry, or unless the particles can be restricted to the surface as in some filter belt mining applications, the risk of premature blinding through pore blockage is always a threat.

7.10 Fabric finishing techniques

As with dust collection filter media, fabric finishing treatments are designed (a) to ensure adequate dimensional stability (b) to enhance cake release, or (c) to increase particle collection efficiency. An additional requirement may also be the removal of the processing aids, which are routinely applied to fibres or yarns and which may subsequently contaminate the products being filtered.

7.10.1 Stabilisation treatments

In addition to heat-setting, hot aqueous treatments are also widely used to induce relaxation from process tensions, thereby alleviating any tendency to shrink during use. In special cases, compressive shrinkage processes

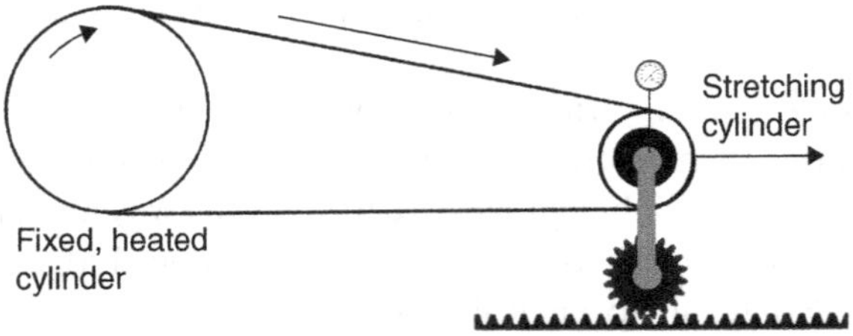

7.11 Filter belt stretching.

have also been employed for this purpose. Conversely, if the fabric were to be used in a filter belt operation, a pre-stretching process would be more appropriate. A traditional stretching operation is shown in the schematic diagram in Fig. 7.11, which is based on paper-machine clothing technology. The cloth to be stretched is sewn into an endless loop, fitted on the stretching cylinders, and then tensioned by outward movement of the smaller cylinder through a rack and pinion system. The larger heated cylinder slowly drives the belt whilst the smaller cylinder gradually 'racks it out' to a predetermined length, the belt tension being carefully monitored throughout the operation. The heat and tension thus has the combined effect of (a) increasing the belt's resistance to stretch, (b) inducing width-way contraction to reduce any further shrinkage during use and (c) ensuring that the belt will track squarely on the filter. Whilst with the aid of additional rollers, this equipment may be adequate for filter belts up to say 100 m in length, for significantly longer belts, as used in more recent tower belt presses, a continuous roll-to-roll stretching operation, (see Wumag) is the only realistic option.

Although the forces applied by the filter are never so great as to stretch the fabric to its breaking point, they can impose sufficient tension to stretch the material beyond the filter's tensioning stroke. As a consequence, the initial modulus of the fabric, and hence its resistance to relatively low stress levels, is the most important consideration in the stress–strain curves (see Fig. 7.12).

7.10.2 Improved efficiency and cake release treatments

As with dust collection media, surface treatments routinely feature singeing and calendering operations to enhance cake release and/or particle capture. On the other hand, were the suspension to contain extremely small particles, a more efficient approach may be needed. Filter aids have already been mentioned, but if for practical reasons they cannot be used, the application of a PTFE membrane may seem a logical choice. Whilst these are used in some applications (e.g. on candle filters in sodium hypochlorite

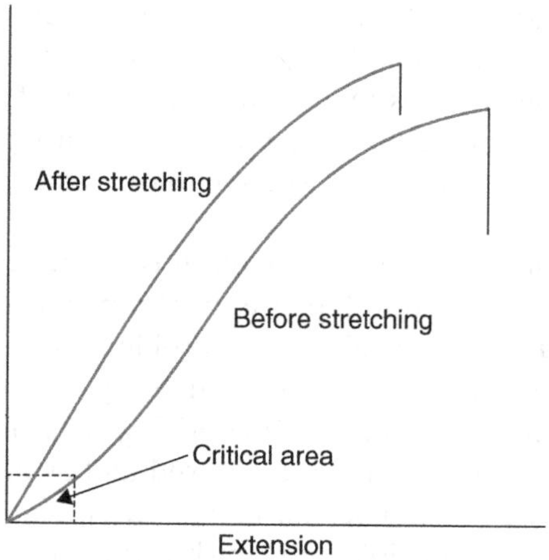

7.12 Stress–strain curves before and after stretching.

filtration), they are both expensive and in many applications easily prone to mechanical damage from abrasive forces. As a consequence, their use is not as widespread as in dust collection. The development of a more durable microporous polyurethane coating, referred to as Primapor™ by the manufacturer Clear Edge, has proved to be a successful alternative, especially where fine pigments are being processed. Details of the operation are not disclosed, although it is claimed that the particle capture is far superior to any woven construction, even after intensive calendering, and second only to PTFE membranes. However, since in very severe applications even this tough surface coating may become damaged, an alternative albeit slightly less efficient treatment, referred to as Azurtex™, also by Clear Edge, was subsequently developed in which the microporous compound is embedded within the body of the structure, thus averting damage from mechanical forces.

7.11 Future trends

In dust collection there is an ever-increasing demand for more stringent control of atmospheric pollutants. Improved media performance will be necessary in order to meet these demands and in conventional dust collection it is expected to see greater use of finer fibres, and fibres of irregular cross-sectional shape, applied in the form of surface layers as previously described. Taken to a higher level, PTFE membranes provide the ultimate in particle collection, and the trend to wider use of this material will almost certainly continue, together with further developments to enhance its toughness and durability.

There is also a growing trend towards the use of cartridge filter elements (see BHA), and it is expected that this will continue, these being installed either in new filtration equipment or as retrofit units in existing filters. As mentioned above, it is also anticipated that demands on the industry will see the wider use of PTFE membranes (together with meltblown and nanofibre webs) in these units.

As more advanced formulations become available, it is further expected that superior chemical finishes will be developed, either for increased fabric protection or to enhance dust release capabilities in problematic areas. For hotter gases than conventional textiles can accommodate e.g. > 300°C, dust collection systems incorporating rigid elements comprising inorganic fibres e.g. alumino silicate fibres will most likely be employed and as municipal waste incineration and kindred plants become more widespread – waste incineration currently processes some 50% of Danish waste – this trend is likely to continue. This technology has seen considerable growth in recent years, and in addition to providing highly efficient particle collection at high temperatures, by impregnating these elements with specific catalysts, the destruction (by oxidation) of flue gas pollutants such as NOx, SOx and volatile organic compounds can be achieved (see Clear Edge TopKat™). It is anticipated that, not least as a result of the environmental benefits that they provide, the future will see a much wider use of such elements.

In liquid filtration, it is considered that in terms of the volume of filter media consumed, future trends will again be driven by environmental considerations coupled with a growing need for more efficient use of water. For example, greater demands are being placed on industry to reduce the suspended solids in wastewater discharges, typically from power plants, and as water is becoming increasingly scarce due to population growth and heavy demands from power and manufacturing industries, there is a constant need to recycle this commodity. Advanced technologies are emerging to deal with this situation, for which purpose textile filter media frequently play a critical part. In similar manner, new technologies are also appearing to ensure maximum use of waste materials, perhaps for use as 'greener' fuel in power plants. Dewatering this waste through filter press cloths will ensure more efficient combustion and, if the waste has to be transported elsewhere for treatment, once again dewatering prior to transfer will ensure that transport costs are kept to a minimum. Apart from environmental issues, filtration is also involved in the clean-up of industrial areas, and this will continue as more contaminated 'brownfield' sites are developed for alternative uses. Close cooperation with equipment manufacturers is vital in all new developments to ensure that the filter fabric manufacturer can deliver what the equipment engineer requires. This is especially important at a time when there is a trend towards larger individual units e.g. longer and wider filter belts, larger filter presses featuring higher feed velocities and heavier filter cakes and a general demand from industry for a higher standard of product quality. To assist in this respect one leading filtration engineering company (Metso) has actually acquired a highly respected filter fabric manufacturer (Tamfelt), though it is not anticipated that this particular initiative will gather pace.

The above developments notwithstanding, in a competitive industry, cost reduction is a constant challenge, and this will be the driver for more competitive raw materials, more efficient processing equipment, and more advanced fabrication techniques.

7.12 Sources of further information and advice

There are a number of textbooks that make reference to textile filter media, albeit usually in the context of filter media in general. This notwithstanding, the following are considered worthy additions to any library: Purchas's Solid/liquid separation technology 1981 and its companion book Solid/liquid separation equipment scale up 1977; Handbook of filter media by Purchas and Sutherland (2002), Filters and filtration handbook by Dickenson (1997), yet another 'Filters and Filtration Handbook' by Sutherland (2008), and a complementary chapter to the present chapter, Textiles in filtration found in the Handbook of technical textiles edited by Horrocks and Anand (2000). Reference to 'Filtration equipment separation modelling and process simulation' by Wakeman and Tarleton (1999) has already been made in the text but is included here for completeness. Finally, although it does not feature filter media as such, a further publication of value is Shreve's Chemical process industries. Edited by Austin, this handbook contains many flow diagrams on industrial chemical processes, providing guidance to, amongst other things, where the filters 'fit in the processes'.

Readers will find useful information on manufacturers' websites, although this will naturally be weighted by 'sales patter' and restricted so as not to give too many clues to competitors. This notwithstanding, the websites of Sefar, Albarrie, Micronics, Gutsche and Andrew Webron are particularly informative and in terms of filtration equipment, Mikropul and Wheelabrator for dust filtration and Larox (part of Outotec) and Andritz for liquid filtration provide useful guides.

Exhibitions are also ideal places to extract and obtain more information on advances in filtration equipment, and filter media in general is one of the most important of which is Achema (Frankfurt), where numerous filter equipment and media suppliers frequently exhibit. The Filtration Society also organises periodic exhibitions and conferences, referred to as Filtech, which are aimed primarily at chemical engineers and students who have a vested interest in the separation of solids from liquids. Short courses often run concurrently with these conferences, which frequently feature an introduction to filter media (textile or otherwise). The same society, which has additional European and American chapters, also publishes its own monthly journal aptly entitled Filtration with occasional contributions from textile media producers. Also widely read are Filtration and Separation and a US publication entitled Filtration News, issued quarterly, is a similar magazine.

7.13 References

Andritz, www.andritz.com

Achema, www.achema.de

Albarrie, www.albarrie.com

Andrew Webron, www.andrewwebron.com

Austin GT editor (1984), *'Shreve's Chemical Process Industries'*, Singapore, McGraw-Hill

BHA, http://www.baghouse.com/products/dust-collector-filters/pleated-filters

BS 6524 (4) (1995), 'Method for determination of the surface resistivity of a textile fabric'. British Standards Institute, London Testrite Ltd, Halifax, UK. www.testrite.co.uk

BS 3321 (1986), 'Method for measurement of the equivalent pore size of fabrics (bubble pressure test)'. British Standards Institute, London

BS EN ISO 9237 (1995), 'Determination of the permeability of fabrics to air'. British Standards Institute, London

Clear Edge TopKat™, http://www.clear-edge.com/cerafil_topkat.Bekaert stainless steel fibres, www.bekaert.com

Clear Edge, http://www.clear-edge.com

Donaldson Torit, http://www2.donaldson.com/torit

Dickenson TC (1997), *'Filters and Filtration Handbook'*, Oxford, Elsevier

Dubey, Savage and Maindola (2005), 'Current trends in membrane fabric filtration for carbon black production', *Carbon black '05 perspective in Asia Pacific*, Suzhou China. See: http://www.gore.com/MungoBlobs/145/600/article_trends_carbon_black.pdf

Epitropic fibres, www.epitropicfibres.co.uk

Filtration News, Grand River, Michigan, U.S.A

Gore, www.gore.com

Gutsche, www.gutsche-mgf.com

Hardman E (2000a), 'Textiles in filtration', in Horrocks AR and Anand SC, *Handbook of Technical Textiles*, Cambridge, Woodhead Publishing Limited, 319.

Hardman E (2000b), 'Textiles in filtration', in Horrocks AR and Anand SC, *Handbook of Technical Textiles*, Cambridge, Woodhead Publishing Limited, 320.

Hardman E (2000c), 'Textiles in filtration', in Horrocks AR and Anand SC, *Handbook of Technical Textiles*, Cambridge, Woodhead Publishing Limited, 348–350.

Hardman E (1994), 'Some aspects of the design of filter fabrics in solid/liquid separation processes', *Filtration and Separation*, **31**(8) 813–818.

Horrocks AR and Anand SC (2000), *'Handbook of Technical Textiles'*, Cambridge, Woodhead Publishing Limited.

ICI Research (1975), 'Chemical properties of Terylene and ICI Nylon', *Industrial Fibres Manual TA1/3*, ICI Fibres Research and Textile Development Department, Harrogate.

Larox, see Outotec, http://www.outotec.com/en/Products--services/Process-equipment/Filters/Pressure-filters/Outotec-Larox-PF-/
Micronics, www.micronicsinc.com
Mikropul, www.mikropul.com
Morden K (1994), 'Dust explosion hazards in white sugar handling systems' *International Sugar Journal*, **96**, 142.
Nexylene®, www.emsgriltech.com
Purchas (1981), *'Solid/Liquid Separation Technology'* Croydon, Uplands press.
Purchas (1977), *'Solid/Liquid Separation Equipment Scale Up'* Croydon, Uplands press.
Purchas DB and Sutherland K (2002), *'Handbook of Filter Media'* Oxford, Elsevier.
Rideal (2005), 'Measuring filtration performance by challenge tests parts 1 and 2, *Filtration News*, 2005, **23** (4, 5).
Sefar, www.sefar.com
Sefar Tetex,
http://techlist.sefar.com/cms/newtechlistpdf.nsf/vwWebPDFs/tetex-dlw_En.pdf/$FILE/tetex-dlw_EN.pdf
silveR.STAT®, http://www.r-stat.fr/uk/index.php
Sutherland K (2008), *'Filters and Filtration Handbook'* Oxford, Elsevier.
Tetratex, www.tetratex.com
The Filtration society, www.filtsoc.org
Wakeman RJ and Tarleton ES (1999a), *'Filtration Equipment Selection Modelling and Process Simulation'* Oxford, Elsevier 27–33.
Wakeman RJ and Tarleton ES (1999b), *'Filtration Equipment Selection Modelling and Process Simulation'* Oxford, Elsevier.
Wheelabrator, www.wapc.com/home.htm
Wumag, http://wordpress.wumag.de/en/

8

High performance textiles for geotechnical engineering: geotextiles and related materials

C. A. LAWRENCE, University of Leeds, UK

DOI: 10.1533/9780857099075.256

Abstract: Geotextiles are high performance materials used to improve soils for the foundations of above-ground structures, and for protection against weather erosion damage. They therefore yield substantial benefits in geotechnical and environmental engineering. This chapter focuses on the structure, properties, and applications of these textiles, which are designed to perform one or more of five functions. The physical mechanisms of these functions are described, along with standard test procedures used to measure related properties. The effect of fabric parameters and other physical factors on the performance of a given geotextile in relation to a particular function is also discussed. The applications of geotextiles cover many areas of geotechnical engineering. Examples described include transportation, fluid containment systems, soil reinforced structures, and erosion control. The chapter concludes by outlining future trends in the development of these materials, aimed at achieving greater multifunctionality with improved cost effectiveness in their applications.

Key words: geosynthetics, geotextiles, geocomposites, structures, properties, functions, multifunctionality, applications.

8.1 Introduction

The prefix geo- means 'relating to the earth';[1] geotextiles are therefore textile materials used in earthworks. The application of fibrous materials in geo-engineering is aimed at making site soils more suitable for a desired end use than they would naturally be. Historical records show that this concept is not new.[2] Indeed, Sweet Track causeway in Somerset, England, was thought to be the oldest engineered road in the world, dating back to 3800 BC,[2] but archaeologists have found that it was built on top of what is known as 'the Post Track', constructed 30 years earlier from ash planks and hazel posts.[3] The employment of reeds to reinforce compacted soils for foundations in the construction of dwellings can be traced back to the first millennium BC, when in biblical times 'The Tower of Babel' was built on a substrate reinforced by river-bank reeds. In the third millennium BC, constructions of reed-reinforced clay were used for erosion control along the banks of the Tigris and Euphrates, and it is believed that around the fifth millennium BC the Persians used compacted soil reinforced with reeds for the construction of dwellings.[4]

The early nineteenth century saw walls and slopes reinforced with brushwood, timber or canvas, and during the first quarter of the twentieth century, cotton fabrics treated with asphalt were used in the USA to reinforce and protect sensitive soils. By the late 1950s geotextiles made from synthetic fibres had been introduced, and used mainly in drainage filtration. In the same decade, woven nylon tapes were utilised by the Dutch as canal-bed protection mattresses. However, in the late 1960s, non-woven geotextiles made of continuous spun filament were gradually and successfully used in roads and railway-track constructions. At about the same time as the introduction of these synthetic-fibre geotextiles, plastic sheet materials, referred to as geomembranes, became available as substitutes for waterproof clay liners, and were developed to become the principal waterproof liner materials for canals, riverbanks, water reservoirs, ponds and, lately, ground storage waste containment sites.

Collectively, these synthetic polymer-based materials are given the generic term geosynthetics, meaning 'artificial ground' and may be defined as:

> mainly sheet products that exhibit a degree of flexibility (from high to low levels) that are placed within or adjacent to soils to enhance the engineering performance of man-made, ground-based structures.[5]

Today there are various types of geosynthetics. They have been classified by grouping those materials that are made to be porous and readily permeable to water and gas as 'geotextiles' and 'geotextile related products', and those made to be impermeable to fluids as geomembranes. This simple classification is enlarged to include combinations of materials, referred to as geocomposites. As indicated earlier, not all geotextiles are made of synthetic polymers. Many natural fibres, in particular bast fibres, are converted into geotextiles products that have an important function in certain ground engineered constructions.

Although this chapter is concerned with geotextiles and their applications in geotechnical engineering, the subject needs to be considered in a broad way to include most of the other classes of materials, since some of the so-called related products are made of textile materials; also, the geocomposites often include one or more geotextiles, and generally the applications of geomembranes require a supporting geotextile. The chapter is therefore structured to initially give a classification of the geosynthetic materials, followed by an overview of their functionality, and consideration of their structure–property relations with regard to the various functions. Examples of

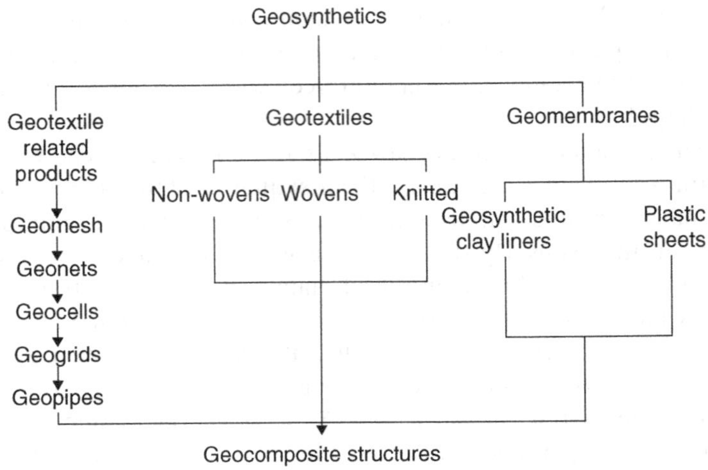

8.1 Classification of geosynthetics.

geotechnical applications for these materials are then presented, and finally possible future developments are described.

8.2 Classification of geosynthetics

Figure 8.1 gives a general classification of geosynthetic materials based on their physical form. As indicated, these materials include solid and perforated plastic sheets, plastic pipes, a range of textile structures, and combinations of these, to form a subcategory referred to as geocomposites. Strictly, the term geosynthetics refers to synthetic polymer-based materials. However, when considering the category of geotextiles, natural fibres as well as synthetic fibres are also of importance, and therefore the subcategory 'natural geotextiles' is 'loosely' incorporated in the general classification.

Most non-textile geosynthetics are used with geotextiles, and therefore it is more informative to consider the functions, structures and properties of all the material types. This should aid the reader's understanding of the role played by geotextiles, whether employed with the other materials or as their substitutes in the various geotechnical applications described later as examples.

8.3 Functions of geosynthetics: filtration

The use of geotextiles in filter applications is probably the oldest, most widely known, and most used function. A geotextile can be structured to be a filter, so that liquids can pass through its thickness (i.e. across the plane of the material) while preventing the passage of soil particles from the upstream

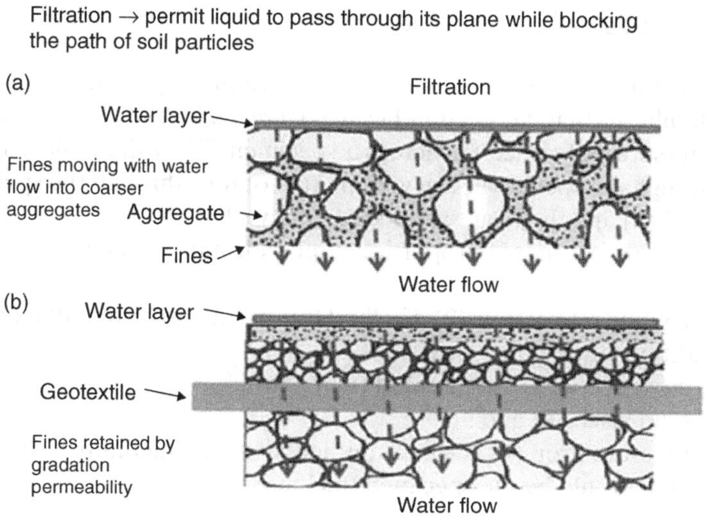

8.2 Filtration function: (a) Without geotextile and (b) With geotextile.

side, as illustrated in Fig. 8.2. This means that the geosynthetic filter must be made to meet two conflicting requirements: it must have a suitable level of permeability to enable the required liquid flow, as well as an average pore size, and pore-size distribution, sufficiently small to prevent all but the finest of particles migrating through its thickness. Generally, textile structures are the only material form that can be readily manufactured to achieve these conflicting requirements.

The basic mechanism of wet filtration is applicable to what we may term geo-filtration. On installation of a geotextile filter, particles finer than the restricting pores of the filter will be either carried through the filter thickness by the fluid flow or lodged within it. As these fines are lost from the soil/filter interface, coarser particles that are blocked at the interface (see bottom illustration of Fig. 8.2) build up into a compact porous layer, small pores resulting from the packing of the particles. This layer of blocked coarser particles then becomes a filtration zone itself by blocking smaller particles, which in turn collectively become an additional filter layer. The net result is a graduated soil filter. The geotextile filter therefore acts as a catalyst to produce a natural filter within the soil. At the dynamic equilibrium stage, there will be a gradation of permeability, the geotextile filter being the most permeable, and the soil furthest from the filter having the lowest permeability. Thus, the fluid flow into the filter will ultimately be governed by the parent soil.

From the above description of the basic geo-filtration mechanism, it becomes evident that the performance of geotextiles as filter media depends largely on two properties: the material's cross-plane permeability and pore-size characteristics.

8.3.1 Cross-plane permeability

The permeability of a geotextile can be determined by using a test arrangement similar to that depicted in Fig. 8.3. Here, an inflow of water (1) under a set pressure difference (4) across a specimen (3) of the material permeates through the specimen. The volume rate of flow through the specimen is measured (2) and the result applied to Darcy's law.

Darcy's law is a simple proportional relationship between:

- the volume rate of flow (2) of a fluid through a porous medium (3), under conditions for laminar flow,
- the cross-sectional area normal to the flow,
- the fluid viscosity, and
- the pressure drop (4) across a unit length of the medium; in the present case this would be the specimen thickness.

For one-dimensional flow, Darcy's law can be written in the form:

$$Q = kA\frac{\Delta P}{\eta t} \qquad [8.1]$$

or

$$v = \frac{Q}{A} = \frac{k}{\eta}\frac{\Delta P}{t} \qquad [8.2]$$

where v = superficial flow rate of the fluid (ms^{-1}), Q = volume rate of flow (m^3s^{-1}), A = area of the specimen through which the water flows (m^2),

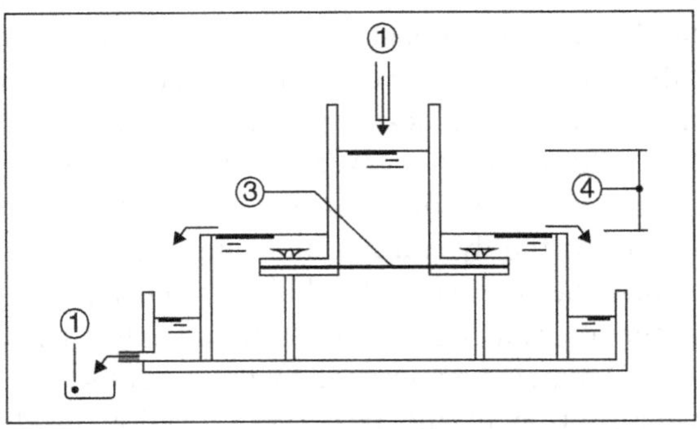

1. inflow, 2. collected outflow, 3. specimen, 4. head loss (H)

8.3 Schematic for cross-plane permeability measurements of geotextiles.

η = dynamic water viscosity (Pa.s), ΔP = differential hydraulic pressure (Pa), t = specimen thickness (m), and k = is the constant of proportionality (m^2).

v is often referred to as Darcy's flux, since it is not the velocity with which the fluid actually travels through the pores of the porous material. The pore velocity would be:

$$v_p = \frac{v}{\Phi} \quad [8.3]$$

where Φ is the porosity of the porous material.

In engineering applications, Equation [8.2] may be rearranged as:

$$v = \frac{k}{\eta}\frac{\rho g \Delta h}{t} \quad [8.4]$$

$$v = K_n i \quad [8.5]$$

where $K_n = k\rho g / \eta$ and is termed the permeability coefficient (ms^{-1}) normal to the plane of the material, also called the cross-plain permeability, ρ = the fluid density (kgm^{-3}), g = acceleration of gravity (ms^{-2}), and Δh = differential hydraulic head (i.e. total head lost) (m), and $i = \Delta h/t$ is termed the hydraulic gradient.

Hence,

$$k = K_n \frac{\eta}{\rho g} \quad [8.6]$$

and k may therefore be called the specific permeability.

An alternative approach to K_n and k is to consider the proportionality between v and Δh,

$$v = \psi \Delta h \quad [8.7]$$

Thus

$$\psi = \frac{K_n}{t} \quad [8.8]$$

where ψ is known as the permittivity of the porous material (s^{-1}) and may be defined as the rate of flow of a unit volume of fluid through a unit area, normal to the plane of the porous material, per unit head of pressure.

The commercial testing of geotextiles for filtration involves the use of one of the two standard methods, namely BSENISO 11058:'Determination of water permeability characteristics normal to the plane without load' or ASTM D4491:'Water permeability of geotextiles by permittivity'. The former is the international standard, the latter is the USA standard.

With BSENISO, superficial flow rates (v) are determined for Δh values of 70, 56, 42, 28, and 14 mm in order to establish the full permeability characteristics of the geotextile. Since ρ and η vary with temperature, the water flow rate values (v) must be adjusted to the equivalent values (v_{20}) for a standard water temperature of 20°C. From Equation [8.4],

$$\frac{v_{20}}{v} = \frac{\rho_{20}}{\eta_{20}} \frac{\eta}{\rho}$$

i.e.

$$v_{20} = vR_T = \frac{Q}{A} R_T \qquad [8.9]$$

where v_{20} = the flow velocity at 20°C in ms^{-1}, and R_T = the correction factor for a water temperature of 20°C

The use of the correction factor enables comparison of v_{20} values for different geotextiles, irrespective of the test laboratories used.

In practice, the units used are $Lm^{-2}s^{-1}$, in which case Equation [8.9] becomes:

$$v_{20} = \frac{Q}{A} R_T 1000 \qquad [8.10]$$

In the ASTM method, the permittivity (ψ) is determined for Δh values from 10 to 75 mm in steps of 5 mm. Hence, from Equations [8.7] and [8.9],

$$\psi = \frac{v_{20}}{\Delta h} = \frac{Q}{A\Delta h} R_T \qquad [8.11]$$

Both the v_{20} and ψ values can be plotted against the Δh values to represent the cross-plane permeability characteristics of the geotextile.

Some manufacturer's literature quotes the permeability coefficient, K_n, which from Equations [8.5], [8.8] and [8.9] can be calculated by:

$$K_n = v_{20} \frac{t}{\Delta h} = \psi t \qquad [8.12]$$

8.3.2 Pore-size characteristics

Several terms may be used to describe the pore size of geotextile materials: apparent opening size (AOS), equivalent opening size (EOS), and filtration opening size (FOS); the most commonly used is the AOS. This term is linked with the symbol O_x, which refers to the apparent pore-size diameter which is greater than x% of the pore diameters contained in the material. Thus, O_{95} would be the 'near-largest' pore diameter in the material. From the earlier description of the filtration mechanism, it can be reasoned that AOS or O_{95} must be considered in relation to the soil particle distribution, and this will be discussed further when dealing with the geotextile structure–property–function relationship. However, it is important to mention here that it is not only for particle retention that AOS values should be determined, but the minimum AOS value in relation to particle size distribution will indicate susceptibility of the geotextile to clogging.

Various methods have been reported for determining the pore-size characteristics of geotextile materials, but the two standard methods are: ASTM D4571 'Determining apparent pore size of a geotextile'; and BSENISO 12956: 'Determination of the characteristic opening size'. Both methods use the progressive sieving of glass beads of known diameters through the pores of the geotextile until a value is reached where 5% or less (by mass) of the beads pass through the material, thereby giving O_{95}. The ASTM method, also termed the dry sieving method, is performed under strictly controlled humidity with the beads and geotextile in their dry state, but this is susceptible to static electricity causing particles to be held by the material. The BSENISO method attempts to circumvent the problem by employing a water spray, and the test is therefore referred to as the wet-sieving method.

8.4 Functions of geosynthetics: drainage

This function is closely related to that of filtration, in that the fluid may enter the geosynthetic normal to its plane, but the geosynthetic is designed for fluid flow to also take place within and along the plane of its thickness (see Fig. 8.4). Generally, when in use, the geosynthetic will be under load from the surrounding soil, so the test for its flow characteristics is performed under loads of from 10 up to 500 kPa normal to the plane of the material. The type of geosynthetics used for drainage are geotextiles, geotextile related products (specifically geonets), and geocomposites (sandwich structures of geotextiles and geonets); these structures will be described later.

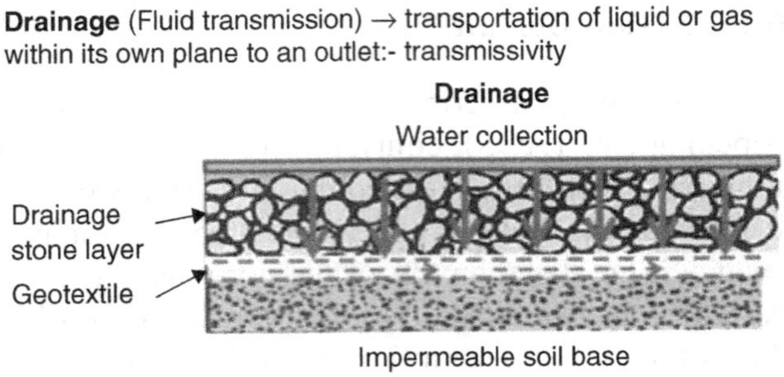

8.4 Geotextile drainage function.

Similar to the filtration function, the requirements for drainage are soil retention, adequate flow capacity, and long term soil-to-geotextile flow equilibrium, in respect of minimum clogging of the pore spaces. Apart from the flow capacity within the plane of the geosynthetic material, the other aspects have been dealt with above. We will therefore now consider the in-plane flow characteristics.

Figure 8.5 shows a diagram of the basic test arrangement for measuring the transmissivity of the above-mentioned geosynthetics. A specimen (4) (usually 300 mm × 300 mm) taken from a sample of the geosynthetic is sandwiched between two impermeable sheets of foam (simulating compacted soil) (7) of similar area to the specimen and placed under a platen (9) to which a specified load (8) is applied. An inlet of water, set to a specified head of flow (10) which corresponds to the required hydraulic gradient (H.G), passes through the specimen along its plane and the volume rate of flow is measured (2). The result is applied to Darcy's law.

For fluid flow along the plane within the thickness of the geosynthetic, Equation [8.5] can be rewritten as:

$$v_{20} = K_p i$$

$$Q = AK_p \frac{\Delta h}{L} \qquad [8.13]$$

where K_p = the permeability coefficient for flow along the plane of the material (ms^{-1}), and the hydraulic gradient is now given by $i = \Delta h / L$, L = flow length along the plane (i.e. specimen length = 300 mm).

Instead of the permittivity, reference may be made to the transmissivity, θ (m^2s^{-1}):

$$\theta = K_p t \qquad [8.14]$$

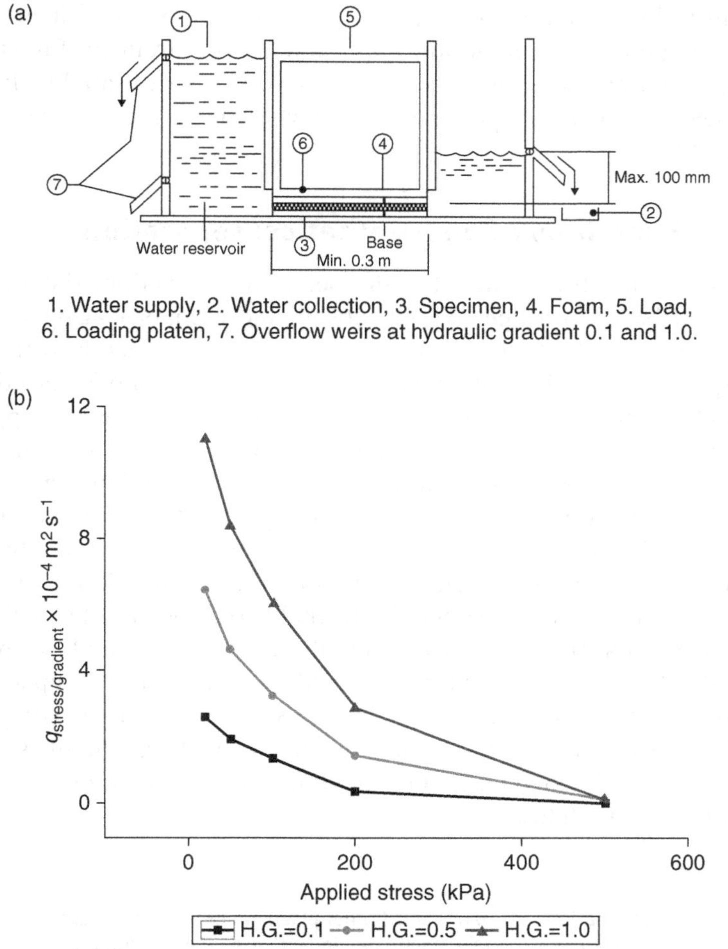

8.5 (a) Schematic for in-plane permeability measurements of geotextiles, and (b) effect of applied stress and hydraulic gradient on flow rate.

The transmissivity therefore may be defined as the flow along the plane, in the length direction, through a unit width of a material of a specified thickness, per unit head of pressure.

One of two standard test methods may be used to measure the in-plane permeability of materials: BSEN ISO 12958: 'Determination of in-plane water flow capacity'; and ASTM D4716: 'Determining the in-plane flow rate per unit width and the hydraulic transmissivity of a geosynthetic using a constant head'. As stated above, a series of loads is applied to the test specimen. At each applied load $Q(\mathrm{lm^{-2}s^{-1}})$, K_p, or θ is determined for three hydraulic gradients within the range $I = 0.05–1.0$. Similar to the filtration functionality, Equations [8.13] and [8.14] with the appropriate dependent variable are applicable to the drainage functionality. Figure 8.5 shows that with increased compression by the applied loads the permeability decreases, often to a point where the effect of H.G. becomes negligible.

8.5 Functions of geosynthetics: separation

Figure 8.6 illustrates that, for the separation function, the geotextile is placed between a fine soil, or subsoil, and a coarser aggregate material to prevent the two from mixing, even when subjected to the action of repeated applied normal loads. If the stone aggregate is placed on the subsoils without a geotextile separator then, when loads are applied to the aggregate, the stone at the interface will be pushed into the subsoil, while concurrently the displaced volume of soil moves up into voids of the aggregate stone. With repeated loading, over time the two materials become effectively a homogenous mix. An example of this is limestone drives, where limestone aggregate is spread to a suitable thickness on a compacted base soil (the subgrade) to enable the passage of vehicle traffic. The idea is that the compression modulus of the aggregate layer prevents the wheels of vehicles sinking into and/ or rutting the compacted soil. However, over time the aggregate layer is seen to decrease in compression modulus as it sinks into the compacted base soil. The geotextile separator, in preventing the mixing, enables the limestone layer to retain a suitable compression modulus.

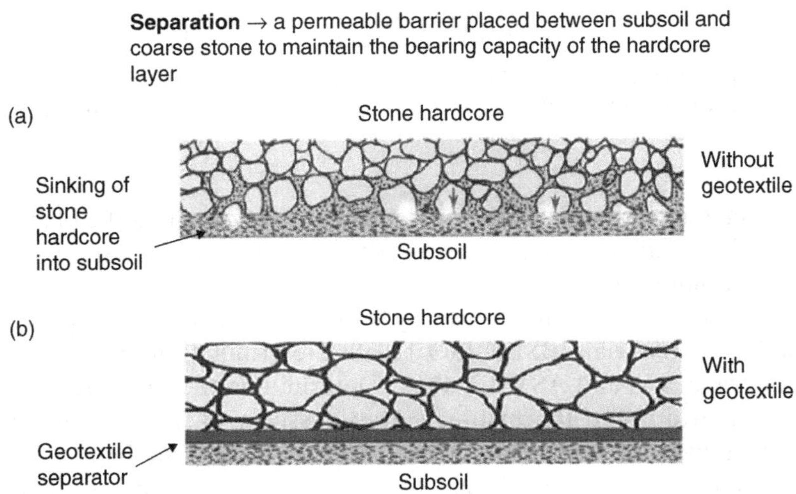

8.6 Separation function: (a) Without geotextile and (b) With geotextile.

The soil retention function of the geotextile separator can be seen as similar to the geotextile filter, in that for a suitable separator the AOS of the geotextile must be related to the particle size distribution and, as stated under the filtration function, this aspect will be considered in more detail later when describing the relationship between structure, properties, and functionalities of geotextiles.

When the aggregate layer is subjected to loads, significant local stresses are created in the separator. The separator must therefore be of sufficient strength to withstand these stresses. The bursting strength, localised tensile strength properties, and puncture resistance are of specific importance to the separation function.

8.5.1 Bursting strength

Consider the mechanism of the intermixing at the stone–subsoil interface as described above. If the geotextile separator is placed to form the interface, the subsoil will push the geotextile into the voids of the aggregate stone. Visualising the cross-section of the aggregate–geotextile separator–subsoil system, Fig. 8.7a can be taken representing the localised situation within a void. Evidently, for effective separation, the bursting strength of the geotextile separator must be greater than tensions, T_p, generated in it by the upward pressure, P, of the subsoil.[6] Hence:

$$T_p = \frac{1}{2} P d_v \left[f(\varepsilon) \right] \qquad [8.15]$$

$$f(\varepsilon) = \frac{1}{4}\left(\frac{4y^2 + b^2}{2yb} \right) \qquad [8.16]$$

where $f(\varepsilon)$ = the strain function of the deformed separator, T_p = the generated tension (kNm^{-2}), P = the upward pressure deforming the separator (kP$_a$), d_v = maximum void diameter at the aggregate interface (m) – assumed 0.33 d_a, d_a – average stone diameter(m), b = width of the void opening (m), and y = height of deformation into the void (m).

The bursting strength of the geotextile can be determined by the ASTM D3786 Mullen burst test method. As Fig. 8.8 illustrates, a specimen of the geotextile sample is clamped over a rubber diaphragm by a ring, and increasing oil pressure is applied to the underside of the diaphragm until the specimen ruptures. The bursting strength, T_R (kNm^{-2}), and deformation, d_R (mm), are thereby obtained.

In respect of the test arrangement, Equation [8.15] can be rewritten as:

$$T_R = \frac{1}{2} P_R d_M \left[f(\varepsilon) \right] \qquad [8.17]$$

where P_R = bust test pressure, and d_M = inner diameter of the clamping ring.

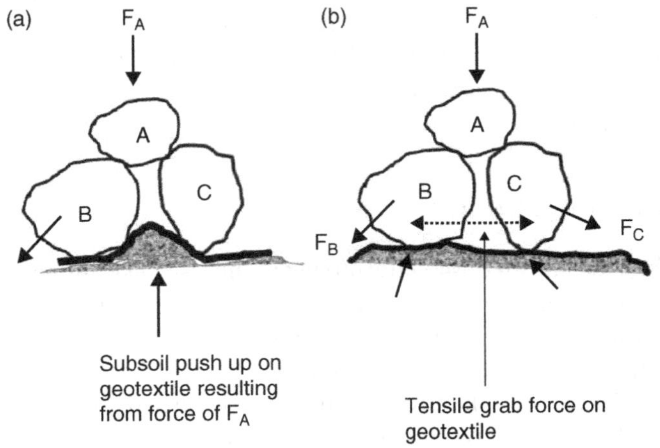

8.7 Schematic illustration of (a) bust and (b) grab strength of geotextiles.

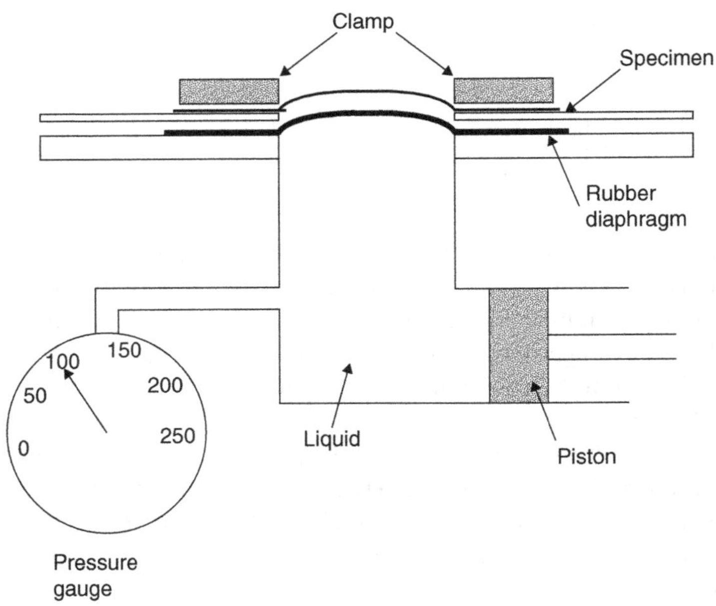

8.8 Schematic of burst strength test.

8.5.2 Localised tensile strength (grab strength)

Not all localised areas of the separator will be subjected to the upthrust of the subsoil. As shown by Fig. 8.7b, the local movement of stones forming the void occurs such where the forces F_A on the upper stone (A) try to push (A) into the space between the two stones (B) and (C) in contact with A. This action would cause forces F_B and F_C to induce a local tensile force, T_G, within the geotextile separator; essentially the stones (B) and (C) respectively grab the separator at the locations B' and C' as they are being pushed apart by (A), simultaneously with (A) being forced into the void. The relationship between the pressure exerted on the stone and the local tensile force generated is given by[6]:

$$T_G = Pd_v^2\left[f(\varepsilon)\right] \qquad [8.18]$$

The test methods 'D4632: Grab Tensile Test' or 'BSENISO 13934-2 Determination of Maximum Force using the Grab Method' are used to determine the capability of a geotextile separator to withstand these localised stresses. Basically, 25 mm width grips, placed 75 mm apart within the loading zone of a tensile tester, are used to hold a specimen (100 mm × 100 mm) along its centre line. The upper grip is then pulled away from the lower grip at a rate of 20 mm per minute until rupture, the tensile force at rupture being the measured grab strength.

8.5.3 Puncture resistance

During the laying and consolidation of the aggregate, the separator must resist being punctured by sharp stones at the aggregate interface or projecting from the surface of the compacted base soil. Two test methods may be used to determine a separator's resistance to puncture. The first, ENISO 13433:2006, the Dynamic Perforation Test (Cone Drop Test), determines the resistance of the separator to penetration by a 1 kg steel cone dropped from a specified height (generally 500 mm, see Fig. 8.9). The specimen is held firmly in a clamping ring system of 150 mm internal diameter, as the steel cone is dropped onto the centre point. The diameter of any hole made by the steel cone, i.e. degree of penetration, is an indication of the puncture resistance when sharp stones are dropped onto the separator during installation; the smaller the diameter the greater the resistance.

The second test, BSENISO 12236 Static Puncture Test (California Bearing Ratio (CBR) Test) is referred to a large-size puncture test and appropriate to the situation of aggregate consolidation. It is also commonly referred to as the CBR test, because it is based on the ASTM D1883 CBR test performed by measuring the pressure required to penetrate a soil sample with a plunger.

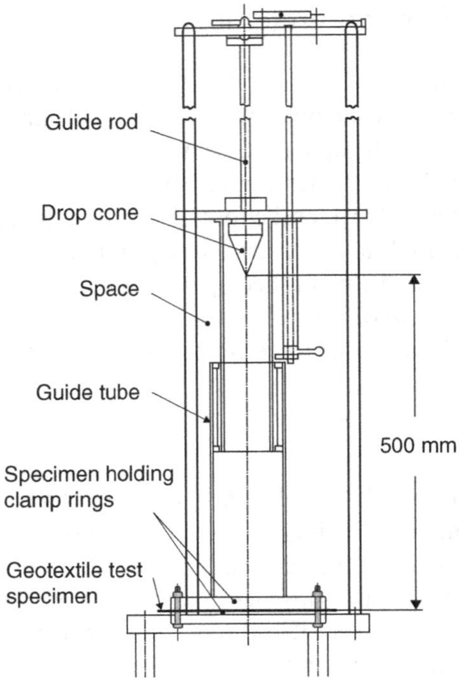

8.9 Schematic of cone drop test rig.

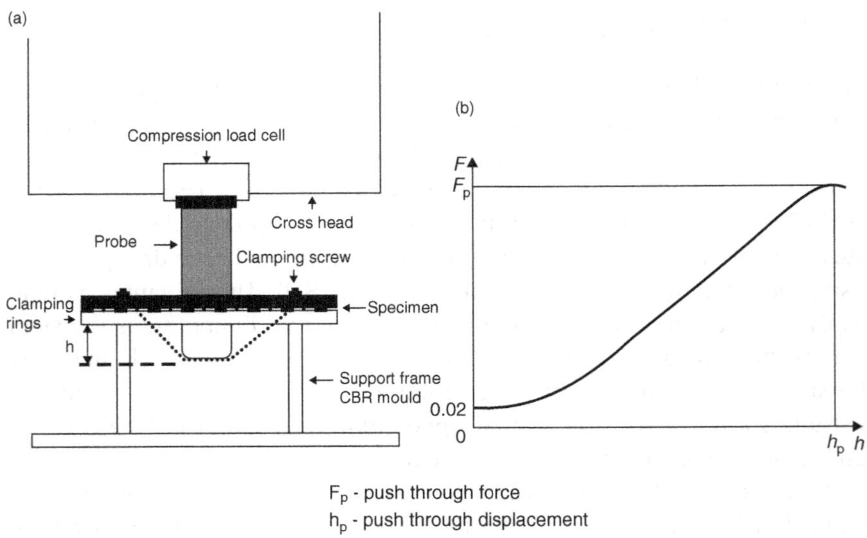

8.10 (a) Schematic of static puncture (CBR) test and (b) force-displacement curve.

Figure 8.10 shows a schematic of the CBR test for a geotextile. The specimen is clamped between the serrated surfaces of two steel rings. A plunger of 50 mm diameter is advanced at a constant rate onto the centre of the specimen and perpendicular to it. The push-through force, and push-through displacement are recorded and a force–displacement curve plotted.

8.6 Functions of geosynthetics: reinforcement

A compacted layer of aggregate generally has good compressive strength but very poor tensile resistance; therefore to give resistance to a compacted soil from breaking up under tensile stresses, a geotextile can be used to reinforce the soil (see Fig. 8.11). Clearly, there are three important mechanical properties of a geotextile used for reinforcement: tensile modulus, tensile strength, and surface friction.

8.6.1 Tensile properties

For certain geotextiles used for reinforcements, mainly non-wovens will be explained later; grab strength, as described earlier, is of importance. However, for all such reinforcements, what are termed the 'wide-width' tensile properties are generally measured according to standards ASTM D4595 1 'Tensile Properties of Geotextiles by the Wide-Width Strip Method' or ISO10319 'Geosynthetics – Wide-Width Tensile Test'. This is essentially the load-elongation test carried out on tensile testing machines. However, whereas, say, in general textile testing a 50 mm width is used (e.g. BSENISO 13934-1 Textiles – Tensile Properties of Fabrics or ISO 9073 Non-wovens – Determination of Tensile Strength and Elongation), for wide-width tests, specimens are cut to 200 mm width. This gives a more representative measure of the geosynthetic's structure and circumvents the Poisson's ratio effect having a major influence on measured values. Similar to general textiles, the tensile stress is calculated as load per unit width, the unit width in this case is metres, and the unit of stress kNm^{-1}. The usual plot of stress vs strain[7] enables the modulus and work of rupture to be calculated.[8]

8.6.2 Surface friction (reinforcement by shear strength)

It can be reasoned from Fig. 8.11 that if the interfacial contact does not fully transfer the shear force to the geosynthetic as a tensile load, then slippage of the top layer of aggregate can occur. This can be particularly important if the reinforcing arrangement is on an incline. A measure of the interfacial behaviour is therefore useful, and this is usually determined by the ASTM D5321 direct shear test.

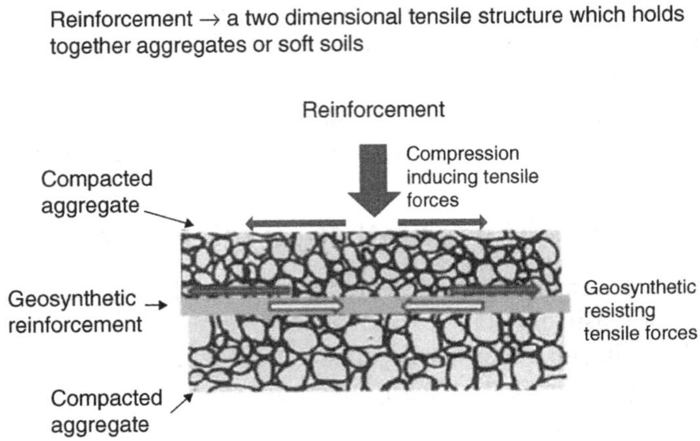

8.11 Geotextiles reinforcing function.

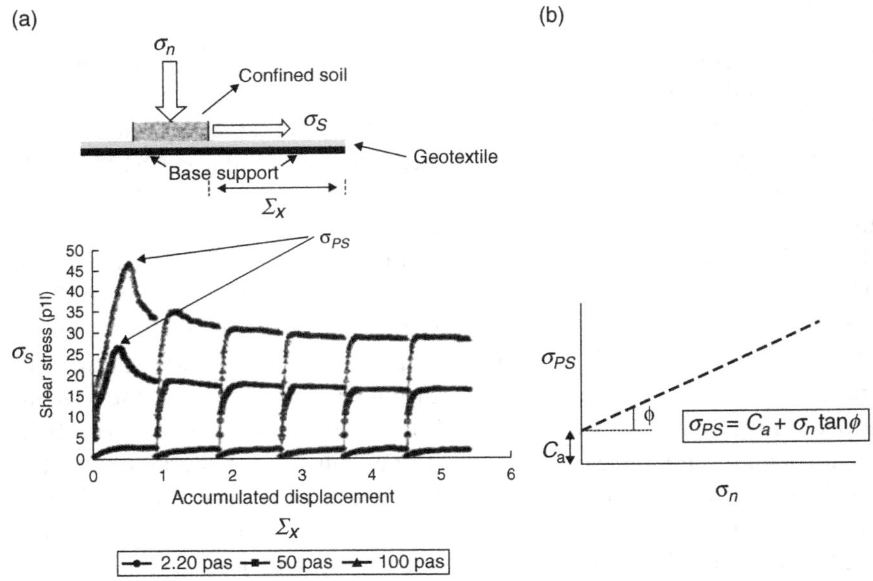

8.12 (a) Principle of the direct shear test (b) Mohr-Coulomb failure curve.

Figure 8.12 depicts the principle of the direct shear test. The geosynthetic is firmly fixed to the bottom half of the test apparatus, referred to as a shear box, and the relevant soil is contained within the upper box (see Fig. 8.12a). A specified normal stress, σ_n, is applied to the upper box and an increasing shear force, σ_S, is also applied until sliding of the upper box occurs. The test is carried out for three normal stresses of increased levels, say low, medium, and high. As illustrated in Fig. 8.12a, the shear stress vs displacement can be plotted for each applied normal load. The peak shear stress, σ_{PS}, is the value that overcomes the static friction of the soil–geosynthetic interface for the corresponding normal stress. The stress at which the curve levels off is termed the residual shear strength σ_{RS}. The Mohr-Coulomb failure criterion for the peak shear stress can be written as (Fig. 8.12b):[4]

$$\sigma_{PS} = C_a + \sigma_n \tan\varphi \quad [8.19]$$

and similarly the residual shear stress can be written as:

$$\sigma_{RS} = \sigma_n \tan\theta + C_R \quad [8.20]$$

where ϕ, C_a and θ, C_R are respectively the interfacial friction angle and interfacial adhesion for the respective shear stresses. Linear plots of σ_{PS} vs σ_n and σ_{RS} vs σ_n provide these friction parameters.

The soil-to-soil direct friction values are considered as the upper limit. Therefore a comparison with the values for geosynthetic–soil friction will give an indication of the interfacial shear strength efficiency of the geosynthetic.

8.6.3 Pullout strength (reinforcement by anchorage)

In certain applications of the reinforcing function, the geotextile has to provide anchorage by sandwiching a sheet of geotextile between two soil layers. The important property is the resistance of the geotextile to pullout from between the soil layers. A test can be devised having a horizontal arrangement of the geotextile embedded in the soil, whereby the applied tension at which the geotextile is pulled from between the soil layers is determined for given normal stresses applied to the upper soil layer, the value obtained being the pullout strength. Clearly the pullout strength is closely related to the direct shear strength, and it is reasonable to use the direct shear test results for both sides of the geotextiles as an indication of the material's pullout strength, although in practice the sum of the direct shear is likely to be greater than the actual pullout strength.

8.6.4 Creep

Creep is the term applied to the time-dependent, continued strain or extension of a material when subjected to a constant load over a prolonged period of time. For the reinforcing function, it is an important requirement that the geotextile has low or preferably negligible creep over tens to hundreds of years of constant loads. ASTM D5262 and ISO13431 describe standard test procedures for creep tests. Essentially, a stress is applied to the geotextile, equal to a specified percentage (commonly 20%, 40% or 60%) of the material's tensile strength. Since results are required as quickly as possible, the time period is important for the results to be meaningful, and usually a period of 1000–10 000 h is specified. Measurements are taken of the material's elongation under the set percentage stress at increasing time intervals, typically 1, 2, 5, 10, and 30 min initially, then 1, 2, 5, 10, 30, 100, 200, 500, 750, and 1000 h and for longer required periods, every 500 h after the 1000 h. Environmental conditions may influence the creep of geotextile materials susceptible to degradation of properties with temperature rise and/or increased moisture content; such factors should be accounted for in testing. To reduce the test period, accelerated creep tests may be used. This involves loading the material at high stress levels sufficient to cause creep rupture within, for example, 100 h. With such creep-rupture testing, indications of environmental effects can be readily determined.

8.7 Functions of geosynthetics: containment (fluid barrier) and protection

The specific objective in containment is to prevent the movement of fluids, mainly liquids, from a collection zone into the neighbouring soil. For this function, the most commonly used geosynthetic is an impermeable plastic membrane, usually high density polyethylene (HDPE), but polypropylene (PP) sheets, polyvinyl chloride (PVC)-coated textiles, and non-woven geotextiles impregnated with montmorillonite clay are also of interest. Since this function is not strictly applicable to geotextiles, no further consideration will be given to it. However, generally where the fluid barrier function is in evidence, a geotextile is present as a protection layer.

Figure 8.13 illustrates the compressibility of a geotextile being utilised for the protective function and, as such, largely thick non-woven geotextiles are used. The main objective is to prevent penetration by stones in the soil base layer causing failure of the geomembrane fluid barrier. Failure can be by an immediate puncture of the membrane or by longer-term localised stress cracking of the membrane. As depicted in Fig. 8.13b, at a localised projection the weight of water contained within a membrane-lined pond can induce localised tensile stresses, T_m, in the membrane, stretching the material as it deforms for the liquid to occupy the void. Over time, this can lead to a type of creep rupture or stress cracking as it is termed. Placing a non-woven

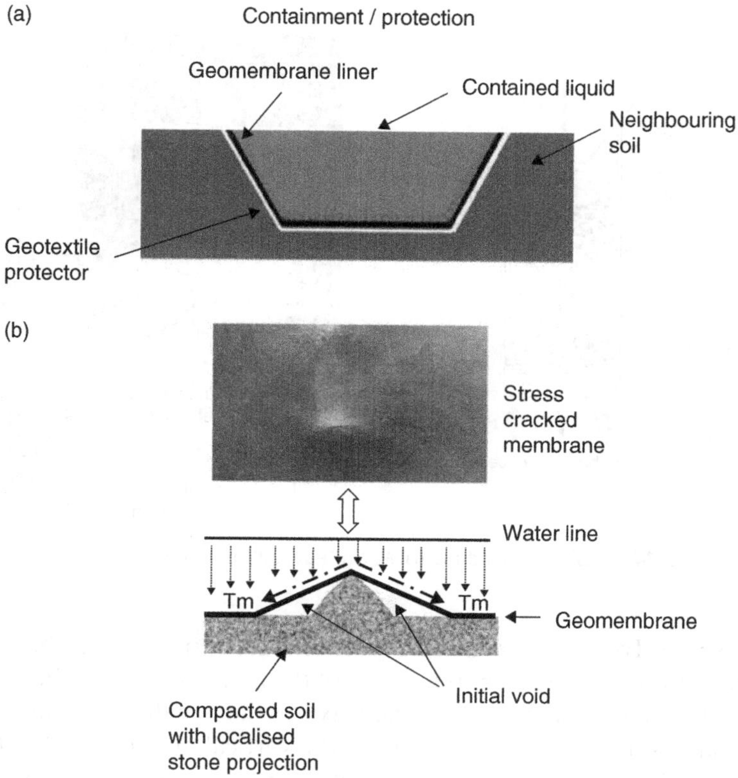

8.13 (a, b) Illustrating protective function of geotextiles.

geotextile between the membrane and the compacted soil would reduce the induced localised stress, and prevent creep rupture.

The protective function of the geotextile can be determined by the dynamic perforation test, commonly called the cylinder test.[9] Figure 8.14a shows the test arrangement. The assembled layers of the geotextile, the geomembrane, the lead sheet and the dense rubber, are all of diameters equal to the internal diameter of the cylindrical container (300 mm) and are placed on a load cell. Approximately a 300 mm thickness layer of 20 mm limestone aggregate is laid on the geotextile surface of the assembly, followed by a separated layer of sand and then a steel plate. The specified pressure, registered by the load cell, is applied to the steel plate for a given period of time, e.g. c.100 days, after which the lead sheet is removed and inspected to assess the strain-induced deformation of the geomembrane, and thereby the cushioning effect of the geotextiles.

Figure 8.14b shows the way in which an identified indentation in the lead sheet, corresponding to an area of deformation in the membrane, is

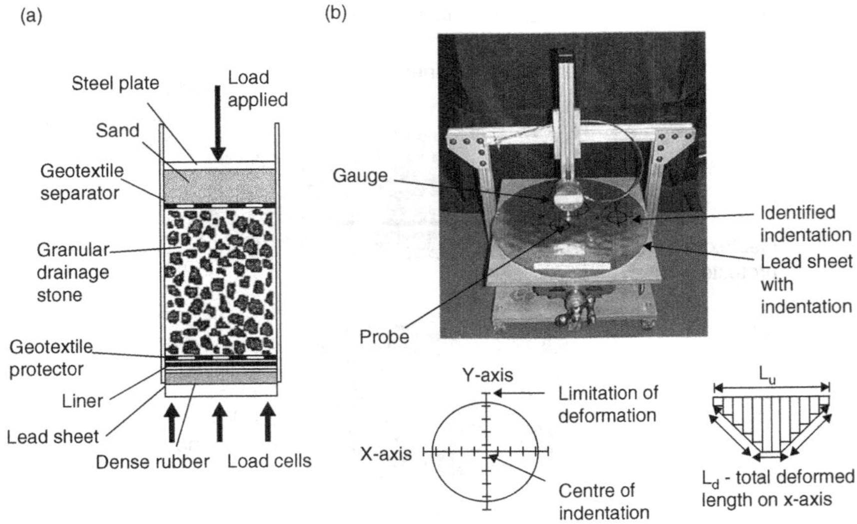

8.14 (a, b) Dynamic perforation test (cylinder test).

measured. Two axes at right angles, X and Y, are marked on the identified indentation. Starting from the centre of the indentation, a probe fitted with a digital gauge is moved along the x-axis in the negative direction (–ve/X) a small distance at a time, say 3 mm. The vertical displacement to the base of the indentation is measured at each 3 mm position using the probe. The edge-limit of the indentation is defined as the point where two consecutive readings taken 3 mm apart have a difference in vertical displacement of less than or equal to 0.06 mm. The process is then repeated for (+ve/X), (–ve/Y), and (+ve/Y). If, say for –ve/X, D_i is the indent depth (mm) at each 3 mm measured by the probe for $i = 1 - n$, where n is the number of measurements, then:

$$L_i = \sqrt{(D_c - D_i)^2 + 9} \qquad [8.21]$$

where D_c is the depth at the centre of the indentation and L_i is the incremental strain calculated from the Pythagoras's triangle.

The total length of deformation, L_d, is therefore obtained by:

$$L_d = \sum_{i=1}^{2n} L_i \qquad [8.22]$$

where $2n$ indicates n –ve/X and n +ve/X measurements, each 3 mm apart.

The membrane strain along the x-axis of indentation is:

$$\% \text{ Strain} = \frac{L_d - L_u}{L_u} \times 100 \qquad [8.23]$$

where $L_u = 6n$ is the estimated undeformed length indicated in Fig. 8.14.

The % $\text{Strain}_{(\text{y-axis})}$ is determined similarly and the average of two percentage strains is taken as the % strain of the indentation. It is recommended that the average % strain should be determined for each of the three largest identifiable indentations in the lead sheet. For effective protection of a geomembrane the average % strain for each of the three selected indentations should be less than 0.25%.[9]

The above test is a direct assessment of the likely performance of a geotextile with limestone aggregate. Other soils can be used for assessment relevant to the installation site. An alternative approach used as a comparative measure of geotextile protectors is to replace the soil layer with a layer of 20 mm diameter steel ball bearings. In this test the geomembrane is omitted from the arrangement in Fig. 8.14a, since the objective is to determine the resistance to a uniformly localised compression.

8.8 Index and performance testing of geotextiles

The tests referred to in the description of the functionalities are generally used in two different ways. Essentially, they are employed to generate data that can be utilised for product comparisons, specifications, quality control purposes, and as an indicator of how a product might survive during the installation process, and of its long term effectiveness.

These tests are called 'index tests'[4] and they include mass per unit area, uniaxial mechanical strength (*grab tensile; load-strain; creep, tear, and seam strength*), multiaxial rupture strength (*puncture, burst, and cutting resistance; flexibility*), and hydraulic tests (*AOS, percent open area; pore-size distribution; porosity; permeability and permittivity; transmissivity*). For indication of long term endurance or durability the materials are tested after being subjected to specific mechanical and/or simulated environmental conditions in order to determine their abrasion resistance: UV stability, chemical and biological resistance, and wet–dry and temperature stability.

A common practice is to subject a geosynthetic product to a required test while it is in contact with the particular soil or granular material in which it is to be installed. Such tests are termed 'performance tests' and are used to obtain a direct assessment of the geosynthetic properties under simulated geotechnical conditions of a site specific installation. Performance tests are not normally used in specifications; rather, a geosynthetic should be preselected for performance testing based on index values, or performance test results should be correlated to index values for use in specifications. Examples of performance tests include in-soil stress–strain, creep, friction/adhesion, dynamic perforation (cylinder test), chemical resistance, and filtration/clogging resistance.

Within the European Union (EU), the Construction Products Directive – 89/106/EEC has resulted in minimum standard test requirements for any geotextile that is to be utilised for the possible functions described above. These are referred to as the CE marking requirements.[10] All manufacturers of building and construction products in the EU are required to comply with the CE marking requirements and provide the appropriate test certificates. This became mandatory for geotextiles in October 2002. The CE mark is therefore seen as a guarantee that, for application of a particular functionality, the properties of a product will match those claimed by its manufacturer.

8.9 Geotextile materials: polymers and fibres

Geosynthetics, and in particular geotextiles, are generically identified by: polymer, type of fibre, type of fabric structure, as well as by the physical, mechanical and hydraulic properties used to describe the material. Most geotextiles are made from fibres of synthetic polymers, but some natural fibres are used to produce what is termed 'natural geotextiles' for specific applications.[11,12] Natural geotextiles are made from vegetable fibres, which consist of the natural polymer cellulose. There are three classes of vegetable fibres:[13] bast fibres, leaf fibres, and seed/fruit fibres. Bast fibres are extracted from stems of plants, such as jute, flax, hemp, and ramie, whereas leaf fibres are obtained from leaves of plants such as sisal, abaca, and henequen, and seed fibres are produced from seed/fruit of plants such as cotton and coir. The most important of these fibres for manufacturing natural geotextiles are jute and coir, owing to their better mechanical properties. Table 8.1 shows these natural fibres have good tensile strength and modulus and low extensibility. However, being highly water absorbent and biodegradable, they are employed only for temporary applications.

The vast majority of geotextiles are made from synthetic fibres primarily because they are water resistant, or have low water absorption, and also have good resistance to biological and chemical degradation. The synthetic fibres used are PP, polyethylene (PE), polyester (PET), and polyamide (PA); in some cases recycled fibres are utilised, which can also be a mixture of the waste of these synthetic fibres. These materials are produced by the 'melt-spinning process'[14] in the form of continuous monofilament or multifilament yarns, or staple fibres which have discrete lengths, typically 20–100 mm. Generally, fibre fineness ranges from 2.2 to 60 dtex. Synthetic polymer yarns

Table 8.1 Properties of jute and coir fibres

Property	Jute fibres	Coir fibres
Fibre length (mm)	180–800	100–300
Fibre diameter (μm)	10–20	10–20
Density (kg/m³)	120–140	1.15–1.33
Tensile strength (MPa)	250–350	140–150
Elastic modulus (GPa)	26–32	4–5
Elongation at break	2–3	15–17.3
Water absorption (%)	25–40	< 10

may also be produced by slitting extruded plastic sheets or films into thin flat tapes, or twisting fibrillated plastic ribbons into yarns.

Since geotextiles made from synthetic fibres are required to have longevity, it is essential that the fibres from which they are made retain their properties when subjected to various environmental conditions. Such fibres are made to have the best thermal stability, resistance to UV radiation, oxidation, and chemical degradation, limited only by the chemical structure of the polymer. PET is inherently stable to ultraviolet light, but susceptible to high pH environments. PP in contrast has excellent chemical and pH resistance, but requires additives for UV stability. During production by the melt-spinning process, various additives, such as carbon black (about 2% w/w), may be incorporated in the polymer to enhance fibre resistance to degradation.

Where very high strength geotextiles are required, high performance fibres such as Kevlar and glass are used. However, because geosynthetics/geotextiles are installed over large areas of ground, material costs need to be kept low, so cost effectiveness is an essential factor in any product procurement. PP and PET are therefore the most used synthetic fibres.

As with all synthetic fibres, mechanical properties depend not only on the chemical parameters of the polymer, in particular molecular weight, but also on the melt-spinning conditions used in their production. The tensile characteristics of these fibres can be tailored to requirements, but generally PET fibres can be made to have much higher strengths than PP fibres (see Fig. 8.15) and much lower creep, both total strain and rate of strain (see Fig. 8.16). In general, PET is more suitable for geotextiles for the reinforcement function and PP for less strength critical applications.

Although woven fabrics can be impregnated and coated in bitumen to give a bituminous geomembrane (BGM), the geomembranes widely used as impervious sheets are largely made from high-, medium-, and low-density polyethylene (HDPE, MDPE, LDPE),[15] but PP, ethylene propylene diene monomer (EPDM) and PVC are used as alternatives for some applications.

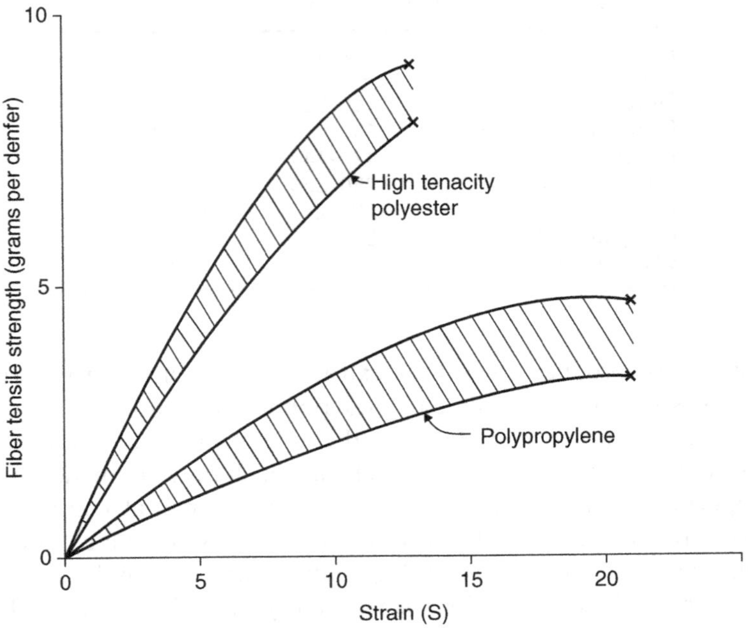

8.15 Comparison of tensile properties of polyester and polypropylene fibres (*Source*: From Reference 57.)

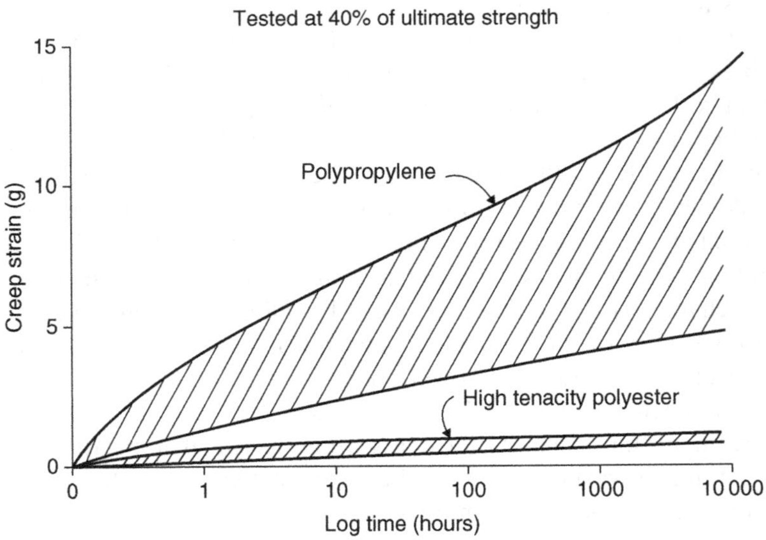

8.16 Comparion of polypropylene and polyester creep characteristics.

These geomembrane materials have differing characteristics which influence their installation, performance, and longevity in respect of resistance to oxidation, temperature stresses, UV radiation, chemical attack, mechanical stress, and microbiological activity. For example, PVC materials are very flexible and when laid, readily conforming to uneven surfaces without becoming punctured. EPDM rubber is also highly flexible, with excellent UV and weathering resistance, but is not suitable for use in long term contact with oils and hydrocarbons. HDPE, on the other hand, has excellent chemical resistance and is highly impervious, but is inflexible, whereas MDPE and LDPE have better flexibility but are not as impervious; all are very susceptible to UV radiation, and should therefore contain about 2% carbon black and not be exposed for prolonged periods, else they will become brittle and fragile and suffer from environmental stress cracking.

The physical properties of PE are mainly determined by its crystallinity. Crystallisation can occur better when the degree of branching is lower. The branching accounts for the differences in physical properties between the different grades of PE. HDPE has a lower degree of branching, which makes it more rigid, but stronger, tougher, better chemically resistant, and impervious to fluids than the low density grades. HDPE usually performs satisfactorily up to 60°C when exposed to most acidic and alkali solutions. However, hydrocarbons severely affect the properties of PEs.

As stated earlier, this chapter is primarily about geotextiles and related materials as a subcategory of geosynthetics. The reader wishing to obtain more information on geomembrane polymers may access References 15–22.

8.10 Geotextile materials: non-woven geotextiles

Figure 8.1 indicates that geotextile structures include continuous sheets of non-woven, woven, and warp knitted fabrics, and stitch-bonded fibres or yarns. The woven and warp knitted materials can be made to have a similar structure to certain of the textile-related products (essentially geogrids) to be used as alternatives. The non-woven geotextiles can be combined with the textile-related structures to make geocomposite products that are multifunctional.

Non-woven geotextiles are fibrous sheets in which the fibres can be almost randomly orientated. These materials may be manufactured from either staple fibres (synthetic or natural) or continuous filaments of PP or PET which are randomly distributed in layers onto a moving belt to form a 'web' of assembled fibres; blends of PP and PET staple fibres are also used.

With staple fibres, the web is formed by a carding machine, usually a roller and clearer card,[23] and is cross-laid onto the moving belt. Continuous filament webs are produced during the melt-spinning process by extruding multifilament yarns to form a swirling pattern of fibres as they deposit onto the moving belt. In staple-fibre cross-laid webs, the fibre directions are semi-random within the two dimensional plane, whereas the swirling pattern of the continuous fibre webs gives, almost, a totally random fibre orientation. Increasing the number of layers and the number of fibres in each layer forming the web, the thickness and thereby the bulk of the assembled mass increases, which will also contain a high volume of open spaces.

To give cohesion and strength to the assembled fibre layers, the fibre lengths are either interlocked or thermally bonded. Interlocking of the fibre lengths is achieved through a process called 'needle punching'.[23] Here, the fibrous web is fed into a machine equipped with groups of specially designed barbed needles. While the web is trapped between perforated plates, the needles punch through the holes of the plate and into the web, re-orientating the fibres gripped by the barbs so that fibre lengths in the upper layers entangle with those in the layers beneath, giving what is termed mechanical bonding. In some cases, the needles may also vibrate or rotate to speed up the entanglement. Figure 8.17 shows a graphic illustration and photomicrographs of a typical non-woven needle-punched structure, and it can be seen that lengths of fibres that were caught by the barbed needles penetrate through the material thickness.

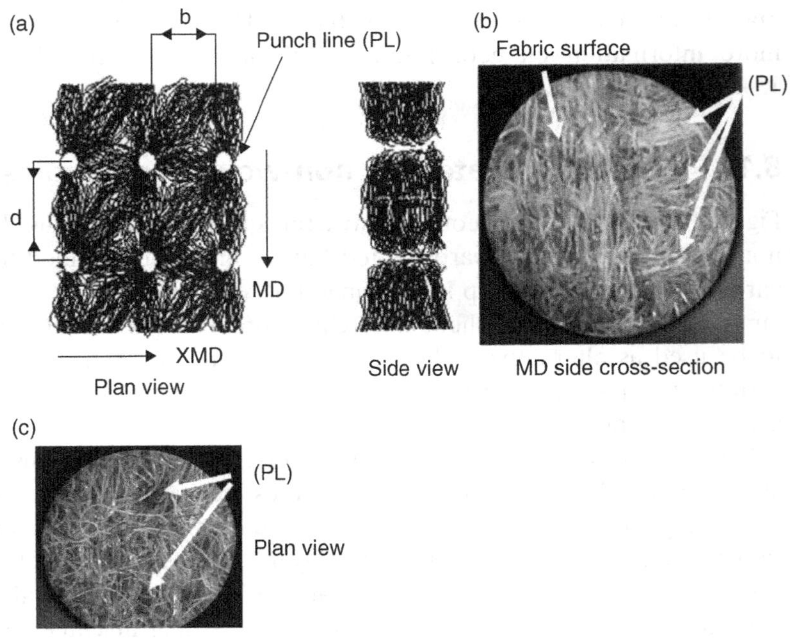

8.17 (a–c) Needle-punched structure of non-woven geotextiles.

Table 8.2 Comparison of needle-punched and thermally bonded non-woven geotextile properties

Property		Unit	Needle punched	Thermally bonded
Mass per unit area		g/m²	200–2000	150–1000
Nominal thickness (under 2 kPa load)		mm	1.2–9.7	1.0–6.0
Porosity (under 2 kPa load)		%	80–85	90–95
Tensile strength	MD	kN/m	5–70	6–50
	CD	kN/m	14–146	6–50
Elongation at break	MD	kN/m	120–140	60–80
	CD	kN/m	80–100	60–80
Grab strength	MD	kN	1.0–6.2	0.5–2.5
	CD	kN	1.0–6.2	0.5–2.5
Trapezoidal tear strength	MD	kN	0.5–2.2	0.11–5.3
	CM	N	0.5–2.2	0.11–5.3
Bursting strength		GPa	–	1.7–5.5
Static puncture test strength (CBR test)		kN	1.3–14.5	$2–25 \times 10^2$
Static puncture test displacement (CBR test)		mm	70–85	–
Dynamic cone drop		mm	2.0–24.0	5–40
Protection efficiency		N	80–1700	40–600
Permeability (normal to the plane) No load		10^{-3} m/s	30–110	5–130
Permeability (within the plane) 20 kPa load at 0.1 H.G	MD	10^{-6} m/s	2.0–8.5	0.1–2.0
	CD	10^{-6} m/s	2.4–8.0	0.1–2.0
Characteristic opening size (AOS)		µm	70–130	70–155

The needle punching action gives compaction to the web and therefore increases the density and reduces the thickness of the non-woven sheet, but a considerable bulk is still retained. Table 8.2 gives a typical range of areal density and thickness for needle-punched non-wovens.

Needle-punched materials are usually cross-laid webs; the fibres tend to lie across the machine direction (CD), and the CD strength is therefore higher than the material's strength in the machine direction (MD) (i.e. throughput direction). Generally, non-wovens of high areal density will be made of a greater mass of fibres, more intensely needled, i.e. a higher number of needle punches per unit area plus depth of needle penetration into the web, and thereby have greater fibre entanglement. The high strength values usually correspond with lower extensions and high areal densities. Greater fibre mass and needling intensities also give good compression resistances, higher values for protection efficiency, and lower figures for the dynamic cone drop. However, high fibre mass and needling intensity result in lower AOS and permeability.

Various methods may be used to thermally bond fibrous webs generally, but for thermally bonded non-woven geotextiles, bonding is performed by passing the web through heated rollers.[24] The heated rollers compress the layers of loose fibres and cause partial melting of the polymer, leading to heat fusion at the fibre cross-over points. The webs processed in this way have their fibres in the form of filaments, deposited in fewer layers than with needle punching. Therefore, thicknesses and areal densities are lower for thermally bonded geotextiles. The almost total random orientation of the filaments results in more isotropic strengths, where CD(strength) = MD(strength), when compared with needle-punched materials (see Table 8.2). In general, both the needle-punched and thermally bonded non-wovens have a wider size distribution of open spaces than other geotextiles. Similar to needle-punched products, thermally bonded materials of greater fibre mass and bonding will have higher values for mechanical strength, but lower AOS and permeability; their in-plane permeability is often much lower than for needle-punched geotextiles.

Needle-punched non-woven geotextiles are the most widely used for ground engineering. Their wide range of AOS, in-plane permeability, and relatively low compressibility, makes them suitable for the drainage and protection functions. Heat-bonded geotextiles also have a wide range of AOS, but being much thinner they will have the better through-plane permeability, making them appropriate for the separation and filtration functions.

The characteristic opening size or pore size, O_{95}, and the pore-size distribution of a non-woven geotextile are determined not only by the quantity of the fibres forming the web and the intensity of bonding used, but also by the fibre diameter, or diameters, and their orientation in the web. Various mathematical models for pore-size prediction have been derived based largely on the open spaces viewed normal to the plane of the deposited web layers.

One approach in developing a pore-size model is to apply the Poisson polyhedron theory.[25–29] Each deposited layer in the build-up to make the web is simulated by a two dimensional network of straight lines, a Poisson line network, forming a random distribution of polygons,[30] as depicted in Fig. 8.18 showing two pore sizes defined as the diameter of largest circle inscribed within a given void.

If T_g is the thickness of the web and T_e the thickness of each randomly stacked elementary layer, then the cumulative probability for a circle of diameter, d, inscribed within a void of the elementary layer greater than d, is given by the function $G(d)$:

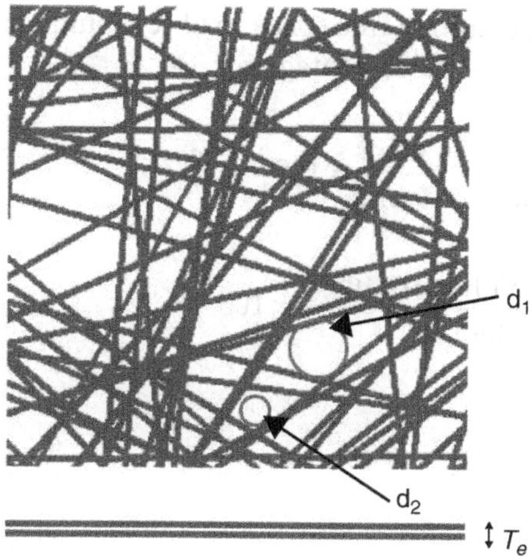

8.18 Simulation of deposited fibre layer in non-woven web.

$$G(d)=\left(1+\frac{2T_e[1-s]}{\pi d_f^2}d\right)^2 e^{-}\left(1+\frac{2T_e[1-s]}{\pi d_f^2}d\right) \qquad [8.24]$$

where d = diameter of inscribed circle (μm), ε = fabric porosity, d_f = fibre diameter (μm), T_e = deposited layer thickness (mm), and T_g = web thickness.

The number of deposited layers comprising the web is T_g/T_e. Therefore the cumulative probability, Q(d), of a pore equal to or greater than d, may be expressed by:

$$Q(d)=G(d)^{T_g/T_e} \qquad [8.25]$$

Hence, the pore-size distribution is the cumulative probability of having an opening, O, smaller than d, i.e.

$$P(0<d)=1-Q(d) \qquad [8.26]$$

It is reasonable to assume that each deposited layer will be approximately equal to the fibre diameter. Then, the pore-size distribution may be obtained by the following equation:

$$Q(d)=\left[\left(1+\frac{2[1-s]}{\pi d_f}d\right)^2 e^{-}\left(1+\frac{4T_e[1-s]}{\pi d_f}d\right)\right]^{T_g/d_f} \qquad [8.27]$$

The porosity ε defined by the ratio of free space to fibre volume in a given material volume, can be expressed as percentage by:

$$\text{Porosity } \varepsilon(\%) = \frac{\text{Volume of voids}}{\text{Volume of fabric}} \times 100$$

$$= \frac{\text{Volume of fabric} - \text{Volume of fibres}}{\text{Volume of fabric}} \times 100$$

$$= \frac{AT - W/p_f}{AT} \times 100$$

$$= \left(1 - \frac{W}{Ap_f T}\right) \times 100 \qquad [8.28]$$

where W = mass per unit area, A, of the given material, p_f = fibre density, and T_g = the material thickness.

Equation [8.28] is applicable not only to the deposited web, but also to the final non-woven fabric, since W will account for the change in the web density with T_g now representing the reduced thickness resulting from the needle-punching or thermal-bonding process. Hence, Equations [8.26] and [8.27] can be used to estimate the pore-size distribution in relation to the structural parameters ε and d_f of a non-woven geotextile, from which O_{95} can be estimated. As ε changes with compression of the material, the change in O_{95} may be estimated for the applied confining pressures.

It is the fabric thickness which most significantly decreases with applied pressure. The reduced thickness means a decrease in O_{95} and pore-size distribution, which in turn influences the fabric's permeability and particle retention capability.

Research studies[31,32] have shown that the compressive characteristic of non-woven geotextiles typically follow a hysteresis curve as depicted in Fig. 8.19, which can be represented by the following equations respectively for compression and recovery.

$$\frac{T}{T_o} = 1 - \alpha \log e\left(\frac{P}{P_o}\right) \qquad [8.29]$$

$$\frac{T}{T_f} = \left(\frac{P}{P_f}\right)^{-\beta} \qquad [8.30]$$

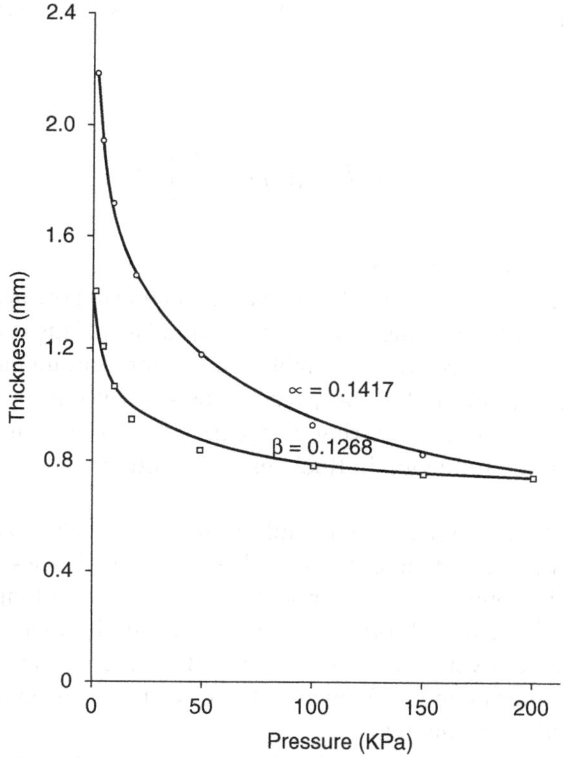

8.19 Typical non-woven hysteresis curve showing compression and recovery.

where T_o = initial thickness, measured under a pressure P_o, T_f = compressed or final thickness, T = thickness under pressure P, α and β are the respective constants for compression and recovery, and are specific to any given non-woven, and can change in value with cyclic loading.

If A is the fabric area under load, then the percentage energy loss is

$$E_L\,(\%) = \frac{E_1 - E_2}{E_1} \times 100$$

where

$$E_1 = \int_{T_o}^{T_f} PA\mathrm{d}T = P_o A \int_{T_o}^{T_f} \mathrm{e}^{1/\alpha(1-T/T_o)}\mathrm{d}T$$

$$E_2 = \int_{T_l}^{T_R} PA\mathrm{d}T = P_f A \int_{T_f}^{T_R} \left(T/T_f\right)^{-1/\beta} \mathrm{d}T$$

This indicates a percentage thickness loss, i.e. an irrecoverable reduction in thickness, given by:

$$
\begin{aligned}
T_L\,(\%) &= (1 - T_R / T_o) \times 100 \\
&= \left[1 - \left\{1 - \alpha \log_c (P_f / P_o)\right\}\left\{(P_f / P_o)^{\beta}\right\}\right] \times 100
\end{aligned}
\qquad [8.31]
$$

where T_R is the recovered thickness.

Experimental findings showed that for a given T_0 increasing the areal density decreases ε, α, and β, resulting in lower $\% E_L$ and $\% T_L$. This is because the greater number of fibres present enables more fibre entanglement in mechanical bonding and more thermally fused cross-over points in thermal bonding to reduce fibre slippage during compression deformation and the associated irrecoverable reduction in thickness. In addition:

- Non-woven geotextiles made from staple fibres have higher $\% E_L$ and $\% T_L$ than the equivalent filament material. Staple-fibre fabrics are usually mechanically bonded and the short lengths that are entangled are more prone to slip past each other than mechanically bonded continuous filaments, since several positions along each filament length will be entangled with other filaments. However, the lowest $\% E_L$ is obtained by the thermal bonding of filaments.
- Coarser/thicker fibres of a given density will have higher stiffness, a higher degree of crystallinity will also give a higher stiffness, and aid in reducing $\% E_L$ and $\% T_L$.
- Lower compressibility and $\% E_L$ and $\% T_L$ are obtained with intensive needle punching, both needling density and depth of penetration, up to the point where fibre breakage occurs. This is attributable to better entanglement of the fibres. Similarly, stronger thermal bonds give better compression resistance.

Clearly, if ε decreases with T then so will O_{95}. Figure 8.20 illustrates this.[33] The graph shows the effect of applied normal stress on the O_{95} rating of a light, a medium, and a heavier weight, needle-punched, non-woven geotextile. The AOS value was measured using a wet-sieve test under load with the glass bead grades denoted A and B in the table.

The resulting graphs of O_{95} plotted against applied normal pressure indicate the initial opening size decreases with increases in T_o and show a similar trend to T plotted against applied normal pressure. There is initially a rapid reduction and then the decrease becomes smaller, gradually levelling out at higher pressures. With the three geotextiles used as an illustration, the application of the confining pressure less than 25 kPa is seen to have a greater

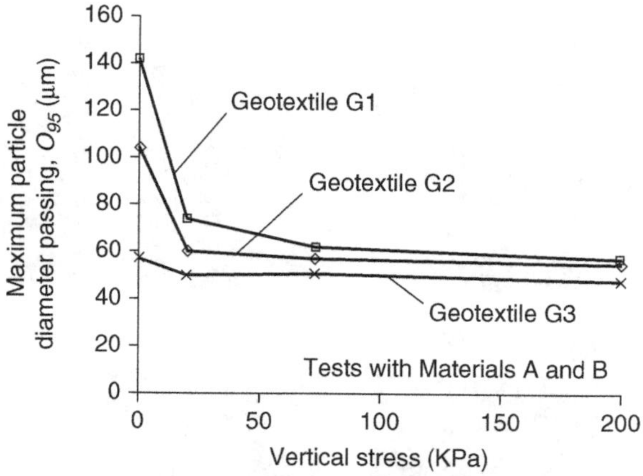

Material	Composition	D_{15} (µm)	D_{50} (µm)	D_{85} (µm)	C_υ	G_5
A	Glass beads	29	48	64	2.1	2.48
B	Glass beads	57	116	140	2.2	2.48

Geotextile	Thickness[(1)] (mm)	Mass per unit area (g/m²)	*FOS* (µm)	k_{Go} (cm/s)
G1	1.9	180	140	0.70
G2	2.7	300	110	0.63
G3	5.6	600	60	0.28

Notes: [(1)] at 2 kPa vertical stress; *FOS* = geotextile pore opening size values from hydrodynamic sieving
K_{Go} = permeability normal to the geotextile plane (no surcharge)

8.20 Effect of confining pressure on pore size.

effect on the thinner, lighter, more open fabrics G1 and G2, than on the thicker, denser fabric G3. However, all curves tend towards a similar value within 50–60 µm. The inference is that the O_{95} and T_o of very dense fabrics should be negligibly affected by the confining pressures used, i.e. up to 200 kPa. The reason for this may be attributed to the tortuosity of pores in the fabric. The most direct path through the fabric for the glass beads would be the needle-punched holes, the initial size of these, and their changes with applied pressure will limit the size of particle passing through the geotextile. Figure 8.21 shows microscopic images of the change in size of the needle holes in the three non-wovens, under a confining pressure of 50 kPa. For G1 and G2, the needle hole had decreased to an equivalent diameter of 60–80 µm, approximately two to three times the fibre diameter, which is close to the corresponding O_{95} values in Fig. 8.20, whereas the needle hole of G3 shows little change from its original size.

Although the above illustration is of needle-punched non-wovens, a similar trend with confining pressures can be reasoned for thermally bonded geotextiles.

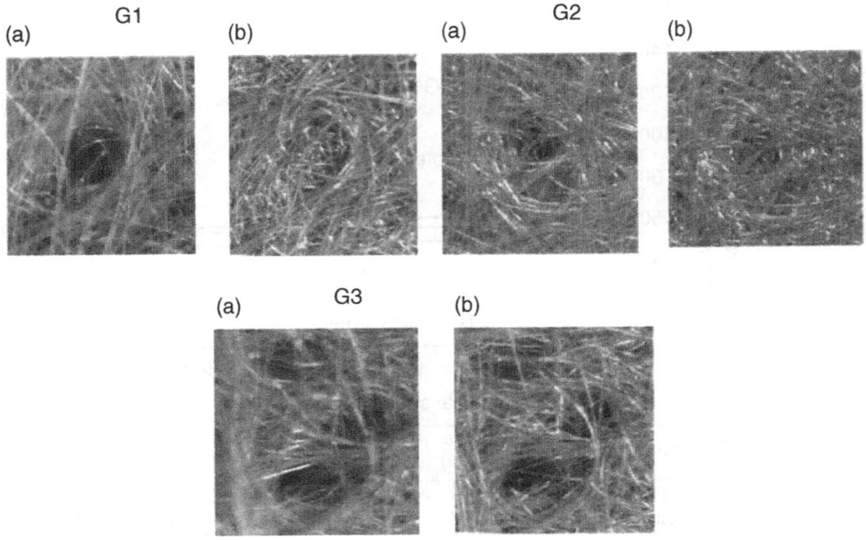

8.21 Effect of confining pressure on needle-punched hole in geotextile fabrics G1, G2, and G3 where (a) no applied stress and (b) 50 kPa applied vertical stress is Magnification 18X.

It should be noted here that the pore size, as identified by O_{95} and measured by the sieving of glass beads of known diameters through the geotextile, does not give the actual void space that a particle passes through (see Fig. 8.18), and therefore is not necessarily useful in relation to fluid flow through that void space. Therefore, prediction of O_{95} may be useful for the prediction of particle retention in respect of the separation and filtration functions, but not for the level of permeability in relation to the drainage function.

Instead of considering the largest particle diameter that can pass through a given void space, the hydrodynamic pore size[57] may be used for predicting the hydromechanics of non-woven geotextiles. If the void is assumed to be equivalent to a circular flow channel, then the hydrodynamic pore size may be defined by its hydraulic radius, $R_{H,}$ where:

$$R_H = \frac{\text{Volume available to flow}}{\text{Surface area exposed to flow}} = \frac{\varepsilon}{S} \qquad [8.32]$$

which, for a unit volume, is the ratio of cross-sectional area of the void to its wetted perimeter.

Using the concept of R_H has enabled a number of mathematical models to be derived for the permeability of a porous material based on its porosity (see Table 8.3). The models were derived from the following three differing theoretical approaches: drag theory, cell theory, and capillary theory.

Table 8.3 Theoretical models for specific permeability

Flow theory	Permeability model	Comments
Drag theory (Iberall model)	$k = \frac{1}{16} * \frac{\varepsilon}{1-\varepsilon} D_f^2 * \left\{ \frac{2 - \ln q\rho D_f / \varepsilon\eta}{4 - \ln q\rho D_f / \varepsilon\eta} \right\}$ [8.33]	
Cell theory (Happel model)	$k_p = \frac{D_f^2 \varepsilon^3 A}{16(1-\varepsilon)^2}$ $A = \left[4(1-\varepsilon) - (1-\varepsilon)^2 - 2\ln(1-\varepsilon) - 3\right]$ [8.34]	K_p = specific permeability of fluid flow direction parallel to cylinders
Cell theory (Happel model)	$k_v = \frac{D_f^2 B}{32(1-\varepsilon)}$ $B = \left\{ -\ln(1-\varepsilon) \left[\frac{1-(1-\varepsilon)^2}{1+(1-\varepsilon)^2} \right] \right\}$	K_v = specific permeability of fluid flow direction vertical to cylinders
Capillary theory (Kozeny-Carman model)	$k = \frac{D_f^2 \varepsilon^3}{k_o (1-\varepsilon)^2}$ [8.35]	k_o = Kozeny constant related to the orientation and tortuosity of the capillary. Depending on ε, k_o has values ranging from 3.0–184[36]

Drag theory considers the physical basis of permeability in terms of the frictional resistance of the walls of the material's pores reducing the free flow of fluid through the material, the degree of resistance being dependent on the viscosity of the fluid.[34] The drag on the fluid by a pore wall is estimated from the Navier-Stokes equations, and the sum of all the resistances of the pore walls is assumed to be the total resistance of the material to the fluid flow. This total resistance would be the reciprocal of the specific permeability coefficient, k, of Equation [8.6].

Cell theory[35] uses the concept of two concentric cylinders as the basis for modelling fluid flow through a porous solid simulated by an assembly of solid cylinders. The central of the two represents the solid cylinder, the outer fluid flow enveloped by a free surface. The relative volume of the annulus to the central cylinder is taken as the same relative volume of fluid to solid in the assembly of cylinders. Navier-Stokes equations are then used to derive the reciprocal of k. Cell theory enables two basic cases to be considered: fluid flow normal to the assembly of solid cylinders, and flow parallel to the axis of the cylinders.

Capillary theory[34] uses the simplest model, whereby pores within a solid material are represented as parallel capillaries of equal diameters in the porous solid. The analogy is between the tortuous pore-system of the solid and the cylindrical pores of the capillaries. The equation for k is then derived from the Hagen-Poiseuille equation for streamline flow through straight circular capillaries taking account of the tortuosity of material's pores. The tortuosity is defined as the ratio of the actual length of the flow channel to the length of the porous medium.

Since ε can be determined for increased confining pressures, the permeability constants may also be obtained for increased applied pressures.

8.11 Geotextile materials: woven and warp knitted

8.11.1 Woven geotextiles

Woven geotextiles are produced from synthetic-fibre yarns, mainly PET or PP, and natural fibre yarns, largely jute or coir, such as the Sulzer projectile loom P7150, used to produce woven fabric in widths of 190–540 cm.[37]

Woven geotextiles have the appearance of two sets of parallel threads interlaced at right angles to each other in the plane of the fabric. 'Warp' yarns lie along the length of the fabric and 'weft' in the transverse direction, i.e. across the width of the fabric. The type of yarns used to produce a woven geotextile may be monofilament, multifilament, a combination of each type, or slit film yarns. Two kinds of slit film yarn can be used, either flat-tape yarns or fibrillated yarns.

Monofilament and multifilament woven fabrics are generally made from PET, the use of monofilament giving the better permeability, whereas multifilament is used for higher strength reinforcement. Slit film, flat-tape fabrics are usually PP materials, which are quite strong but they form a fabric that has relatively poor permeability. Alternatively, fabrics made with fibrillated-tape yarns have better permeability and more uniform interstice openings than flat-tape products.

In weaving, the warp and weft yarns are commonly referred to as picks and ends. The interlacing of the picks and ends gives a coherent structure, and the repeating pattern of the interlacing is referred to as the weave of the fabric. Woven geotextiles are commonly plain weave, but twill weave and leno weave are also used.[38]

Plain weave is the simplest interlacing pattern, as shown in Fig. 8.22 for jute and coir woven geotextiles; in the length of the fabric a warp yarn crosses over alternate wefts, and in the fabric width, a weft yarn crosses alternate warps. The frictional contact of the yarns at the interlacing points prevents each yarn slipping from its woven position, and it also increases the stiffness of the fabric. The interlacing causes the warp and weft yarns to have a wavy configuration in the fabric structure. This waviness is called the yarn crimp.[39] The amplitude and frequency of the sinusoidal waviness of a warp yarn will depend on the yarn diameter, the number and spacing of the weft yarns across a unit length of the fabric width, and similarly for a weft yarn in the direction of the fabric length. Quantifying the warp and weft crimp is important, because it indicates the length of a specified yarn needed in each warp and weft to produce a certain length of fabric of a required width. To circumvent having to determine the yarn diameters of the warp and weft and their frequency per unit length, the crimp is calculated as the percentage difference between the unit length of yarn and unit length of fabric in the warp and weft directions.

Thus,

$$c = \frac{(L_y - L_f)100}{L_f} \qquad [8.33]$$

where L_y = uncrimped yarn length and L_f = fabric width.

The following expression can also be used:

$$\left(1 + \frac{c}{100}\right) = \frac{L_y}{L_f} \qquad [8.34]$$

and is called the crimp ratio.[40]

The warp and weft yarns are usually specified in mass per unit length, termed the count of the yarn.[39] In the SI system of units, the tex of a yarn is used and is defined as the number of grams per 1000 m. Thus, knowing the tex of the yarns in a given fabric and the crimp ratios, the fabric areal density, W_F, can be calculated:

$$W_F = W_p + W_c \text{ (units gm}^{-2}\text{)}$$

where W_p and W_t are the mass of the warp and of the weft yarns, respectively, given by:

$$W_p = T_p\left[1 + \frac{C_p}{100}\right]10^{-3} \qquad [8.35]$$

$$W_t = T_t\left[1 + \frac{C_t}{100}\right]10^{-3} \qquad [8.36]$$

The areal weights of woven geotextiles can be from 70 g/m^{-2} (for lightweight synthetics) to 1400 g/m^{-2} (heavy weight natural geotextiles).

Woven constructions produce geotextiles with high strengths and moduli in the warp and weft directions with low elongations at rupture. When a woven fabric is placed under increasing load or a constant rate of strain, say in the warp direction, the effect will be to initially straighten the sinusoidal waviness of the warp yarns, reducing their crimp height and lengthening the warp direction of the fabric, by only a relatively small load, before the filaments in the yarns undergo strain to break. The percentage crimp can therefore cause a small reduction in the initial modulus. Importantly, as the warp crimp reduces, the weft crimp increases, narrowing the width of the fabric in the weft direction. The converse occurs with increased loading in the weft direction. Hence, flat-tape yarns should show negligible fabric narrowing. The woven construction and the filament yarns used can be varied so that the finished geotextile has equal or different strengths in the warp and fill directions.

At the points of the warp and weft, increased pressure is exerted on the contact surfaces when increasing tension is applied to the woven fabric, and this pressure tends to prevent filament slippage. This effect is known as fabric assistance. Fabric assistance can therefore under certain conditions result in the fabric having a higher strength than the sum of the individual filament strengths in the particular warp or weft loaded direction. When woven geotextiles are pulled on a bias, the modulus decreases, although the ultimate breaking strength may increase.

Table 8.4 gives an example of the range of tensile properties for the different types of woven geotextiles. Fabrics made from multifilament yarns have usually the highest strengths and therefore the best reinforcing function. Compared with synthetic fabrics, woven geotextiles made of natural fibres are heavier (400–1400 gm^{-2}), thicker, and have low elongation at break. This is because the natural staple fibres are coarse and have low extension, and consequently so are the yarns produced for weaving. In contrast to non-wovens, woven synthetic geotextiles are made only from continuous filaments.

Woven synthetic geotextiles usually have higher strengths and lower breaking extension than non-wovens geotextiles of the same areal weight and polymer type,[37] as illustrated in the stress–strain graphs of Fig. 8.23. As shown, a woven fabric weighing 100 gsm would have the same strength as a non-woven weighing almost 300 gsm. This difference in strength is also reflected in their CBR values. The reason for the higher strength is self-evident from a comparison of two fabric structures. In the woven fabric structure, the warp or weft threads will be considerably straighter and more parallel to each other in the direction of axial loading compared with the fibre orientation in a non-woven fabric. However, the curled and

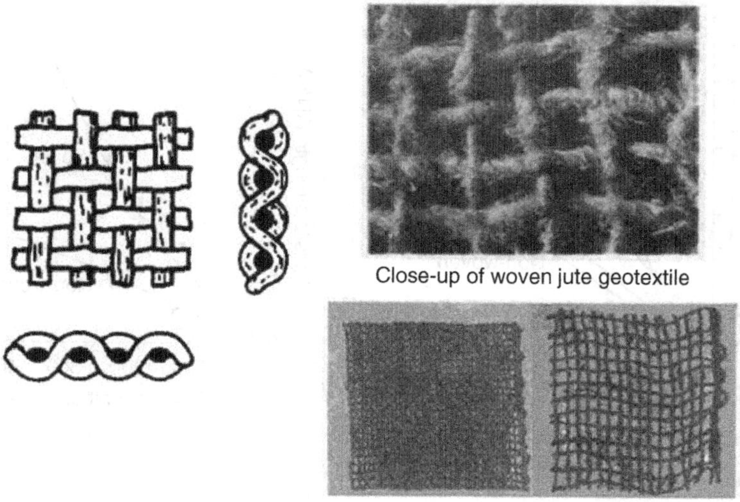

8.22 Woven jute and coir geotextiles.

looped fibre configurations as well as their orientations, generally give non-wovens much higher extensibility.

Woven geotextiles have a simple pore structure. This can be seen in Fig. 8.22 for the plain-weave structures of jute and coir geotextiles. The interstices or open spaces per unit area of the fabric appear as a regular equilateral rectangular pattern (a square in this case) defined by the inter-yarn spacing of the warp and weft, and the diameter or thickness of the yarns; the situation is similar with synthetic fabrics. As with non-woven geotextiles, it is the pore structure of the woven fabric that governs the O_{95} value and the permeability, and thereby the fabric's separation, filtration, and drainage functionalities.

Staple, slit film and multifilament yarns used to make a fabric will have open spaces between the fibres, but these are very small in size and may be neglected when considering the geotextile functionalities.

The interspacing and diameter/thickness of the yarn is often used to calculate what is termed the fabric's 'cover factor', meaning the actual area the solid part of the fabric covers when the fabric is laid on a surface. From a more practical sense, this is the fraction of a fabric surface area that comprises the fibres/filaments. Neglecting the very small interfibre interstices, Fig. 8.24 depicts the projected areas seen in this way for a plain-weave structure, which can be used to obtain a calculated estimate of the fabric porosity.

Let d_p and d_t = the width of the warp and weft yarns as projected, and P_p and P_t = the pitch of the warp and weft yarns as projected.

Then, considering the unit cell ABCD, the warp and weft cover fractions are given by:

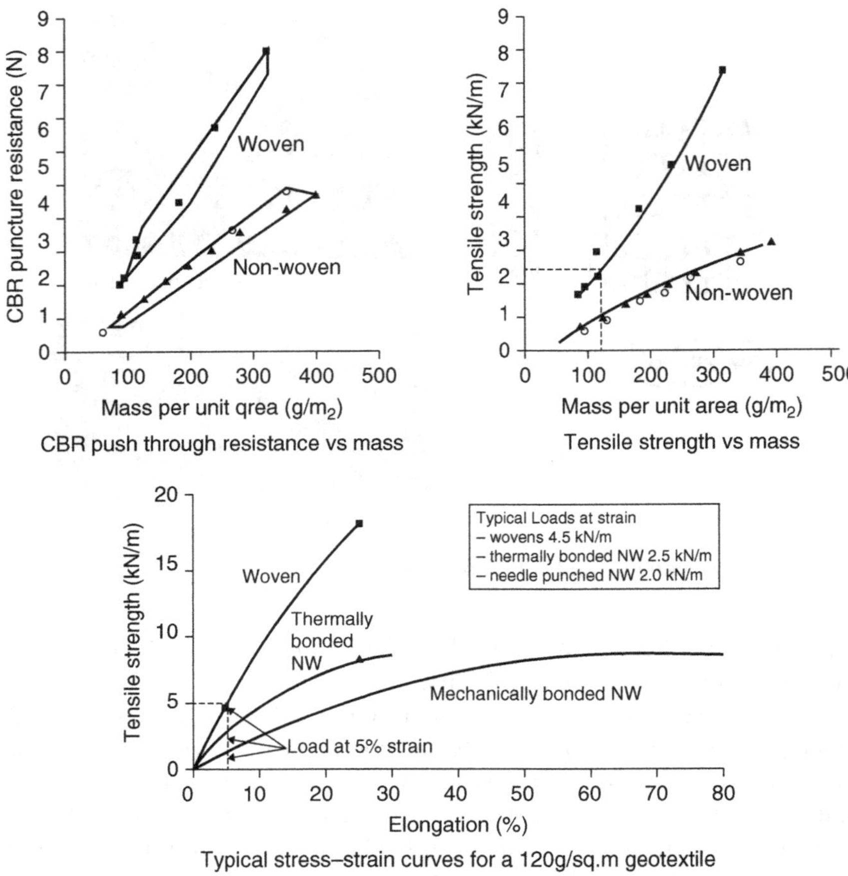

8.23 Comparison of woven and non-woven geotextile tensile properties.

$$C_p = \frac{d_p}{P_p} \quad [8.37]$$

$$C_t = \frac{d_t}{P_t} \quad [8.38]$$

The fabric cover factor, C_{fab}, is then

$$C_{\text{fab}} = \frac{\text{projected area of yarns}}{\text{area of fabrics surface}}$$

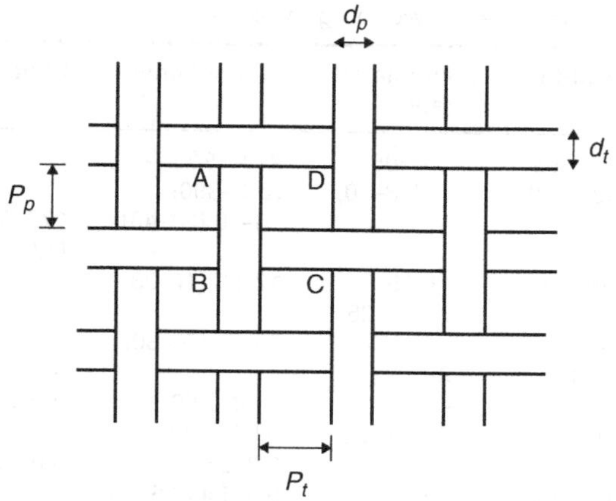

8.24 Cover factor of balanced plain weave.

$$C_{\text{fab}} = \frac{(P_p - d_p)d_t + d_p P_t}{P_p P_t}$$

When expanded it becomes

$$= \frac{d_t}{P_t} + \frac{d_p}{P_p} - \frac{d_p}{P_p}\frac{d_t}{P_t} \qquad [8.39]$$

or

$$C_{\text{fab}} = C_t + C_p - C_t C_p \qquad [8.40]$$

The fabric porosity ε_{fab} is simply

$$\varepsilon_{\text{fab}} = (1 - C_{\text{fab}})100 \qquad [8.41]$$

Since the tex of a yarn are usually referred to, then C_{fab} may be estimated in terms of the yarn tex.

If N is the tex of a yarn, A its cross-sectional area, and ϱ its density in gcm^{-3}, then

$$\varrho = \frac{N10^{-5}}{A}$$

For a circular cross-section,

Table 8.4 Examples of the properties of woven geotextiles

Properties	Polymer	Flat-tape	Fibrillated tape	Monofilament	Multifilament
Tensile strength, warp (weft) kN/m	PP	38–103 (38–93)	18–205 (18–50)	210–390 (210–390)	–
	PET	–	–	26–50 (24–40)	175–664 (175–420)
% Ext at break	PP	18 (18)	11–25 (11–25)	12–23 (12–23)	–
Warp (weft)	PET	–	–	10–30 (10–30)	11–15 (11–15)
CBR (kN)	PP	3–8	3–8	–	–
	PET	–	–	0.30–0.80	2.0–28.0
AOS (O_{95}) (mm)	PP	0.15–0.30	0.15–0.30	0.04–0.07	
	PET	–	–	0.3–0.6	0.07–0.5
Tear strength (kN) warp (weft)	PP	0.5–1.4 (0.5–1.4)	0.5–1.4 (0.5–1.4)	0.5–1.4 (0.5–1.4)	–
	PET	–	–	0.3–0.7 (0.3–0.7)	0.5–3.0 (0.5–3.0)

$$d = \sqrt{\left[\frac{4N}{\pi \varrho}\right]} = \Phi\sqrt{N} \qquad [8.42]$$

where Φ is dependent on the density of the polymer. Equations [8.39] and [8.41] can then be used to estimate C_{fab}.

Table 8.4 gives examples of AOS values for woven synthetic geotextiles made from the various yarn types. Generally flat tapes give a low percentage open area (POA) suitable for the separation function. Monofilaments facilitate a uniform opening size and a relatively high POA, typically AOS > 0.21 mm and POA > 4%, which is advantageous in filtration function as it lowers the probability of blockage and aids the formation by soil of a natural filter cake. Fibrillated-tape yarns and multifilament yarns fall within the two extremes.

It was shown above that the cover factor is directly related to the POA, and it is therefore reasonable to assume that the relation extends to the fabric permeability normal to the plane of the fabric. However, the in-plane permeability is more complex, as it concerns the cross-section area of the fabric.

One approach[41] for deriving a mathematical relationship between a woven fabric in-plane permeability and its porosity involves applying a modification to Peirce's model[42] for the unit cell of the fabric geometry of a balanced plain-weave geotextile, taking account of yarn flattening when under load. Figure 8.25 illustrates that cross-sections of the flattened warp yarns may be considered elliptical with a major axial width a_p and a minor axial width b_p. Let P_p be the warp yarn spacing and t the fabric thickness, b_t the minor axial width of the flattened weft yarn, and AA*, BB* the geometrical parameters shown.

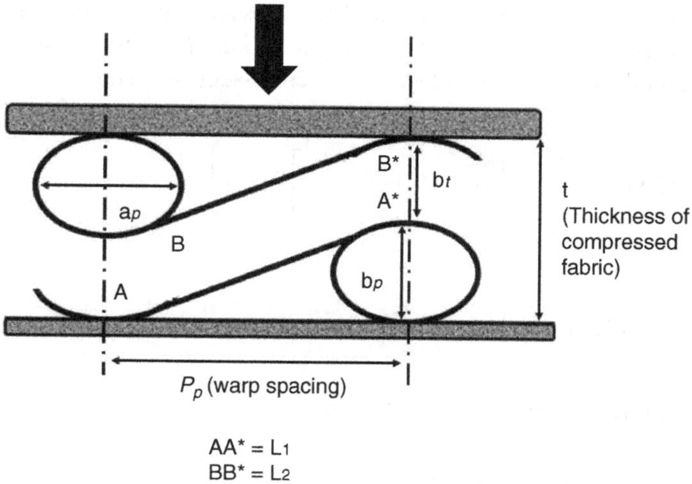

8.25 Unit cell of a balanced plain-weave geotextile under load.

Assuming laminar flow through the plane of the fabric, then Poiseuille's equation can be written for the rate of flow, Q:

$$Q = \frac{R_H^2 \Delta P A}{8\eta L}$$

where ΔP = pressure difference over L, L = length of the fabric, A = cross-sectional area of the pore space for the flow through the plane of the fabric, and R_H = the hydraulic radius

If A_{fab} is the fabric cross-sectional area, then $A = A_{\text{fab}}\, \varepsilon_{\text{fab}}$

Considering the specific permeability, k, the above equation can be rewritten as:

$$Q = \frac{k \Delta P A}{\eta L}$$

where

$$k = \frac{R_H^2 \varepsilon_{\text{fab}}}{8}$$

For a unit volume of fabric, R_H is given by;

$$R_H = \frac{\text{cross-sectional area of pore (CAP)}}{\text{wetted perimeter (WP)}}$$

CAP is the numerical difference of the cross-sectional area between the central vertical axes of flattened warp yarns (i.e. $t.P_p$) and the space within this occupied by the warp and weft yarn cross-sections. Thus:

$$\text{CAP} = tP_p - \left\{ \frac{\pi b_p^2}{4} + 2b_p \left[\frac{a_p}{2} - \frac{b_p}{2} \right] + L_1 b_t \right\}$$

and

$$L_1 = P_p \left(1 + C_p\right)$$

WP is the length of pore boundary, hence:

$$WP = 2\left[P_p + L_2\right] + \pi b_p$$

$$L_2 = L_1 - \left[a_p - b_p\right]$$

Combining,

$$R_H = \frac{tP_p - \left\{\left(\pi b_p^2 / 4 + b_p\left[a_p - b_p\right]\right) + \left(b_t P_p\left[1 + C_p\right]\right)\right\}}{2P_p\left(2 + C_p\right) + \pi b_p - 2\left(a_p - b_p\right)} \qquad [8.43]$$

8.11.2 Warp knitted geotextiles

Knitted geotextiles are a special type of warp knitted structure made with synthetic filament yarns and referred to as a directionally orientated structure (DOS).[43] Figure 8.26 illustrates the multiaxial warp knitted DOS structure. The term is somewhat self-explanatory, in that the load-bearing filament yarns are kept straight and parallel to each other, and aligned with the fabric's load bearing directions. These yarns are placed in the fabric structure in four different directions, warp, weft and diagonally, to give the fabric multiaxial strength. One set of the knitting machine operations lays down sheets of the multidirectional reinforcing yarns, and these are then passed into the knitting zone where they are held together by the knit loops, termed stitches, of a third filament yarn (knitting or stitching yarn) at their cross-over points to produce a coherent structure. DOS fabrics therefore have the advantage that the fabric modulus is effectively equal to the load bearing yarns, since yarn crimp is absent, and these reinforcing yarns enable the fabric to withstand loads from various directions.

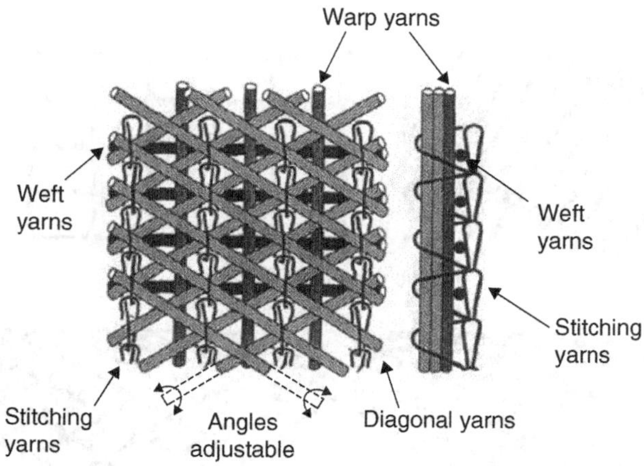

8.26 Warp knitted multiaxial DOS structure.

DOS fabrics are mainly made for special reinforcement. The load-bearing yarns used are high strength PET of 37–400 dtex, and the DOS fabric strengths range from 35 to 110 kN/m in the warp, and from 30 to 80 kN/m in the weft and diagonal directions. The warp inlay yarns absorb the axial tensile forces, say for example, down a slope, while the weft and diagonal inlay yarns give frictional resistance and transfer forces across the fabric to the cross-over points so that loads are uniformly induced into the warp yarns.

8.12 Geogrids, geonets and geomembranes

8.12.1 Geogrids

Geogrids are primarily made for reinforcing soil or aggregate. They can be manufactured by extruding polyolefin sheets (HDPE or PP) that are subsequently hole-punched, then heated, unidirectionally or bidirectionally stretched, and cooled to give a grid structure with large openings or apertures that enable the interlocking of the structure with the soil or aggregate to provide the reinforcing function (see Fig. 8.27). These grids have tensile strengths up to 86 kN in the reinforcing direction, but they are usually stiff; the ribs of the grid are susceptible to facture during installation. Geogrids are also made by weaving (leno weave) or warp knitting (biaxial DOS fabrics) and are flexible and less susceptible to installation damage. These geotextile grid structures, see Fig. 8.27, comprise either PP or high tenacity PET filament yarns and can be made to have directional strengths of 35–110 kN/m. After the structure is formed, the fabric is given a protective coating, which binds the filaments together in the structure. Application methods include spread coating with a knife or roll, dipping, and spraying.

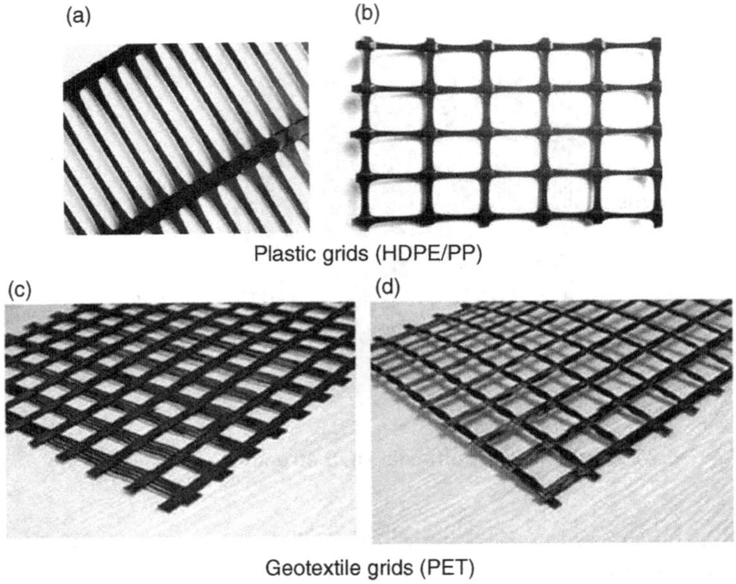

8.27 (a–d) Uniaxial and biaxial grids.

Typical coatings may be bitumen or PVC, acrylic-based or an ethylene/vinyl acetate (EVAc) copolymer coat, all containing carbon black for UV protection. The coating used must also protect against chemical and biological degradation of the underlying filaments, and in the case of PET hydrolytic degradation. Therefore, good adhesion to the filament yarns is essential to prevent wicking into internal voids. PET geogrids usually have PVC coatings that are typically in the form of plastisols, i.e. the resin dispersed in a plasticiser (dioctyl phthalate plasticiser). The purpose of plasticisers is to modify the resin, reducing its glass transition temperature, T_g, and making the polymer a more flexible and ductile coating. The formulations would also contain heat and light stabilisers, pigments, and biocides. The coating thickness is usually greater than 150 μm, as thinner coatings do not give adequate long term protection.[44–47]

8.12.2 Geonets

Geonets are stiff criss-cross, open grid-like sheet materials formed by two sets of coarse, parallel, extruded plastic strands intersecting at an acute angle (see Fig. 8.28). The network forms in-plane flow channels, making these sheet materials suitable for coupling with sheets of non-woven geotextiles to produce drainage geocomposites. Nearly all geonets are made of PE. The molten polymer is initially extruded through slits in counter-rotating dies to produce the intersecting plastic strands in the form of a

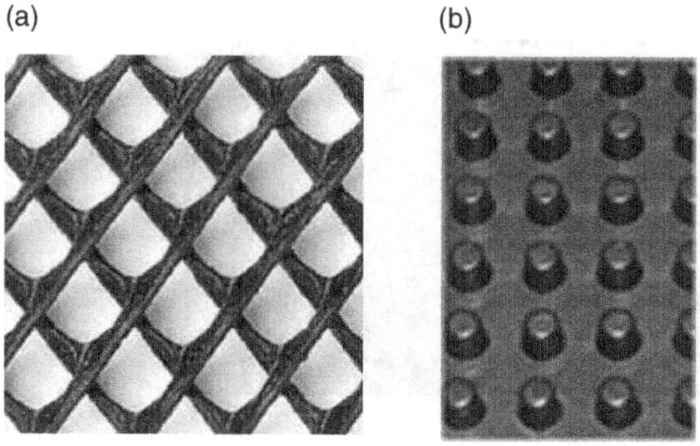

8.28 (a) PE Geonet and (b) PP cuspated sheet.

tubular mesh where one layer of strands is overlaid at the acute angle by a second layer. The tubular mesh is then slit along its length to create the 'bi-planar' grid-like sheet. A third layer can also be extruded to give a 'tri-planar' sheet, having increased thickness and, thus, increased flow capacity. A further development of a plastic sheet material with high in-plane flow capacity is the cuspated drainage sheet, also shown in Fig. 8.28. This is essentially a PP sheet, hot pressed to give parallel rows of truncated cones protruding from the plane of the sheet. The rows define the flow channels, which are usually more widely spaced than the strands of the geogrid, enabling a greater in-plane flow capacity. If made from a plastic sheet of adequate thickness, the rows of cone-like protrusions can give a cuspated drainage material of greater compressive strength than a geogrid.

8.12.3 Geomembranes

A geomembrane is usually extruded as a flat sheet or as a large tube to be slit along the length of one side and then opened into a flat sheet. Being a continuous, highly impermeable plastic sheet, suitable for liquid or gas containment or as a vapour barrier, a geomembrane generally has no macrostructural features. However, where its use with, say, a geotextile requires enhanced interfacial friction between the geomembrane and the geotextile, the membrane surface can be textured (see Fig. 8.29).

A typical uniaxial stress–strain behaviour of an HDPE geomembrane is also shown in Fig. 8.29. While maximum elongation is several hundred per cent, yield occurs at 10–15% strain.[48] A safety factor of 1.5–2.5 is usually taken into account when employing geomembranes, which brings any allowable strain down to 5–6.5%. Geomembranes are viscoelastic materials and their stress–strain behaviour is time dependent, so over the long term, applied stresses could lead to their creep failure.

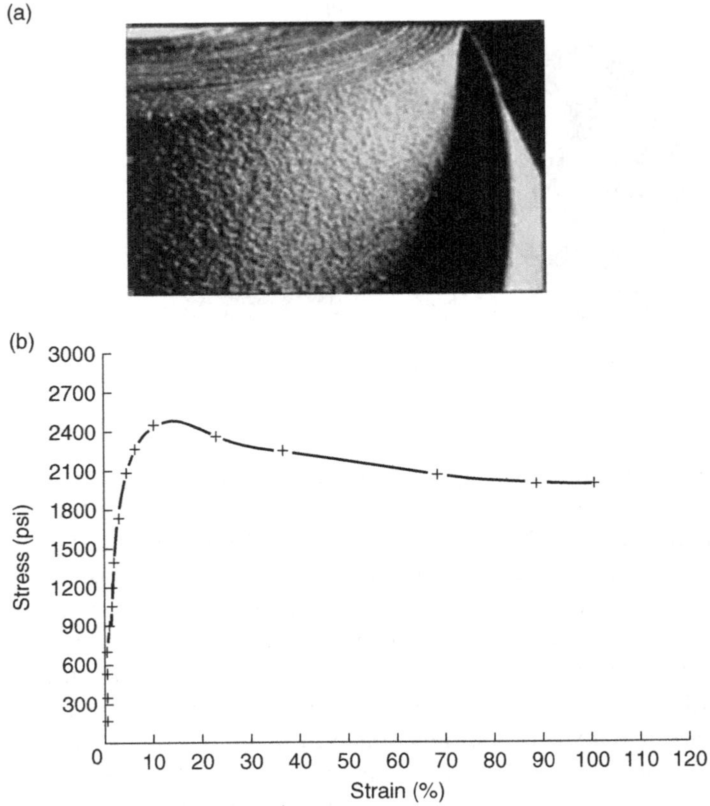

8.29 (a) Textured geomembrane and (b) stress-strain behaviour of HDPE geomembrane.

When subjected to a sustained stress in a chemical environment at an elevated temperature, the membrane will yield and undergo creep, becoming less ductile with time through strain hardening, resulting in brittle failure or rupture. Figure 8.30 schematically illustrates creep curves for stresses σ_1–σ_5. The initiation of crazing and stress whitening, which represents the start of micro-cracking and voiding, is also indicated. These show that at some time before rupture is irrecoverable, damage may be sustained by the membrane as it supports the load. The curves also indicate that the higher the stress or supported load, the faster is the time to failure.

Figure 8.30 also shows the change in yield stress of an HDPE membrane for accelerated life tests of up to 100 h duration.[49] The downturns of the lines in the graphs are associated with the transition from ductile to brittle behaviour.[18, 50] Depending on temperature, there is a 40–50% reduction in the yield stress over the time period, with a transition from ductile to brittle failure within the 100 h. Extrapolating to 1000 h at normal operating temperatures, the yield would be reduced by 70%.

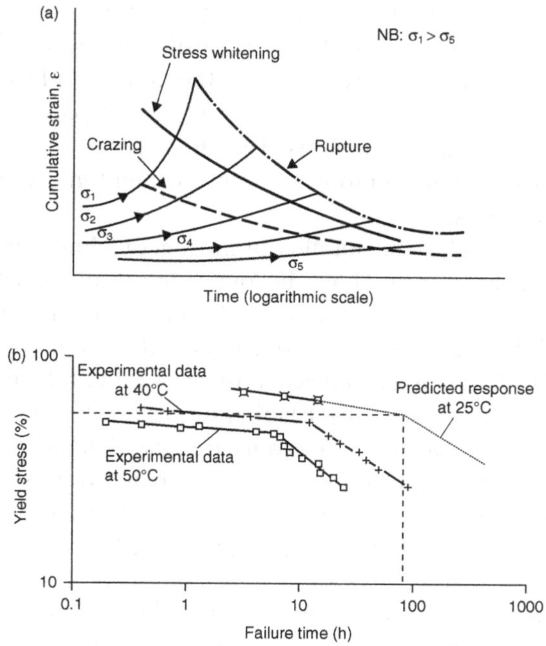

8.30 Creep characteristic of geomembranes: (a) schematic behaviour of geomembrane under constant load at prolong periods, and (b) notched constant load test on HDPE geomembrane (immersed in solution of 10% IGPAL/90% tap water).

The downturns which mark the transition to brittle behaviour can prematurely result from defects or localised damage in the membrane sustained during construction or protrusions during contact with soils. The shape and size of the defect and the presence of a chemical environment which promotes stress cracking will affect the time at which the transition occurs. Similarly, intermittent or cyclic loading can promote a transition to brittle behaviour due to the initiation and growth of fatigue cracks

It can be understood, then, that geomembranes are not intended to support heavy loads for sustained periods and deformation giving rise to high localised stresses should be prevented. Therefore, in most applications geomembranes are supported by a geotextile to provide tensile strength and enhanced tear and puncture resistance.

8.13 Geocomposites

Geocomposites are basically combinations of two or more different types of geosynthetics. As most of the individual components are thermoplastic they can be thermally laminated, but adhesive bonding and needle punching are also used. Examples include: geotextile–geonet, geotextile–geogrid, geonet–geomembrane, or what is termed as a geosynthetic clay liner (GCL). There is almost no limit to the variety of geocomposites that is possible, and the development of these materials results from the anticipated usefulness of their multifunctionality and the opportunity for more rapid installation than by using the individual components. The three main geocomposite material types are: drainage geocomposites, reinforcement geocomposites, and fluid barrier geocomposites.

Control of water is critical to the stability of most geotechnical constructions, and drainage geocomposites have become important materials for such a requirement. Common configurations of drainage geocomposites are of a geonet sandwich between two non-woven geotextile filters (termed a blanket drain – see Fig. 8.31), or a sandwiched thick or thin preformed core (panel drain, edge drain or wick drain). Blanket drains are commonly used as liquid collection–removal layers; edge drains are placed adjacent to road foundations to collect lateral and remove seepage, panel drains are used to reduce hydrostatic pressures on buildings, and wick drains are vertically embedded deep into the ground, providing drainage paths to expedite the consolidation of saturated soils.

Reinforcement geocomposites are structures in which a spun-bonded or melt blown non-woven web is incorporated into a knitted geogrid by the stitch-knit action of holding yarns, or bonded by needle punching to one or both sides of a woven or knitted geogrid. The non-woven adds separation and filtration functions to the geogrid reinforcement to give the multifunctionality of the geocomposite.

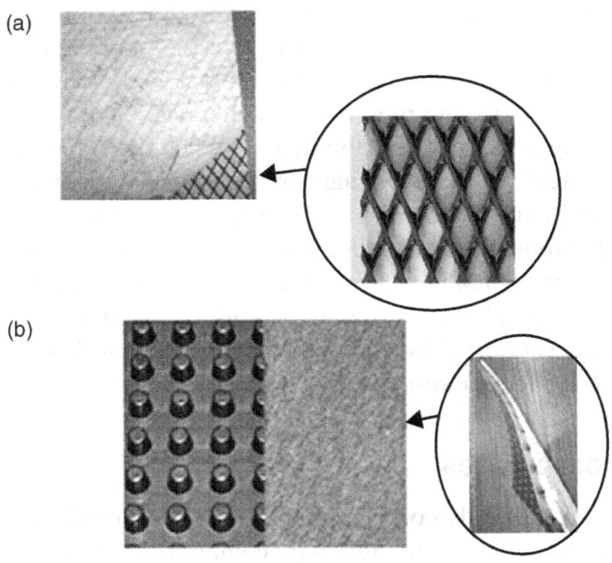

8.31 Drainage geocomposites: (a) geonet and (b) cuspated. Inset: magnified portion of the geocomposite.

Both woven and non-woven geotextiles can serve as moisture barriers when impregnated with bituminous, rubber–bitumen, or polymeric mixtures. Such impregnation reduces both the cross-plane and in-plane flow capacity of the geotextiles to a minimum. However, for liquid containment applications, what is referred to as a GCL will be more effective.

GCLs are geocomposites that are typically prefabricated with a sodium bentonite clay layer sandwiched between two geotextile layers: two needle-punched non-woven layers, or one needle-punched non-woven and one woven layer. The process commonly involves attaching the top and bottom layers by stitching or needle punching through the bentonite core which also gives the structure its internal shear resistance. When hydrated the bentonite core swells and becomes an effective barrier to liquid or gas. The hydraulic conductivity of most sodium bentonite GCLs ranges from 1.0×10^{-11} to 5.0×10^{-11} m/s[51] and they are also more resistant to cracking by freeze–thaw cycles and desiccation than compacted clay liners.[52–54] However, GCLs need to be chemically compatible with the liquid or gas against which it is used as a barrier. This is because sodium bentonite tends to maintain its swelling capacity only within the pH range 2–13. Therefore, GCLs are not compatible with strong acids ($pH < 2$) or with strong bases ($pH > 13$). There is also the need to minimise the chances of ion exchange of the monovalent sodium ions, Na^+, with divalent cations such as calcium, Ca^{2+}, and manganese, Mg^{2+}, since these can lead to significant increases in a GCL's hydraulic conductivity, owing to the divalent ion decreasing the swelling capacity of the bentonite, which in turn results in higher porosity. GCLs are therefore commonly used in conjunction with a geomembrane as secondary barriers, but can be used in as a primary moisture barrier for less demanding applications.

8.14 Factors affecting the separation and filtration mechanism of geotextiles

Geotextiles are often used as separators to prevent fine-particle subgrade soils from migrating into a permeable, granular stone layer placed on top of the subgrade. The geotextile simultaneously prevents the stone layer penetrating into the underlying soft subgrade. In this function, a geotextile is required to act as a barrier to the undesirable mixing of two materials of different particle sizes, thereby maintaining the thickness and load-bearing capacity of the stones. The migration of an amount of fine or soft subgrade equal to only around 20% by weight of the stone layer will substantially reduce the bearing capacity of the stone.

A notable problem, which readily accelerates the mixing of materials of stone and subgrade, is the movement of water when large loads periodically pressurise the surface of the stone layer. A pumping action occurs and the water movement erodes the subgrade and accelerates the migration of the fines by lifting them into the voids of the stone layer. To prevent this from happening, the geotextile separator must allow water to move through it and away to drainage while retaining the soil fines. The permeability and AOS of geotextile, as well as the grain size distribution of the subgrade, are therefore key factors, and in this regard woven geotextiles made from tape yarns are well suited to the separation function.

The above factors are also applicable when geotextiles are used for filtration. However, with filtration the intention is to form a natural filter cake at the fabric/soil interface that will retain the fines; consequently the pore-size requirement is different than that for the separation function. For filtration, the AOS of the textile must be smaller than the larger sizes of soil particles, but of sufficient dimension for the small soil particles to pass through, otherwise the geotextile will become clogged or blind and its permeability adversely affected. The fabric must have a pore-size distribution with a large number of larger size pores, so that a relatively unhindered water flow is maintained where some of the smaller size pores become blocked.

Flow conditions affect the way in which soil particles interact with one another and with the filter. As a result retention requirements, given in terms of the AOS value and in relation to particle size distribution parameters, should be determined for unidirectional flow or multidirectional flow conditions by using a wet-sieving performance test. In considering the soil parameters, the particle size distributions of typical soil types can be placed into the three zones shown in Fig. 8.32.

Zone 1 soils consist of more than 85% clay and silt (particles smaller than 0.075 mm) and often have sufficient cohesion and low permeability to allow

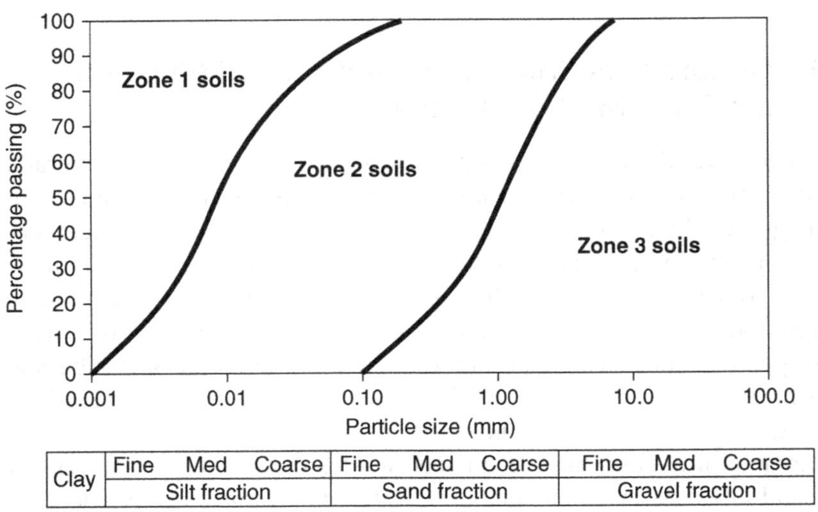

8.32 Soil zones particle size distribution.

Table 8.5 Summary of soil retention requirements for geotextile filters

Soil zones	Flow conditions	Retention criteria
Zone 1	Unidirectional Multidirectional	150 μm < O_{95} > 300 μm $O_{95} < d_{90}$ $O_{95} < 10\ d_{50}$ O_{95} > 80 μm
Zone 2	Unidirectional Multidirectional	150 μm < O_{95} > 300 μm $O_{95} < d_{50}$
Zone 3		Requirements for Zone-2

for the use of filters of more open dimensions. Table 8.5 gives examples of the relationship between the required O_{95} of such a geotextile filter and the d_{90} and d_{50} of soils, for unidirectional and multidirectional flow. If the soil is dispersive, then, a 100–150 mm layer of sand placed between the soil and geotextile will aid the filtration.

Zone 2 soils consist of coarse silts and fine sands (d_{85} > 0.075 mm and d_{50} < 0.6 mm) and these exhibit an increased tendency for the movement of fine soil particles. They are considered to be problem soils with regard to geotextile filters. Table 8.5 shows that the AOS requirement for retention has a smaller range than for Zone 1 soils.

Zone 3 comprises granular soils (d_{85} > 2.0 mm), and the hydraulic filter efficiency can often be easily maintained using a geotextile which meets the retention requirements for Zone 2 soils.

The retention criteria presented in Table 8.5 are given as a systematic way of viewing the AOS–particle size relationship. However, there are a number of other reported retention criteria[55] which refer to specific types of geotextile structures. These tend to suggest that non-woven geotextiles can be made to have better soil particle retention than woven geotextiles of equivalent % open space, because of the wider pore-size distribution of the non-woven structure. Also, that there is no significant difference in efficiencies between needle-punched non-woven structures (porosities 80–90%) and thermally bonded (porosity 71% and above). However, when long term performance is taken into account, and the potential risks of impediment to water flow are included, the situation changes.

There are essentially four ways by which water flow through geotextile filters can be impeded: by blinding, blocking, particulate clogging, and biological clogging or chemical/biochemical clogging.[56]

Blinding occurs when soil particles form a thin impervious layer at the geotextile surface, whereas blocking occurs when the soil particles obstruct the filter opening, and clogging happens when there is a build-up of trapped soil particles within the geotextiles. Blinding is most detrimental and clogging is least, but in making a suitable filter structure to resist clogging, blinding and blocking would also be alleviated.

Particulate clogging takes place when the fine particles moving with the water flow through the pores of the geotextiles become trapped and build up to a level where they slow and eventually inhibit the flow. The risk of trapping is greater with non-woven geotextiles than with woven geotextiles, because of the larger number of smaller pores in their pore-size distribution and the tortuosity of these pores. However, the thickness of the non-woven geotextile is important because of the corresponding increase in the number of pores with thickness. The more open spaces of thicker geotextiles will accommodate a higher fine-particle loading before the permeability decreases significantly.

When accounting for particulate clogging in the retention criteria, consideration must be given to the total % of open area and the faction of open area that has small pore space. For the total open area, the principle is that a geotextile should always be more porous than the equivalent soil filter. However, because non-woven structures are more susceptible to clogging than woven structures, it is recommended that non-woven geotextiles used for filtration should have a porosity of 40–60%,[57,58] whereas woven geotextiles may have a total open area of only 4–6%.[57,59] With regard to the fine pore space, the recommendations are mainly applicable to non-wovens owing to their structural characteristics. Consideration has been given to the O_{95}, O_{90}, O_{50}, and O_{15} in relation to d_{85}, d_{50}, d_{15}, and the ratio d_{60}/d_{10}, i.e. the coefficient of uniformity, C_u, of the soil particle size distribution. The recommendations for the minimum pore sizes are:[60–62]

$$O_{95} > 3d_{15} \text{ for } Cu > 3$$

$$O_{90} < 10 \times d_{50}$$

O_{95} (geotextile) $< B$ D85 (soil) ; $B = 0.5$–2 depending on the type of soil

$$O_{15}/d_{15} > 0.8\text{–}1.2$$

$$O_{50}/d_{50} > 0.2\text{–}1$$

The relationship of O_{95} with d_{15} is used because the clogging resistance criterion is generally smaller (for silt/clay size particles) than the O_{95} value for soil retention.

When a geotextile filter is to be used for environmental applications, it becomes necessary to design the filter with resistance to biological clogging. Biological clogging, initiates with the development of a network of 'biofilms'. The biofilms adhere to the geotextile fibres and along with micro-organisms will grow, reaching a size where they can cause clogging of the small openings in a non-woven geotextiles. A second aspect of biological clogging is the development of encrustations. The growth of the biofilm does not lead to complete clogging of the filter so as long as no encrustation occurs. Encrustations occur when bacteria alter the pH of contaminated ground water, which can lead to the dissolution of metals (decrease of pH) such as calcium and iron, or precipitation of carbonates and sulphides from metal ions (increase of pH). When fully developed, encrustations can completely clog the interstices of a geotextile.

Studies show that the most important factor for biological clogging resistance is the open area or opening size of the filter. Reportedly, biological clogging readily reduces the efficiency of non-woven geotextile filters,[62] and in time will spread throughout the entire thickness of the filter. Therefore, where there is a high risk from such clogging, in preference to a non-woven geotextile, use should be made of a monofilament woven geotextile with a minimum O_{95} of 0.5 mm and an open area of 30%. In critical applications, performance testing should be conducted to evaluate the behaviour of potential geotextile filters with a given soil.

Additional to the criteria for particle retention, the permeability requirements of the geotextile filter for various soil types must be considered. The permeability of a geotextile filter in severe soil and hydraulic conditions can be reduced by an order of magnitude over its life. To ameliorate this, it is important that the geotextile maintain or exceed its index permeability while under load i.e. any re-orientation of the fibres should not decrease or increase the permeability. Criteria against clogging have also been recommended for the permeability as well as for the retention properties of a geotextile filter.

The principle behind all permeability criteria is that the flow of water will not be impeded at the soil/geotextile interface, and this means the permeability of the geotextile ($k_{gtextile}$) must be greater than the permeability of the soil (k_{soil}), i.e.

$$k_{gtextile} > B * k_{soil}$$

where, for a reduced risk of clogging, $B = 10–100$ and the k_{soil} is determined by laboratory or field measurement. Although dependent on soil conditions and the hydraulic gradient likely to be present, the rule of thumb is that the minimum allowable $k_{gtextile}$ is $B = 10$.[60]

Table 8.6 gives values for the d_{50} particle sizes and permeabilities of typical underlying soils. If the soil permeability is 10^{-5} m/s then the geotextile must have a permeability $>10^{-4}$ m/s and for retention $O_{90} < d_{50}$ i.e. < 0.30.

Table 8.6 Particle d_{50} values and related permeabilities

Soil type	Filtration d_{50} (mm)	Permeability k_s (m/s)
Clayey silt	0.02	1×10^{-9}
Sandy silt	0.02	1×10^{-7}
Fine sand	0.30	1×10^{-5}
Coarse sand	0.50	1×10^{-4}
Mixed sand and shingle	2.00	1×10^{-3}

8.15 Factors affecting the drainage mechanism of geotextiles

Drainage may be defined as the action of intercepting, collecting, transporting, and disposing of unwanted waste water/liquids. Although a geotextile filter is not necessarily designed to perform these actions, it may be used as a component of a drainage system. Figure 8.33 shows examples of the type of drainage which may be used with a geotextile filter. The subsurface drain system uses perforated pipes placed at the bottom of a trench. The pipe is wrapped in a geotextile filter to prevent blockage of the pipe's perforation and the assembly is back filled with granular material. French drains, also called 'blind drains' are without geopipes, and are used as lower cost alternatives.

As explained earlier, when a geotextile or geocomposite is designed to function as a drainage layer, its in-plane permeability is a key property. However, a properly functioning drainage layer must retain the surrounding soil while readily accepting water from the soil and removing it from the

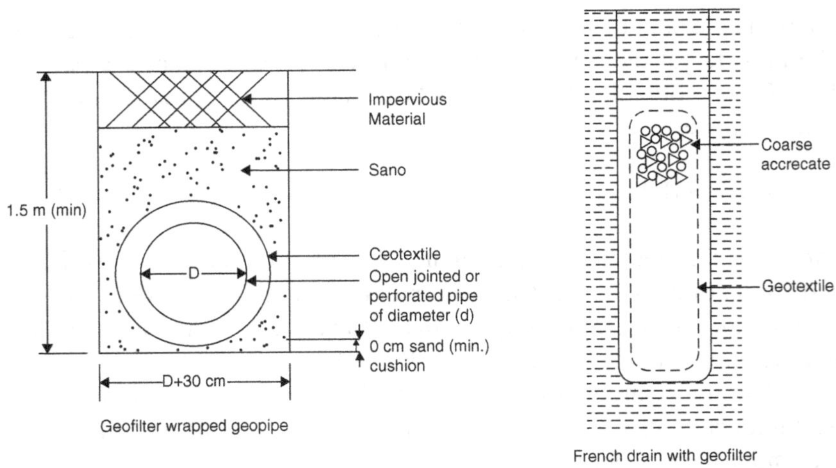

8.33 Subsurface geotextile drainage.

area. The principle of soil retention is similar to the filtration function, so similar criteria will apply.

Most drainage applications utilise the geocomposite structure. The drainage geocomposite is placed above the groundwater table to aid the rapid transport to waste of rising water and/or water moving down through the top soil. The weight of the top soil will cause compressive creep reduction in the thickness of the geocomposite structure, and protruding stones from the top and underneath soils can intrude into the geocomposite structure to constrict some of the flow channels. These factors in addition to the various modes of clogging mean that the estimated long term in-plane permeability of a drainage geotextile must take account of these reducing factors.[63]

Even though a geotextile for separation/filtration, or a drainage geocomposite, may have been made with adherence to the relevant criteria, in order to project for long term performance, their measured properties, i.e. the index test values, must be moderated to take account of likely decreases over their period of use. Reducing factors are therefore applied to obtain property data more applicable to the geotechnical conditions when installed. The index test results are considered the ideal condition values, while their reduced values pertaining to geotechnical conditions are termed the realistic or allowable values.[5] Hence, the general formula applies where,

$$\text{Allowable value} = \frac{\text{Index value}}{\text{RF}} \qquad [8.44]$$

where RF is the total reduction factor and is equal to the product of the numerical values for the individual reducing factors extrapolated from experience or research data.

Thus for separation, filtration, and drainage applications the cross-plane and in-plane permeability values obtained by index testing would be moderated according to:

$$v_{20\text{allow}} = \frac{v_{20\text{index}}}{\text{RF}} \qquad [8.45]$$

where

$$\text{RF} = \text{RF}_{\text{SCB}} \cdot \text{RF}_{\text{CR}} \cdot \text{RF}_{\text{IN}} \cdot \text{RF}_{\text{CC}} \cdot \text{RF}_{\text{BC}} \qquad [8.46]$$

The individual reduction factors being

RF_{SCB} = for possible particle clogging and blinding

RF_{CR} = for reduced pore size caused by mechanical creep of material
RF_{IN} = for reduced pore space due to adjacent materials intruding into the geotextile
RF_{CC} = for chemical clogging of pore spaces
RF_{BD} = for biological clogging of pore spaces

In general the published numerical values for these reduction factors fall within the range of 1.5–20; the more likely the severity of the geotechnical conditions, the higher the number that would be used.[64]

It should be noted that moderating the index test results for the mechanical strengths (wide-width tensile, bursting, and grab-strength tests) is also carried out. Although moderating the index test results gives a more realistic indication of long term performance, a margin of safety, or a safety factor, FS, is nevertheless applied when relating the moderated value to that specified as the required figure for a particular application.[64] Thus:

$$\text{Specified value} = \frac{\text{Allowable value}}{\text{FS}} \qquad [8.47]$$

8.16 Factors affecting the protection mechanism of geotextiles

For the protection functionality, the geotextile used is usually placed immediately over or under the geomembrane (or both) to act as a stress-relief layer. The aim is to prevent wear, due to abrasion, and puncturing of the membrane by sharp stones in the subgrade or overlying layer, by reducing point stresses and ensuring that local deformations in the geomembrane remain below a permissible limit of 0.25%.[9] The suitability of a geotextile to give adequate protection depends on its ability to withstand the intended weight of the overlying layer. In this regard, the key factors are fabric thickness and structure.[65–67] Consideration must also be given to the type of stone involved, in that the larger the stone size used, the greater is the chance of membrane damage; graded stone (i.e. stones with a range of sizes) is less damaging than single-sized stone; angular or crushed stone is more damaging than round stone; and the likelihood of stone damage increases with the hardness value of the stone.[68]

Needle-punched non-woven geotextiles are the most widely used as protective cushions. They are made from PP or PET fibres of 2–40 decitex fineness, and have an areal density within the range of 300–4000 gm^{-2}, depending on the type of stone to be used. The surface and cross-sectional images of the needle-punched non-wovens shown in Fig. 8.17 indicate that the needled regions are sections of fibre lengths re-orientated by the needling process to form vertical columns of grouped fibres through the thickness of the fabric, while the remaining lengths lie between the columns, crossing each other in an entangled state. These columns of vertical fibres provide the cushioning effect, absorbing stresses generated by the overlying layer, and thereby minimising the penetrating force of sharp stones overstraining or puncturing the membrane.

Researchers have shown that the structural properties of non-woven geotextiles are greatly influenced by the physical characteristics of the fibre type used.[68] Besides the use of a dynamic perforation test, mainly for performance testing, CBR test values are used as a guide to the likely protective effectiveness of a geotextile. Reportedly,[68] continuous filament non-wovens give better resistance properties than their staple-fibre counterparts. However, the majority of non-woven membrane protectors are made from staple fibres. Both fibre linear density and fibre length influence CBR strength, but the effect is more pronounced with linear density, in that the finer the fibre the higher is the CBR value, which can be explained by the higher fibre-to-fibre cohesion and fabric compactness obtained with finer fibre linear densities.[69] The CBR increases with increasing fibre length but only for the mid to lower end of the fineness range.

Often the installation of a geomembrane involves a sloping surface. In such situations, the interfacial friction of the non-woven and geomembrane, as well as the soil and geomembrane, is important to the stability of the combination. The surface texture of the geomembrane is the dominant factor. Studies[68] of the effect of smooth and textured geomembranes have found that for textured sheets the peak and residual friction angles are within the respective ranges of 20–26° and 10–14°, whereas smooth sheets show much lower corresponding values of 8–10° and < 6°.

In certain circumstances, mainly when heavy membrane protection is required, GCLs – needle-punched bentonite mats – have been used in addition to non-woven geotextiles.[68] When hydrated, the GCLs are generally found to be efficient in dissipating localised strains. They also have useful very low cross-plane permeability coefficients and thereby aid the barrier function of the geomembrane. Other geocomposites used for added membrane protection include non-woven–sand–non-woven structures and three-dimensional geotextile mats containing mineral filler combinations of sand and clay.[70] Dynamic perforation tests, using 16/32 mm gravel and a loading of 1350 kN/m^2 for 1000 h duration, showed that mineral-filled geocomposites keep the geomembrane strains to very low levels, within the range of 0.01–0.2%, compared with 0.6–2.5% for thick non-woven of heavy areal densities subjected to the same test conditions.

8.17 Factors affecting the reinforcement mechanism of geotextiles

The reinforcing function of geotextiles is one of the most used in geotechnical engineering. Soils have much lower resistance to tensile forces than they do to compressive forces. Embedding a suitable geotextile within the soil, so that the geotextile is both within the plane and aligned with the direction of the tensile forces, enables the soil to transfer such forces to the geotextile, utilising its axial strength and obtaining soil stability by the reinforcement.

For the geotextile to provide an effective reinforcement function, it should have not only a high tensile strength, but also a high tensile modulus so that its resistance to tensile loads generated within the soil occurs at sufficiently small strains to prevent excessive movement of the reinforced soil structure. It is self-evident that decreases in these properties with time (i.e. creep behaviour) must be low, and that the polymers used should have resistance to degradation by the soil.[71–73] An estimate of the anticipated reduction in strength can be determined from an analysis of creep strain versus time plots for various stress levels and a suitable reduction factor applied.

For the majority of reinforcements, fabrics are made from PP or PET filament, although for very high strength applications, requiring strength of the order of 400 kN/m, para-aramid (e.g. Kevlar), glass or basalt filament may be used. PET is usually the filament of choice for applications where strength/cost ratio is the most important factor. However, due to its susceptibility to chemical attack in high pH environments, it cannot always be utilised. PP is more resistant to chemical degradation, but has much poorer long term creep characteristics than PET. As such, where long term strength is important PET fabrics with a protective coating are used. In addition to this, sample pieces of the fabric may be buried with the main fabric and removed for analysis at periods during the life of the structure to assess any degradation to the integrity of the geotextile. Allowance must, nevertheless, be taken to account for reduction in tensile properties caused by chemical and biochemical degradation.[72–78]

The robust nature of earthwork constructions is likely to make material damage a more significant factor for soil reinforcement than for other applications of polymer materials, and therefore consideration should also be given to possible strength loss caused by installation damage.

Similar to the other functions, the combination of the above reducing factors should be accounted for to obtain the allowable tensile strength, T_{allow}. Thus:

$$T_{allow} = \frac{T_{index}}{RF} \quad [8.48]$$

$$RF = RF_{ID} \cdot RF_{CR} \cdot RF_{CD} \cdot RF_{BD} \quad [8.49]$$

where RF_{ID} = reducing factor due to installation damage

RF_{CR} = reducing factor due to creep
RF_{CD} = reducing factor due to chemical degradation
RF_{BD} = reducing factor due to biological degradation

A most important additional requirement for the reinforcement function is that the geotextile is made to have a good bonding strength, T_b, with the soil, so as to resist slippage of the soil over fabric surface or the pullout of the fabric from the soil. Evidently $T_{allow} >> T_b$ in order to ensure effective reinforcement.

The geotextile's bond strength comes from its interaction with the soil through frictional or adhesion forces to resist tensile or shear forces. Geogrids, straps, and woven or warp knitted geotextiles, are the structures that readily do so without alteration to their tensile behaviour, and consequently are mostly used for reinforcement applications. Some high density non-wovens are provided for moderate level reinforcement, but generally the stress–strain behaviour of non-woven geotextiles is significantly altered by soil confinement, making them unsuitable for the vast majority of reinforcing applications.[79]

Clearly the ability of the geosynthetic to interface with the soil and be anchored by it is a key element in the transfer of the tensile force and hence performance of the reinforced soil structure. Thus soil–geosynthetic interaction, or interface friction, becomes critical as it governs the soil–fabric bonding strength. The required embedded length or bond length is determined by this parameter.

Starting from the free end of the buried reinforcing material, the axial tension generated in the fabric increases up to a limiting value equal to fabric's bond strength with the surrounding soil, provided $T_{allow} >> T_b$. The maximum bond strength per unit area for the reinforcing material may be expressed in terms of the product of the bond coefficient (f_b), the soil angle of friction (tan φ being determined from shear test results), and the normal applied stress (σ_n')[80, 81]:

$$\tau_b = \sigma_n' f_b \tan\phi \qquad [8.50]$$

For a reinforcement element of length (L_r) and width (W_r), the maximum bond strength is given by the expression:

$$(P_r)_{max} = 2L_r W_r \sigma_n' f_b \tan\phi \qquad [8.51]$$

There are three mechanisms by which the interaction between the soil and the reinforcing geotextile can resist induced tensile loads:

- shear friction between the soil and the surface of the geotextile.
- the embedding of soil within the large open spaces of the geotextile structure. This gives the additional resistance of soil-to-soil shear friction at these open spaces.
- the resistance of the soil embedded within the large open spaces to the movement of the geotextile by pullout forces, i.e. the bearing resistance of the soil. Usually, the bearing resistance is much higher than the friction resistance.[82] Only the first mechanism applies to woven geotextiles, whereas the latter two also apply to grid structures.

The bond coefficient between the reinforcing material and soil may be generally defined as:

Bond coefficient = resistance between soil and reinforcement resistance between soil and soil.

Consider the case of a grid structure. This can be a woven or knitted grid, or one made from welded straps. The expression for the bond coefficient may be written as:[80]

$$f_b = \alpha_s \frac{\tan\delta}{\tan\phi} + \frac{1}{2}\left(\frac{\sigma_b'}{\sigma_n'} \cdot \frac{B\alpha_b}{S} \cdot \frac{1}{\tan\phi}\right) \quad [8.52]$$

where δ is the direct shear friction angle between the soil and the reinforcement material surface, and σ_b' / σ_n' is the ratio of bearing stress to normal effective stress in the soil. B and S are the parameters of the grid shown in Fig. 8.34, α_s α_b and may be obtained from published data relating σ_b' / σ_n' with the soil friction angle (φ')[81, 83–94] from which grid bond coefficients may be calculated.

For a geotextile, the coefficients are $\alpha_s = 1$ and $\alpha_b = 0$, and Equation [8.52] may be simplified to give:

$$f_b = \frac{\tan\delta}{\tan\phi} \quad [8.53]$$

The above analysis assumes that:

- the normal effective stress (σ_n') is unaffected by soil spread on shearing;
- for woven geotextiles $f_b < 1.0$ in non-saturated soil, meaning that the surface profile of the woven fabric cannot give better bonding than that of soil-to-soil. This may not be the case with geogrids, where values of 1–1.2 apply.

Besides calculating bond strength, the effectiveness of a reinforcing fabric can be judged by considering the parameters, C_a; tan φ and C; tan δ, of the Mohr-Coulomb failure criterion obtained by Equations [8.19] and [8.20] for the fabric–soil and soil–soil shear tests, respectively. The ratios of these parameters give the adhesion and friction angle efficiencies for the geotextile.

$$E_g = \frac{C_a}{C} 100 \quad [8.54]$$

$$E_{ga} = \frac{\varphi}{\delta} \quad [8.55]$$

With woven geotextiles, the absolute minimum value for the bond coefficient or adhesion efficiency would be for soil shear over a smooth solid sheet of the parent polymer at normal effective stresses which would not cause indentation of the polymer surface. The filaments, tapes or yarns from which a geotextile is made give the fabric a much rougher surface for bonding than a smooth polymer sheet. Studies have shown that the type of fabric construction can significantly influence the interfacial bonding strength with soils. Published information for smooth polymer sheet give values of f_b from 0.55 to 0.65, depending on soil type, whereas for woven geotextile f_b falls within the range of 0.37–1.0.

Geogrids have the mechanisms of both interfacial friction and bearing resistance. With the ribs (i.e. the bearing members) of the geogrids spaced to give a ratio of spacing to bearing member thickness (S/Bα_b, see Fig. 8.34) in the range 9–15, just the bearing resistance of the material alone would act like a fully roughened fabric sheet to give good bonding.[81]

The significance of the bond coefficient for geotextiles in reinforcement applications is that the higher the f_b the shorter the required bond length, and comparisons have been published for various textiles including woven flat-slit tape and multifilament, heavy needle-punched non-wovens, and geogrids.[95]

As may be expected, the moisture content of soils will tend to reduce the interfacial friction of reinforcing geotextiles. Several studies have shown significant reduction in both friction angles and adhesion efficiencies, to less than 2° and 0.5° respectively, for near-saturated cohesive soils.[96–98] Overcoming such reduction requires the use of a material with good in-plane permeability, which means the use of an added non-woven layer.[99–101] With the presence of the added non-woven layer in the saturated soil, the Mohr-Coulomb failure envelope effectively becomes a curve as opposed to a straight line, the deviation being at the lower range of the normal stresses. The curved shape indicates a continuous transition from saturated soil to drained soil behaviour, as the value of the normal interfacial stress is increased;[102] the non-woven facilitates the draining of the water from the soil–fabric interface expressed by the increasing

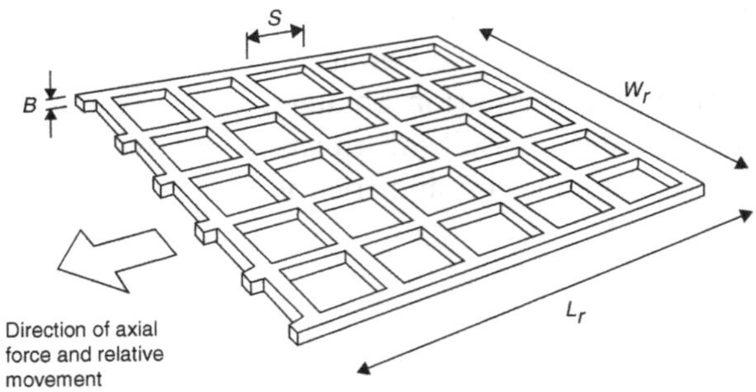

8.34 Geogrid parameters.

normal stress. In saturated soils there is therefore great potential benefit from the use of dual-function geotextiles, i.e. drainage-reinforcing geocomposites.

8.18 Applications: road construction

The applications of geotextiles cover almost all areas of geotechnical engineering. Examples fall within transportation, such as roads, airport runways, railways and embankment and bridge constructions, liquid and gas containment structures, and erosion control including inland, river bankings, and coastal defences. Generally, in these applications the geotextiles used fulfil several functions. Therefore, when considering the applications of these materials, the multifunctionality each provides will be described.

With road construction the following functions are incorporated into the engineered ground structure:

- separation and filtration
- drainage
- reinforcement
- moisture barrier.

The base of a road is likely to consist of a compacted subgrade soil on which a thick layer of coarse aggregate is placed for primarily bearing traffic loads. Based on the earlier description of the separation and filtration functions, a geotextile would be needed to prevent pumping failure where repetitive, cyclic traffic loads would push the coarse stone aggregate into the subsoil (see Fig. 8.35a). As this happens, the bottom stones of the aggregate spread while simultaneously moving the soil fines up into the void spaces of the aggregate, i.e. the aggregate layer becomes contaminated with fines from the subgrade. Rising ground water table and rain, in the case of untarmacked (unpaved) roads, accelerate the fines migration, leading to signs of rutting on the road and ultimately full

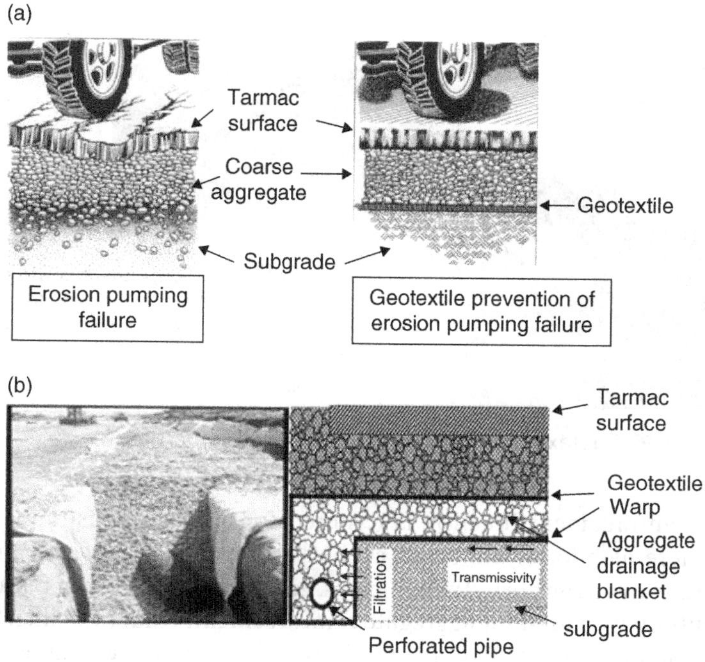

8.35 Separation filtration (a) and drainage functions (b) of geotextiles in road construction.

deterioration of the surface. An aggregate base layer, which is saturated 10% of its time, will have its service life reduced by as much as 50%.[103] Thus, eliminating saturation of the base course is an important requirement. The geotextile placed at the interface of the subgrade and aggregate has to provide, in addition to separation and filtration, the drainage function in order to retain the strength characteristics of the aggregate material. This is achieved by making part of the bottom layer of aggregate a drainage blanket, encapsulated within the wrap of the geotextile. The in-plane permeability of the geotextile-aggregate system must be effective in removing rising ground water and, where necessary, surface water ingress, to a subsurface perforated pipe drainage at the road side; with tarmac/pavement roads, this is referred to as 'a pavement edge drain' illustrated in Fig. 8.35b.

The type of geotextile which provides the best combination of the three functions is a non-woven fabric with suitable through-plane and in-plane permeability, as well as adequate puncture resistance, CBR, and bursting, grab and tear strengths to resist the aggregate penetration under cyclic loads. Where ground water is substantial, geocomposite blanket drains have been used to enhance road base drainage.

8.36 Geotextile reinforcement and moisture barrier.

The reinforcement function is required for aggregate restraint, particularly where high loads will occur, such as motorways (highways) and, especially, runways at airports. In such cases, geotextiles made from para-aramid filaments may be used for aggregate reinforcement. Under loading, the total road structure behaves like a beam in bending. The upper region is under compressive stress while the bottom region experiences tension. Since the aggregate layer has no useful tensile resistance, the bottom region of an unreinforced base course will spread allowing the upper region to progressively collapse downwards. Although the non-woven geotextile used for the separation, filtration, and drainage functions will impart some degree of reinforcement to resist the spread, it is more effective to use a geogrid, woven or warp knitted geotextile within the aggregate layer when high loads are anticipated. The fabric would be placed at a depth where it is likely to coincide with what would be the failure plane in the absence of the geotextile (see Fig. 8.36). The encapsulation effect of the aggregate by the non-woven geotextile gives a supporting role by enhancing the CBR of the drainage blanket.

The use of a base course reinforcement is also seen as a requirement for road construction on soft cohesive soils, weak collapsible soil, and in areas having underground cavities, whether of natural (Karstic phenomena) or anthropic origin (e.g. closed mines).[104]

On weak substrates the geotextile is likely to extend the service life of a road by a factor of 2.5–3.0 times compared to a non-stabilised construction.[105] Roads built in closed mining areas can be susceptible to sink holes suddenly occurring, owing to the collapse of old mine shafts resulting from sub-erosion processes. Road constructions in mining areas are, therefore, reinforced with uniaxial or biaxial geogrids.

Besides the above technical advantages, there is a significant economic advantage in using geotextile reinforcements. Local soils can be used for the construction, dispensing with the need to transport stronger mineral materials to replace weaker local soils, and thereby eliminating the added material costs and the cost of transportation. In addition, because the geotextile reinforcement greatly improves the load-bearing resistance of the road, a reduced thickness of aggregate layer may be used compared to that for an unreinforced structure, thus further reducing costs.

The vast majority of roads and airfields in the built environment are constructed with asphalt/tarmacadam surfaces or pavements, and the modern method is to incorporate a geotextile fabric interlayer, called a pavement geotextile (see Fig. 8.36). Tarmacadam, more correctly referred to as bituminous macadam or 'bitmac' for brevity, is a layer of small stones bound together by bitumen. The term 'Tarmac™©®' is the name of a publicly listed company, although used in the vernacular 'tarmac' to mean tarmacadam. Asphalt, or more correctly asphaltic concrete, according to British and European (CEN) definitions, is a mixture of bitumen and coarser mineral aggregates.

Traditionally laid surfaces have a shorter service life than those with a pavement fabric. Besides flexural fatigue by the dynamic loading from traffic, natural ageing through changes in weather conditions of temperature and moisture will cause deterioration of road surfaces. These factors cause differential expansion and contraction between the various material layers of the constructed road, resulting in small horizontal or vertical movements generating cracks which eventually propagate to the road surface. Water entering surface cracks can penetrate deep into the pavement layer and severe temperature changes can cause water within the cracks to freeze and damage the bitmac layer, leaving pot holes and ruts. Repairing by just placing a new surface overlay often results in the cracks from the existing pavement propagating up through the new overlay surface (termed reflective cracking). The inclusion of a paving geotextile interlayer, in both new road constructions and rehabilitation work, significantly retards the potential for such deterioration of the road surface, and notably increases the pavement service life.

Commonly, a PET non-woven geotextile of 135–200 gm^{-2} is used. When laid, it becomes impregnated with bitumen to form a moisture-barrier flexible membrane, relatively impermeable to cross-plane and in-plane liquid flow, and minimising ingress of water deep into the pavement structure.[106] Although the primary function of the paving geotextile is to prevent penetration of surface water, it may also improve the bitmac layer performance by acting as a stress-relieving/absorbing membrane interlayer. PET is preferable to PP because of its higher thermal resistance, since bitmac is normally heated to temperatures of 150°C–180°C for onsite 'workability'.

However 'doping' the bitumen with a light oil as a solvent (often kerosene, or creosote), known as 'cutback', can render the bitmac workable at lower temperatures.

8.19 Applications: rail track construction

Geotextiles were introduced into railway construction in order to improve track support, for both the laying of new lines and rail track rehabilitation. Subgrade soils make up the main foundation of a rail track on which layers of granular materials are placed, and subsequently the sleepers and rail lines. The role of the aggregate layers is essentially that of a load-supporting intermediary between the rail lines and the subgrade; the aggregate layers are subjected to a repeated cyclical stress as the wheels on each axle of a rolling stock traverses the line above a sleeper. For example, Fig. 8.37a illustrates the complex loading profile on any one sleeper caused by passenger trains running on an assumed perfectly straight, level track at a maximum speed.[111] The graph shows three parts to the profile. The first section is the preceding increase in load as the wheel leaves the previous sleeper but before reaching the sleeper of the considered loading profile. The main hump occurs when the wheel passes over this sleeper and the last hump, the tail, occurs when the wheel leaves the sleeper and passes onto the next. The time index is the fraction of the time it would take to travel the distances between the sleepers. The load index is the percentage of the total static load on that particular train axel. For track longevity, a major requirement is to use appropriate layers of granular material for bearing capacity to support the wheel loads. In heavily trafficked areas, crushed stone or gravel is normally placed on top of low-cost uncrushed aggregate.

Though economical, the use of an uncrushed stone layer can cause a decrease in the stability and the holding capacity of crushed granular cover. To circumvent this, a geogrid reinforcement is incorporated within the construction. However, further problems can still arise with the longer-term stability of the railway track.

Similar to roadway construction these problems are associated with 'erosion pumping failure'. Over time, the track stability becomes greatly weakened by the ingress of water and the movement of 'fines' from the subgrade into the ballast layers. Water ingress, along with train effluent (waste water and diesel/oil/grease), occurs from the environment above the track; water ingress can also result from a rising water table and run-off from the side embankments of, say, a railway cutting. The movement of the fines is much amplified by the presence of water. In addition to the fines from the subgrade, stone damage due to the physical pounding of the ballast by the concrete sleeper and the frictional wear of the stones can give wet contaminants of the aggregate. The movement of the fines and muddying of the

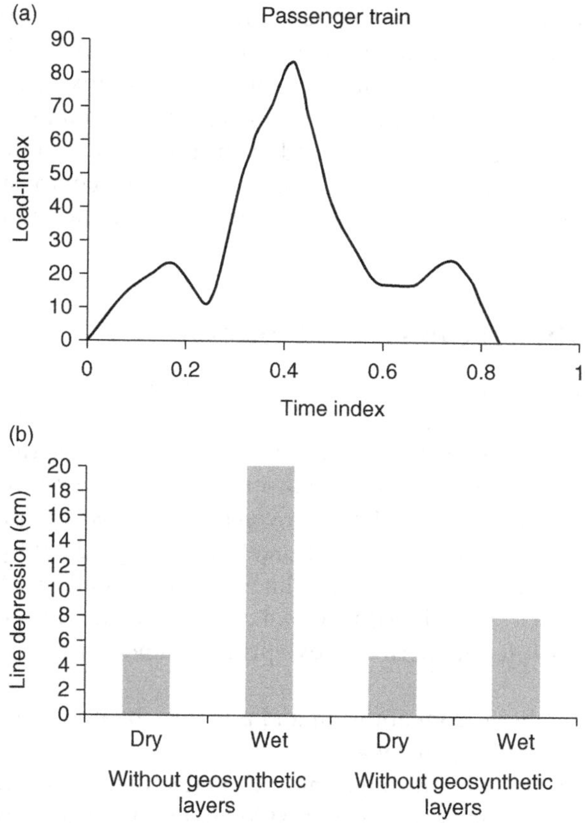

8.37 (a) Loading profile and (b) line depression of rail sleeper.

initially separate layers in the rail bed leads to uneven settlement of the line. Figure 8.37b compares experimental results of the line depression above a sleeper in dry and wet conditions with increasing cyclic load on the sleeper. The graphs show that water ingress can result in significant changes in track movement. Lateral and vertical movement of the track leads to displacement of the rail lines away from their initially 'optimal positions'. The consequences would be the speed reduction of the line (i.e. the speed at which it is safe for rail traffic to proceed along it) and eventually the removal of the line from service while the track undergoes repair. Drainage is therefore an important factor in maintaining track stability, and various geotextile materials can be used for this purpose.

For example, the ingress of a rising water table can be restricted by a cross-slope drainage geocomposite with a GCL layer or geomembrane placed above to act as a capillary break in the water flow. A second drainage geocomposite above the impermeable layer would then remove surface water penetrating the aggregate. Figure 8.37b also shows experimental results which indicate the effectiveness of a drainage geocomposite comprised of the following sequential layers: non-woven geotextile/geomembrane/geonet/non-woven geotextile for improved track stability in wet conditions; the line depression was reduced by around 80% with the use of the geosynthetics.[107] The geotextiles used in such drainage composites also provide the functions of separation and filtration, similar to the situation for road construction.

8.20 Applications: tunnel construction

Modern tunnel design and construction involves engineered lining systems that provide effective drainage and a long term moisture barrier for the tunnel walls. After the excavation of the tunnel, the rock is sprayed with concrete (shotcrete) to smooth out the unevenness of the rock face. Either a geocomposite or needle-punched non-woven geotextile is then placed on the sprayed concrete and anchored using steel bolts coupled with polymer fastening discs (see Fig. 8.38). The tunnel liner (HDPE membrane) is then welded to the discs. After weld testing a final concrete layer (ring concrete) is put in place. The geotextile performs two primary functions:

- protection of the liner from abrasion and puncture by protrusions from the base surface of the tunnel wall.
- drainage of water seeping through the rock and sprayed concrete layer. The in-plane permeability of the geotextile enables the water to flow to a subsurface drainage collection system, where it is directed away from the tunnel.

The three primary applications of soil reinforcement using geotextiles are:

- reinforcing the base of embankments constructed on very soft foundations,
- increasing the stability and steepness of slopes, and
- reducing the earth pressures behind retaining walls and abutments.

In the first two applications, geotextiles permit construction that otherwise would be cost prohibitive or technically not feasible. With the case for retaining walls, significant cost savings are possible in comparison with conventional retaining wall construction.

Figure 8.39 shows the three ways in which embankments constructed on soft foundations can fail. The application of high strength woven or knitted geotextiles, or geogrids can enhance the bearing capacity. The geotextile can be made sufficiently strong to prevent rotational failures, and lateral

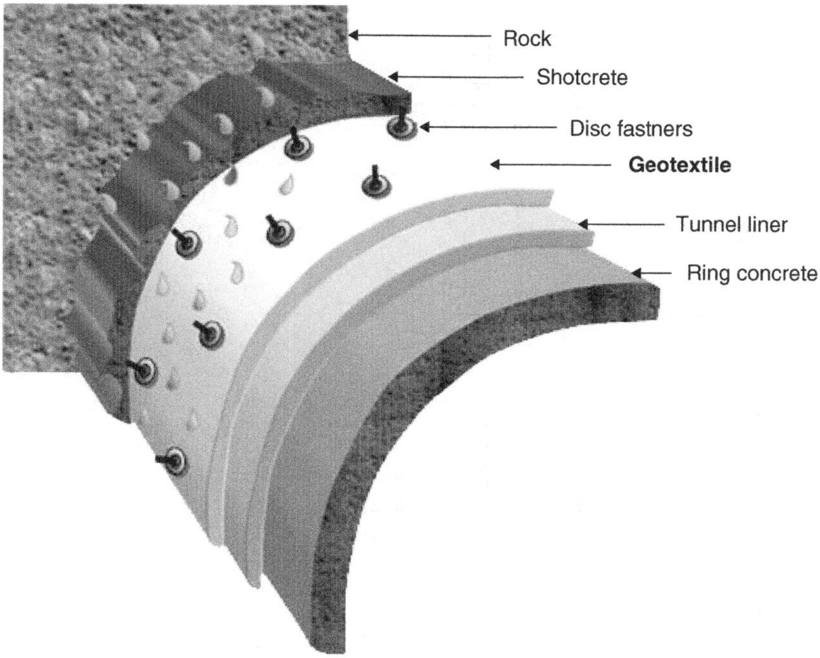

8.38 Tunnel lining system.

spreading failures can be prevented by the development of adequate shearing resistance between the base of the embankment and the reinforcement.[108–110] For some very soft ground, the use of geogrids may require a lightweight geotextile separator to provide filtration and prevent contamination of the material used for the embankment, i.e. the embankment fill. Also, a drainage layer may be needed to ensure ingress of water does not adversely affect the direct shear properties.

Slope reinforcement is also employed in the construction of embankments, and as an alternative to some retaining walls. To provide slope stability, multiple layers of geogrids, or woven/knitted geotextiles, may be placed during construction in the fill material used to make the slope.

Reinforcement in a slope contributes to stability in two ways. Firstly, the reinforcement directly improves the shear resistance of the soil to resist the shear loading caused by the steep face. Secondly, the reinforced zone acts to hold the unreinforced soil mass of the interior in equilibrium without overstressing the underlying foundation soils.

Geosynthetic properties required for reinforced slopes are similar to those for other reinforcements. Allowable tensile strength, T_{allow}, soil and geosynthetic friction, E_g and E_{ga}, and creep strength are the most important. The anticipated half-life of geotextile reinforcements is between 500 and 5000 years;

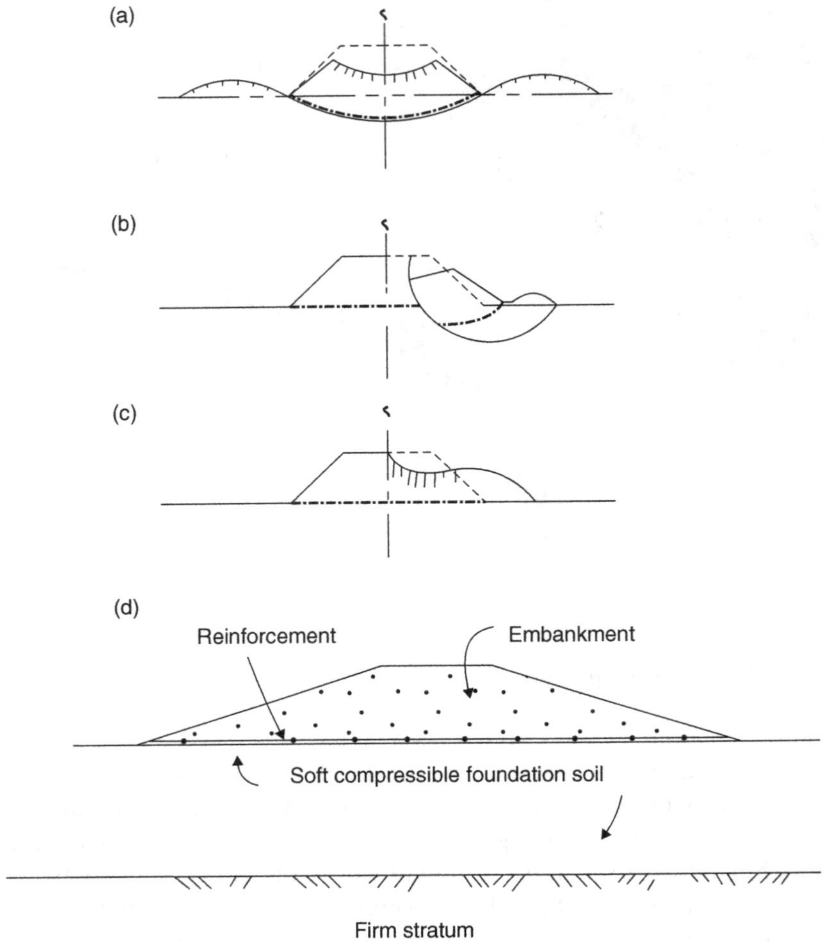

8.39 (a) Bearing failure mode, (b) rotational failure mode, (c) lateral failure mode, and (d) reinforcement.

therefore, the specified values must be adjusted by safety factors covering potential degradation in the specific environmental conditions.[111,112]

A major cost advantage of geotextile reinforcements is that a wider range of slope and compacted cohesive fill-soil types can be used. Usually a granular fill is preferred, especially for the first few lifts above the base geotextile layer. However, various other soil fills can be used, depending on the slope angle required and the type of geotextile to be used.[113] Similar to other situations of reinforcement, filtration, separation, and drainage functions may be needed and a suitable non-woven geotextile with in-plane drainage capabilities would be incorporated.

Retaining walls are generally required where a soil slope is uneconomical or not technically feasible. When compared with conventional retaining

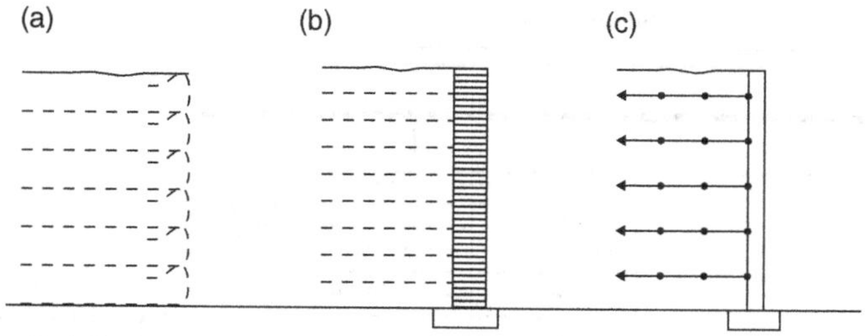

8.40 Geotextile reinforced retaining wall systems : (a) woven fabric wrap-around facing, (b) geogrids with outer edges gripped within segmental or modular concrete block facing, and (c) geotextiles with ends embedded in precast concrete wall facing.

structures, walls with geotextile reinforced backfills offer significant advantages. They are very cost effective, especially for high walls. They are also more flexible than conventional earth retaining walls, such as reinforced concrete cantilever or gravity walls. They are particularly suitable for sites with poor foundations, and for seismically active areas. With conventional soil reinforcement the backfill materials are usually granular. In the case of geotextiles, however, several successful applications have demonstrated the potential of using cohesive soils as backfill materials.[114–116] A further significant benefit of geotextile reinforced retaining walls is that they permit a wider variety of wall facings to be used than traditional systems, resulting in more aesthetic and economic options.

Figure 8.40 illustrates three typical geotextiles reinforced retaining wall systems: (a) woven fabric wrap-around facing, (b) geogrids with outer edges gripped within segmental or modular concrete block facing, and (c) geotextiles with ends embedded in precast concrete wall facing.

There are two possible failure modes for such geotextile systems: rupture of the geotextile itself, and pullout of the material from the soil. Similar to other geotextile reinforcing applications, the corresponding reinforcement properties of importance are the geotextile tensile strength, creep, surface friction, and pullout resistance. To circumvent the last case, the geosynthetic reinforcement must extend some distance behind the assumed critical length so that it will not be pulled out of the backfill. Other important considerations include drainage and potential seismic loading.

Backfills for geosynthetic reinforced walls should be free draining, because unexpected failure of these structures can be caused by lack of control of groundwater or seepage. An appropriate geocomposite can therefore be incorporated horizontally and/or vertically to provide adequate drainage.

8.21 Applications: reinforced soil structures

The seismic behaviour of reinforced soil structures has been studied by 'shaking table tests', where a near full-size geotextile reinforced wall is fitted

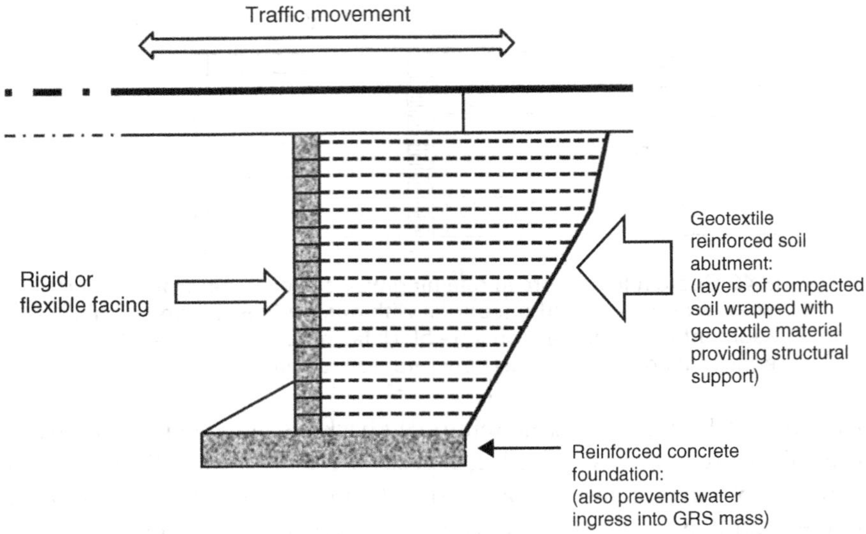

8.41 GRS bridge abutment.

with sensors to measure strain, pore-water pressure and seismic acceleration.[117–120] Results have shown significant benefits compared with conventional retaining walls. With conventional walls there is strong amplification of the bedrock vibration, which is transmitted to the top of the wall. However, geotextile reinforced walls have a damping effect, owing to multiple layers of fabric–soil friction (i.e. an interlayer Coulomb-type frictional damping) in addition to the visco-elastic damping effect of the geotextile itself.[121]

A geotextile reinforcing supporting (GRS) bridge abutment is essentially a GRS mass with a facing (see Fig. 8.41), and is therefore similar to a geotextile reinforced retaining wall. The facings used are of two types: 'rigid', and 'flexible'. A 'rigid' facing is of continuous reinforced concrete, either precast concrete blocks or cast-in-place. A 'flexible' facing GRS system typically takes the form of natural rocks or gabions. Although the facings give little structural aid, a 'rigid' facing, in contrast to a 'flexible' facing, will offer some degree of 'global' bending resistance along the entire height of the facing.

Similar to geotextile reinforced walls, these structures have distinct advantages over their conventional counterparts. They are more flexible (hence more tolerant to differential settlement), adaptable to low-cost backfill, easier to construct, require less overall excavation, and are more economical to erect.[122–125]

8.22 Applications: fluid containment systems

Fluid containment in the present context means fresh water, liquid waste including that leached from solid waste, and biodegradation gases such as methane. Geosynthetic lining systems for water reservoirs, dams, canals, ponds, and irrigation channels can conserve millions of gallons of water by preventing seepage losses. These lining systems can eliminate 70% of the water losses that occur between the storage and usage points. Additionally, for small scale storage units such as ponds, lining covers can prevent contamination, control evaporation, and prevent chlorine loss. Generally, therefore, facilities capable of storing 25 000 m^3 of water and above have some form of synthetic lining system as an alternative water barrier to conventional clay liner systems.

Various plastic liners may be used such as butyl rubber,[126] PP, PVC, and HDPE.[127, 128] When these types of liner are employed, it is necessary to utilise the protection function of a needle-punched non-woven geotextile above and below the plastic sheet, with a layer of gravel or other erosion resistant mineral material on the upper geotextile. A GCL may be used as a secondary barrier liner beneath the plastic liner in addition to the non-woven geotextile protector. For non-critical water storage systems, such as architectural ponds, and recreational ponds, GCLs are used as alternatives to plastic membranes owing to their self-sealing ability as puncture repair.

One of the most important applications of lining systems is landfill. A lined landfill is a carefully engineered excavation in the ground in which waste is contained. At the bottom of the excavation would be a geomembrane liner, called a basal liner, to minimise ground and surface water contamination by the waste. There are several categories of waste: municipal solid waste (MSW) comprising consumer and household items; hazardous waste (e.g. toxic chemicals and flammable, radioactive, or biological substances); industrial waste (i.e. waste chemicals, paints, building demolition residues); agricultural wastes, such as animal manure and crop residue; and mining wastes, such as slag heaps and coal refuse piles. At mine sites, basal liners are used for liquid containment (drainage waters, process solutions, treatment ponds), as well as tailings impoundments, heap-leach facilities, and other solid waste amenities. The principal difference between the landfill constructions for these categories of waste is that MSW landfills generally have one basal liner whereas the others have multiple basal liners. The upper (primary) basal liner (or liners) performs the principal barrier function to contamination of the neighbouring ground environment and surface water, while the lower liner acts as both a secondary barrier and leak-detection system. If local conditions require it (e.g. high water table) a double liner system may be used in MSW landfills.

The contamination risk is usually caused by leachate, which is the liquid that is drained from the landfill basal layer. Leachate forms during the operation of the landfill site mainly by rain water percolating through the decaying solid waste. With the body of the waste of an active landfill average temperatures can be between 40°C and 50°C, reaching up to 70°C in the basal area. Therefore, as the water moves through the waste, the thermal

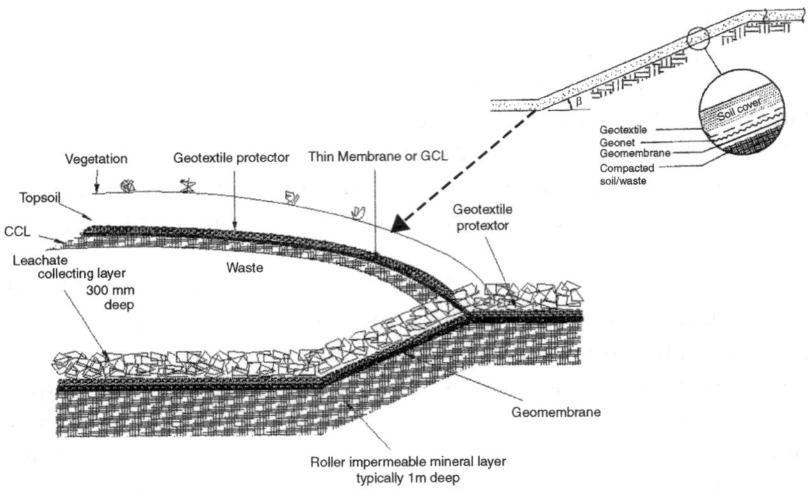

8.42 Cross-section of basal liner and capping in MSW landfill systems.

energy can be sufficient for it to slowly dissolve metals, organic compounds, and other contaminants, collecting to form toxic liquid.

Figure 8.42 gives a cross-sectional diagram of a typical MSW landfill construction. As shown, an HDPE membrane is placed on a compacted soil layer (compacted clay); optionally, a non-woven geotextile protector, a drainage geocomposite (for leakage detection) or a GCL may be present beneath the plastic membrane. This type of basal layer is referred to as a single composite (compacted clay plus geomembrane) liner.

A needle-punched non-woven geotextile protector (typically 1200–2000 gsm) is placed on top of the membrane, and on top of this is laid a 300 mm thick stone drainage layer (usually a deposit of 20 mm semi-angular lime stones), through which the leachate can be tapped off from the waste. The non-woven protects the membrane from installation damage and from stress cracking due to long term compression by stone projections at the drainage interface. Sometimes other materials are used for the protective function, such as sand-filled geotextile blankets.

A second, but thin, non-woven geotextile is placed above the drainage layer as a separator/filter to prevent waste particles, particularly fine waste, caught in the downward flow of the leachate, entering the granular drainage layer. Typical geotextiles used above the drainage system are spun-bonded non-woven PET, 271–542 gsm or 7–32% open area PP wovens. It is often proposed to replace the granular drainage layer and geotextile filter with a drainage geocomposite layer, but this is not a widely used option.

The granular drain system for leachate removal must remain continuously effective. This means the geotextile separator/filter must have a suitable cross-plane permeability to give a $v_{allowable}$ value that will accommodate potential clogging. Some clogging of this filter media may be expected. However, clogging must not build up and eventually block the flow to the underlying granular layer. Otherwise the leachate will pool into the waste – referred to as 'perched leachate' – and may accumulate to reach a level where it threatens the stability of the landfill.[129] As explained earlier, filter efficiency can decline through particle, biological and chemical/precipitation clogging, and the reducing factors (RF_{SCB}, RF_{BD}, RF_{CC}) should be determined by testing.

On the slopes of the landfill, a geonet may be used for reinforcement of the basal liner against the angled load of the waste. Analyses of past slope failures have shown that liner-induced failures occurred at the geomembrane interface with under- or over-lying materials. The geonet is therefore placed above the non-woven geotextile to avoid slippage of the geonet on the membrane. Texturised geomembrane can improve the resistance of the non-woven to slippage.

Figure 8.42 also depicts the capping structure of the landfill. The primary aim of landfill capping is to control and minimise leachate generation by restricting water ingress into the landfill. A low permeability domed cap is therefore constructed over the whole site to achieve the highest possible surface water run-off. The cap is also constructed to facilitate landfill gas control and collection. Landfill gas can be as polluting as leachate and might also be dangerous, if not controlled, causing risk of fire and explosion. Geopipes are usually installed to collect and remove the generated gases from the waste.

The capping of an MSW landfill is usually done by incorporating a non-woven geotextile separator/filter above the waste, and then geopipes enveloped by gravel with a second non-woven geotextile filter above the gravel. This combination is covered with a layer of compacted clay, on which is placed a geomembrane, followed by a needle-punched non-woven, a granular drainage system or a substitute drainage geocomposite, and then a geonet, and finally soil for vegetation. The cap must be built to withstand agricultural machinery, drying and cracking, plant root penetration, burrowing by animals and erosion.

The steepness of capping system slopes ranges from 4:1 (14°) to 2.5:1 (22°), and the interface friction angles between adjacent geosynthetics can be from 8° to 25°.[130,131] The soil load, with the combination of low interface friction and relatively steep side slopes, creates potential slip surfaces that can cause instability of the soil cover and tension cracks in the membrane. Cap failures can therefore occur as a result of surface sliding on the geomembrane and by vertical collapse of decomposing underlying wastes.[132] Similar to situations for basal slopes, geogrids may be used to provide structural reinforcement to soil layers of closure caps and relieve tensile stress on the underlying geomembrane.[133] Generally, therefore, geotextiles in landfill construction have the functions of protection, filtration, drainage, reinforcement, and stabilisation via interface resistance.

8.23 Applications: erosion control

Erosion can result in major damage to coastal areas, river banks, and ground slopes, particularly where the vegetation hold is poor or there is an absence of vegetation. The basic principle of erosion control[134] is to prevent or limit soil movement by the erosive forces such as moving water and driving wind. The control techniques typically involve the use of armoured protection with geotextile support for tidal situations (i.e. coastal and rivers) that occur either naturally or through water movement induced by marine transport. Geotextiles are also employed for the containment of silt and the retention of soil cover, and revegetation on steep slopes. With the former, only synthetic materials are effective as long term solutions, whereas with the latter both natural and synthetic materials are used for erosion control.

8.23.1 River embankment and coastal erosion

Protection from tidal and wave erosion commonly involves the use of revetments constructed from rock armour or precast concrete units on the bank or shoreline (see Fig. 8.43), the size of which is determined by the anticipated wave action. A filter layer is required beneath these constructions in order to prevent the progressive removal of the underlying soil by the infusion of water, which would result in the collapse of the revetment into the increasing void and the loss of its effectiveness against the tidal action. Traditionally, granular mineral filter layers were used, one metre thick and graduated so that progressively larger grain sizes are nearer the rock or concrete material, as illustrated in Fig. 8.43. Installation of such filter layers is difficult and time consuming, particularly if this involves inter-tidal working. The filter layers must allow the free movement of water in both directions, i.e. repeatedly into the shore or banking and then out again, in accordance with high and low tide, at the same time preventing the leaching of the underlying soil. Without the ability to facilitate this action over the entire life of the revetment, there is the potential for the armour to be undermined, as beach material is progressively eroded, or for a build-up in hydrostatic pressure, which then leads to the failure of the revetment. Besides the cost and construction difficulty, the traditional filters do not provide good long term performance. The use of geotextile filters as replacements for traditional filters, see Fig. 8.43, enables better restriction of movement of the underlying soil, and thereby its retain-

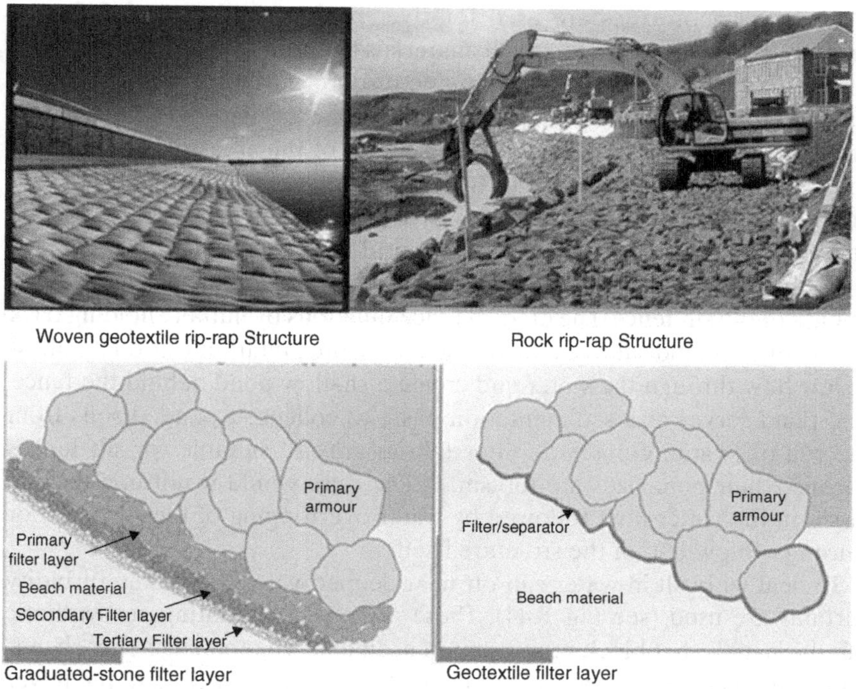

8.43 Application of geotextiles in coastal erosion protection.

ment.[135] The repeated reversed action of the water flow reduces the chances of the filter blocking, and so gives the geotextile long-time performance.

In addition to good filtering efficiency, the geotextile is required to have a high level of stress absorption and high resistance to abrasion. The rock weight, its angularity, and the drop height all contribute to puncture loads. The geotextile must therefore be able to withstand the puncture loads imposed during installation and in service. The fabric should be of sufficient thickness to cushion potentially penetrating point loads and also to provide a lateral drainage path (in-plane permeability) around any compressed areas. The most suitable material is a needle-punched non-woven with almost equal tensile properties in the machine and cross-machine directions, as this would uniformly distribute any tensile loads.

Fabric formed revetments (FFR) can be used as an alternative to rock rip-rap structures, as shown in Fig. 8.43. They are constructed by pumping a very fluid fine-aggregate grout (i.e. cast-in-place concrete) into a woven fabric envelope consisting of two layers connected by tie-chords or by interweaving. FFRs can be engineered to perform as impermeable or permeable wave barriers.

Although geotextiles are used for river banking and coastal defences, there is also a need to use such materials to protect near-shore waters, streams, rivers, lakes, and other aquatic resources from contamination by silt and sediment forming debris. Silt fences are a common solution to such problems and are placed along the perimeter of the areas of the polluting source; typically, they are 1 m (3 ft) high and 30 m (100 ft) long. A PP slit film or flat-tape woven geotextile, which is permeable, is usually used by attaching it to supporting posts that are firmly inserted into the ground. The fabric is usually wider than the wood post length to allow burial at the toe or base of the silt fence. The erected fence functions by initially holding back sediments (silt and sand particles) carried by precipitation run-off, to allow a slow flow through the water and create a shallow pond behind the fence. The pond serves as a sedimentation basin to collect suspended soils from the run-off water. Geotextile filter fabrics are not suitable for silt fences because their pore sizes are too small. Their use would result in a backup of silt in the water flow, followed by water overtopping of the silt fence or causing a 'blow-out' of the structure itself.

To deal with silt in water run-off near deeper water shorelines, turbidity curtains are used (see Fig. 8.44). These are reusable floating geosynthetic tubular panels that block movement of sediment. The top edge of each curtain contains floats, and a cable or chain. Weights are attached to the lower edge of the curtain to keep it vertical in the water. Posts, piling, or anchors hold the curtains in place. Generally, they are intended for use with currents no greater than about 1.5 m/s and to depths of no more than 5–8 m.

8.23.2 Soil erosion

Inland soil erosion follows a process of detachment and transportation of soil particles by wind and rain water under the influence of gravity. Wind erosion is a problem largely in open areas (lack of shelter) – exposed slopes of non-cohesive (loose and dry) soils, or exposed smooth bare surfaces with

8.44 Geotextile turbidity curtain.

little, if any, protective vegetation. Wind erosion occurs when air turbulence imparts sufficient energy to initiate movement and impact of smaller size gains on heavier size soil particles. In geology this action is termed saltation; saltation grains are 0.05–0.5 mm dia. in size, but can move heavier sized particles of up to 200 times their own weight by force of impact, giving what is referred to as surface creep – i.e. slow movement of the larger soil surface particles of 0.05–1 mm dia. Some of the salting particles may become suspended in the airflow to be eventually deposited large distances away from their original grounds.

Ground erodibility or susceptibility by wind, and by water, is dependent on the type of soil, i.e. the particle size, the clay content, and other factors as the existing moisture level, density, compaction and, therefore, permeability. Wind erodibility is low for particles sizes of 0.84–6.4 mm dia or larger, e.g. coarse gravel, but high for silt and fine sands. Soils largely comprising particles < 0.43 mm will be highly susceptible to erosion; at 0.42–0.84 mm dia the susceptibility is significantly lower. However, erodibility decreases with low void ratios and high antecedent moisture content, i.e. with increasing clay and organic content.

Water erosion is generally more destructive than wind erosion. The movement of falling or running water begins the erosion process by moving soil particles. Rain drops will, through conversion of kinetic energy to impact energy, loosen soil particles in the surface layer of the ground, enabling them to travel with the rainwater flow. Figure 8.45 illustrates the process of rain erosion of a slope. For a certain length at the top of the slope, there is no erosion. Beyond this initial zone, water starts picking up soil particles along the slope and the erosion begins. The concentration of soil material in the water increases as the flow travels down the slope until some location is reached (*b-b′*) beyond which the erosion of the lower part of the slope length may be partially offset by material deposition from the flow. If the flow continues to the lowest point of the slope, most of the eroded mass within the flow will enter the basin at the bottom of the slope, which may be a stream, a river or part of a shoreline.

Rain intensity can range from 2l/sqm/h for drizzle, with a 0.9 mm droplet and intensity of < 1 mm/h,–4000 J/sqm/h for a cloud burst of 4 mm droplet and 100 mm/h intensity. The higher viscosity and density of water enable low velocity flows to start erosion by rolling semi-round and round soil particles down any sloping surface. If the flow is concentrated, then small rivulets initially form and, as the water becomes greater in volume and more prolonged, the rivulets will get larger. With the repeat of significant amounts of rainfall over a period of time, gullies will form and ultimately the eroded surface may be turned into a ditch, which widens and deepens in the process. If not arrested in the initial stages, the sequential chain of erosion can impair ground slope stability, where the formation of gullies leads to catastrophic failure.

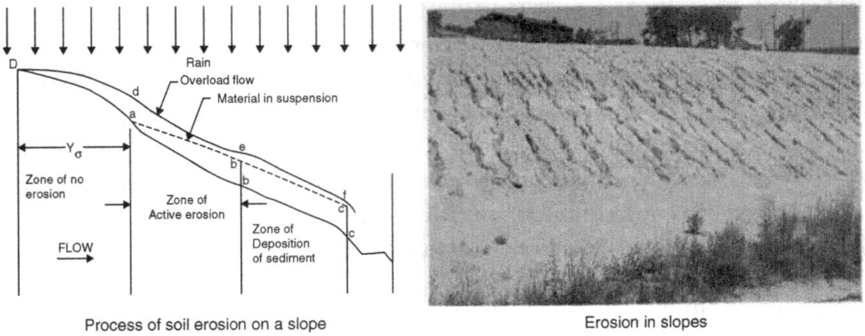

8.45 The process of rain erosion of a slope.

The solution is to create a cover for the soil that keeps the water and wind from eroding the soil particles. In many cases vegetation alone will not provide an effective long term erosion protective cover, particularly over steepened slopes. During initial water flow loose vegetation would be removed by the drag force of the flowing water. Locally, flow can slowly scour soil away from the roots of a plant, thereby weakening its anchorage until the plant itself is removed by the drag force of the flow.

It is evident from the above descriptions that the problems of wind and water erosion vary depending on location, soil conditions, as well as with the physical parameters of the wind and rain. A range of geotextile materials has therefore been developed as solutions for such situations as: bare soil, where there is a need for revegetation; the support of vegetation in erodible soil; temporary biodegradable support for new seedlings; and long term permanent non-biodegradable support.

Where biodegradability is a requirement, jute and coir are the most used natural fibres. This is because, as the vegetation growth becomes established, the degraded by-products from these fibres are beneficial to the plants. The ability of jute and coir fibres to absorb water aids in attenuating soil movement by water run-off and ameliorates high variation of soil moisture content. This is in addition to the regulating effect of soil temperature variation that erosion control geotextiles in general have. For more longer-term permanency (defined as retaining 75% of its original strength after 10 years life) the usual synthetics, PP, PET, and PA are used, the fibre being made with additives for UV resistance.

Geotextile erosion control products may be classed as erosion control nets (ECN), meshes (ECM), blankets (ECB), and turf reinforced mats (TRM). These textile structures are required to meet two conflicting requirements. They must have good flexibility to conform to the ground undulations, thereby giving good ground cover, and also have rigidity to prevent soil movement. Conflict of these requirements results from the fact that good flexibility is associated with high stretch characteristics, and that high stretch is a disadvantage in preventing soil movement during large volume water flow rates over steepened slopes. Such flow rates would tend to stretch flexible structures, loosening their hold on some soil particles with potential for progressive erosion to occur. However, if the stiffness of the fabric inhibits good ground conformability, water flow will scour beneath the fabric where it is not in contact with the soil. Needle-punched non-wovens, and plain open-weave and mesh structures are widely used to provide flexibility while retaining good elastic resistance. Care must nevertheless be taken in respect of the degree of openness because of the reduced cover factor for protection against rainfall impact on the soil.

ECN and ECM are open-weave structures whereas ECB are needle-punched non-wovens. They are the most commonly used temporary, degradable systems. An improvement on these structures is organic fibre-filled blankets (OFB), which are essentially jute or coir fibres sewn into or placed between synthetic nettings. They give increased ground cover and better resistance to rainfall impact. The installation life of OFBs is from 4 to 24 months. Their biodegradability depends on the choice of nettings, fibre type, and chemical nature of the rain and soil.

A further development of the OFB concept is erosion control and revegetation blankets (ECRB). These are a combination of PA or PP staple fibre bound within two PP or PA nets. They have similar ground cover characteristics to the OFBs. However, the use of ultraviolet stabilised fibres provides for a much longer installation life than the organic blankets.

TRM structures are composed of fused polymer nettings, made of randomly laid monofilaments, or yarns woven or tufted into an open, 3-dimensionally stable but flexible mat. Soil filling is carried out during installation and the weight of the soil fill ensures the mat conforms to undulations in the soil surface. They are designed to retain seed and soil cover, and therefore to entangle with the root and stem of the plants as the vegetation becomes established.

A variation of the TRM is a synthetic matting called an erosion control and revegetation mat (ECRM). This is made of a plasticised PVC monofilament that is continuously extruded and formed into a non-woven fabric. ECRMs are usually 6 mm thick with a porosity of about 85%. Stiffness of the ECRM is significantly reduced by using a monofilament which enables good ground surface conformity without soil filling. Seeding of the soil occurs before the ECRM is installed, although hydro-seeding can be done after mat placement.

8.24 Future trends

With the continuing growth of the world's population, there will be continuing demand on land space for infrastructure development and the built environment. This will conflict with the need to safeguard the natural environment and its resources, while accommodating changing weather patterns through flood defences and erosion control. Geotextiles, along with other geosynthetics, will play a major part in the future application of geotechnical engineering to meet these anticipated challenges ahead. Future developments of geosynthetic materials need to have enhanced functionality but yet a reduced carbon foot-print, enabling them to be used in even wider range of applications.

The current trends can be seen in the development of material structures to give better performance in certain specific functionalities. For example, new challenges for non-woven geotextiles in application areas such as filtration require media designed to meet particular small pore-size specifications in the 3–20 micron range. To achieve small pore sizes requires small diameter fibres of the order of 1–10 microns, which are difficult to control during conversion into bulk structures. Studies have shown that bicomponent melt-spinning fibre technology can be adapted to enhance the geotextile filtration function. The particular bicomponent fibre spinning method known as 'island-in-the-sea' produces relatively coarse fibres (1–5 denier), each made up of two polymers. Within each coarse fibre, one of the polymers forms a multiple of small diameter fibres while the other polymer forms a matrix binding them together to produce the coarse fibre. A cross-sectional view of the coarse fibre diameter (90° to the fibre axis) shows the multiple of small diameter fibres, as 'islands' and the matrix as the 'sea'. The coarse fibres can be processed as continuous filaments to make spun-bonded non-woven geotextiles or chopped into staple fibres for carded and needle-punched products. In either case, the sea polymer is eventually removed by dissolving it with a solvent or other means, leaving behind the small diameter fibres, enabling the resultant fabric to achieve targeted pore diameters and porosity.

Coarse and fine fibres have also been used in a two-layer media to be used as high performance filters in rock rip-rap structures for coastal protection. The fine diameter fibres provide the pore size which retains soil particles, while the coarser fibres act as a protective/reinforcing layer against mechanical stresses form rock movements. Such multifunctional filter systems can be a two-layer non-woven combination or the coarser fibre layer could be a monofilament woven geotextiles. Melt blowing technology enables the production of a fine fibre layer on a coarse fibre ground fabric. However, the on-going developments in nanofibre technology may facilitate two or three-layer geotextile filter media with further improvements in performance.

The approach of multilayering of coarse and fine fibres has also been applied for enhanced drainage functionality. One development is a three-layer non-woven geotextile which is claimed to give a better performance than drainage geocomposites with a geonet core.[136] The three-layer non-woven structure is a sandwich of thick non-woven core made from coarse fibres with top and bottom non-woven layers made of finer fibres. The improvement in performance results from the higher in-plane permeability obtained under increased normal load and the greater resistance to soil intrusion into the drainage channels. Again, variations on this three-layer structure could lead to other novel geotextile composites where coarse woven monofilaments in a pile-type structure is heat welded and sandwiched between woven finer filament layers or non-woven layers.

Important to filtration and drainage application is the minimising of clogging, in particular biological clogging. Although the above multilayered structures may be designed to accommodate this problem, there is interest in the possibilities of modifying the surface chemistry of fibres to retard the onset of biofilm formation which may extend the operating life of filters and drainage systems.

Environmental applications of geotextiles are likely to grow in importance, not only in areas of erosion control but also in geotechnical reclamation, pollution control, and prevention. Developments are on-going for better ECB and TRMs involving natural fibres and 3D structures of woven, warp knitted or non-woven synthetic scaffolds. Plantable reinforcement geonets have also been developed, combining jute or coir non-woven geotextiles with high tenacity PET geotextile grids to offer not only reinforcement for wall construction but also for the construction of walls with plantable facings to initiate rapid vegetation cover. Geotextile structures are also being designed to assist plant growth in arid climates where intense evaporation draws saline water to the top soil from the water table by capillary action; the subsequent evaporation increases salt concentration and precipitation leaving salt in the soil. The geotextile is made to disrupt the rapid capillary flow and reduce the rate of evaporation.

The potential development of electrokinetic geotextiles offers an opportunity to increase the redevelopment of derelict land as a means of reducing the pressure for construction within the natural environment. This electrically conductive material enhances the filtration and drainage functionalities by electro-osmosis and ion migration. Various types of conductive component can be incorporated in the geotextile structure to achieve this effect, but the possible use of activated carbon fibre or similar additives with pollution sorptive surface properties can aid land recovery of polluted areas.

Geotextiles with activated carbon in combination with titanium dioxide (TiO_2) are also being developed.[137] These materials are aimed at pollution control, and are claimed to be effective for the degradation of organic pollutants in gases and liquid effluents.[138] The activated carbon absorbs the pollution while the TiO_2 oxidises the organics by photocatalysis, their full oxidation resulting in CO_2 and H_2O. Photocatalysis degrades the majority of organic pollutants in gaseous or liquid effluents such that geotextile materials can be used in numerous applications for protecting local environments, e.g. covering of lagoon ponds, composting stations, water cleaning stations, agricultural or food industry effluents.

Geotextile installations involve very large surface areas and the possibility of incorporating sensory devices into these textile structures has attracted much attention. Sensory geotextiles are being designed, which can be used simultaneously for separation, filtration, protection, drainage and, after their installation, permit the pinpointing of failure places in their structures. For example, in reinforcement applications these materials can monitor movements and stresses via the sensors incorporated within their structures. Any registered significant changes will also give the location of the faults. The types of sensors used are polymeric optical fibres, based on the principle of optical time domain reflectometry (OTDR). Short light pulses are emitted into one end of the fibres and the intensity of the light reflected back by a fault is detected and measured, along with the time of reception. The precise location of the fault can then be determined from the time interval between the pulse transmission and the reception of the reflected light signal, and the speed of light in the fibre. Typical structures that are so far considered suitable for such sensory devices are geogrids, in particular geocomposite non-woven–warp-knit–grid structures

It can be seen from the above descriptions that future developments of geotextiles will be wide and varied, driven not only by the demand for improvements in performance, but also by the desire of added multi-functionalities for solutions to new geotechnical problems, particularly with the on-going changes to global climate.

8.25 References

1. Pearsall, J. (Ed) (2001), *New Oxford Dictionary of English*, Oxford University Press, New York, USA.
2. Copplestone, T. (1963), *World Architecture*, Hamlyn, Essex, p. 178.
3. Somerset Historic Environment Record, Somerset County Council, Retrieved 19 June 2010.
4. Ingold, T. S. Dr (1994), *Geotextiles and Geomembranes Manual*, Elsevier, Oxford.
5. Koerner, R. M. (1997), *Designing with Geosynthetics*, Prentice Hall, New Jersey, USA.
6. Giroud, J. P. and Noiray, L. (1981), Design of geotextile reinforced unpaved roads, *Journal of Geotechnical Engineering Division, ASCE*, **107**, No. GT9, pp. 1233–1254.
7. Saville B. P. (1999), *Physical Testing of Textiles*, Chapter 5, Woodhead Publishing Limited, Cambridge England, p. 145.
8. Morton, W. E. and Hearle, J. W. S (1962), *Physical Properties of Textile Fibres*, Chapter 13, Butterworths, London, UK.

9. Shercliff, D. (1998), A Methodology for Cylinder Testing of Protectors for Geomembranes, A performance test to determine the protection afforded to a geomembrane by a geotextile or other protective materials, in specific site conditions, UK Environment Agency, March 1998.
10. Construction Products Directive – 89/106/EEC.
11. Datye K. R. and Gore, V. N. (1994) Application of natural geotextiles and related products, *Geotextiles and Geomemhranes,* **13**, 371–388.
12. Lekha K. R. and Kavitha, V. (2006), Coir geotextile reinforced clay dykes for drainage of low-lying areas, *Geotextiles and Geomembranes,* **24,** 38–51.
13. Mauersberger, H. R. (1947), *Matthews Textile Fibres: Their Physical, Microscopical and Chemical Properties*, Chapters 9, 10, 11, John Wiley & Sons, London, UK.
14. Fourne, F. (1999), *Synthetic Fibres: Machines and Equipment, Manufacture, Properties*, Hanser Publishers, Munich, Germany.
15. Heger, F. J., Chambers, R. and Dietz, A. G. H. (1984), Structural Plastics Design Manual, *ASCE Manual on Engineering Practice No. 63,* Task Committee on Design of the Structural Plastics Research Council, ASCE, New York.
16. Ogorkiewicz, R. W. (1947), *Thermoplastics Properties and Design,* John Wiley and Sons Ltd, Chichester.
17. Truss, R. W., Duckett, R. A. and Ward, I. M (1981), Effect of hydrostatic pressure on the yield and fracture of polyethylene in tension, *Journal of Material Science,* **16**(6), 1689–1699.
18. Gotham, K. V. (1972), Long-term strength of thermoplastics: the ductile-brittle transition in static fatigue, *Plastics and Polymers,* April 1972, **40**, 59–64.
19. Lockett, F. J. (1982), *Engineering Design Basis for Plastics Products,* National Physical Laboratory, HMSO, London.
20. Heger, F. J., Chambers, R. and Dietz, A. G. H. (1984), Structural Plastics Design Manual, *ASCE Manual on Engineering Practice No. 63,* Task Committee on Design of the Structural Plastics Research Council, ASCE, New York, p. 1176.
21. Kay, D., Blond, E. and Mlynarek, J. (2004), Geosynthetics durability: a polymer chemistry issue. *57th Canadian Geotechnical Conference/5th Joint CGS/IAH Conference*, Quebec City, Quebec, 24–27 October 2004. Session 4D, pp. 1–14.
22. Lupo, J. F. and Morrison K. F. (2005), Innovative geosynthetic liner design approaches and construction in the mining industry. *2005 GeoFrontiers Conference*. Austin, Texas. 24–26 January 2005.
23. Lawrence, C. A. and Collier, B. (2005), *Natural Geotextiles in Biodegradable and Sustainable Fibres*, Chapter 11, Ed. Blackburn, R. S., Woodhead Publishing Limited, pp. 343–365.
24. Fourne, F. (1998), *Syntheic Fibres: Machines and Equipment, Manufacture and Properties*, Chapter 5, Hanser Publishers, Munich, Germany, pp. 554–564.
25. Communications in Statistics: Series C Stochastic Models, **12**(2), 199–214, 1996.
26. Simmonds E., Bomberger, J. D. and Bryner M. A. (2007), Designing non-wovens to meet pore size specifications, *Journal of Engineered Fibers and Fabrics*, **2**(1), 1–15.
27. Lombard, G., Rollin. A. and Wolff, C. (1989), Theoretical and experimental opening sizes of heat-bonded geotextiles, *Textile Research Journal*, **59**(4), 208–217.

28. Faure, Y. H, Gourc, J. P. and Gendrin, P. (1990), 'Structural Study of Porometry and Filtration Opening Size of Geotextiles', Geosynthetics: Microstructure and Performance, Peggs, L. D., Editor, ASTM Special Technical Publication 1076, pp. 102–119.
29. Lifshutz, N. (2005), On the mean flow pore distribution of microfiber and nanofiber webs, *International Non-wovens Journal*, **14** (1), 18–24.
30. Miles, R. E. (1964), Random polygons determined by lines in a plane. *Proceedings of the National Academy of Sciences USA*, pp. 901–907 and pp. 1157–1160.
31. Kothari, V. K and Das, A. (1992), Compressional behavior of non-woven geotextiles, *Geotextiles and Geomembranes*, **11**, 235–253.
32. Kothari, V. K and Das, A. (1993), Compressional behaiour of layered needle-punched non-woven geotextiles, *Geomembranes*, **12**, 179–191.
33. Palmeira, E. M. and Fannin, R. J. (1998), A methodology for the evaluation of geotextile pore opening sizes under confining pressure, *Geosynthetics International*, **5**(3), 347–357.
34. *Scheidegger Physics of Flow Through Porous Media* (1957), University of Toronto Press.
35. Happel, J. (1959), Viscous Flow Relative to Arrays of Cylinders, American Institute of Chemical Engineers, 5, June, 174–177.
36. Davis, C. N. (1973), *Air Filtration*, Academic Press Inc., London, UK.
37. Wildhaber, J. (1999), Geotextiles – attributes applications and manufacture, *Non-wovens and Industrial Textiles*, No. 4, 42–46.
38. Swan, R. H. (1987), The influence of fabric geometry on soil/geotextile shear strength, *Geotextiles & Geomembranes*, **6**, 81–87.
39. The Textile Institute (1995), Textile Terms and Definitions, 10th Edition, p. 79.
40. Robinson, A. T. C. and Marks, B. (1973), *Woven Cloth Construction*, The Textile Institute, p. 25.
41. Anadurai, S. P., Potluri, P. and Whyte, I. L. (1999), Modelling the in-plane permeability of woven geotextiles, *Textile Research Journal*, **69** (5), 345–351.
42. Peirce, F. T. (1947), Geometrical principles applicable to the design of functional fabrics, *Textile Research Journal*, **17**, 123–134.
43. Rankilor, P. R. and Raz, S. (1989), *The Karl Mayer Guide to Geotextiles*, Karl Mayer, Textilmaschinenfabrik Gmh.
44. McMahon, W. (1959), Degradation studies of polyethylene terephthalate. *Journal of Chemical & Engineering Data,* **4**(1), 57.
45. Davis, A. and Sims, D. (1983), *Weathering of Polymers,* Applied Science, New York, USA, pp. 179–186.
46. Grassie, N. and Scott, G. (1985), *Polymer Degradation and Stabilization,* CUP, London, UK, pp. 213–216.
47. Schneider, H. and Groh, M. (1987), An analysis of the durability problems of geotextiles. *Geosynthetics Conferences Proceedings.* Industrial Fabrics Association International, MN, USA, pp. 434–441.
48. Giroud, J. P. (1984), Analysis of stresses and elongation in geomembranes. In *Proceedings of International Conference on Geomembranes,* Vol. **II**. Industrial Fabrics Association International, St Paul, Minnesota, pp. 481–484.
49. Heger, F. J., Chambers, R. and Dietz, A. G. H. (1984), Structural Plastics Design Manual, *ASCE Manual on Engineering Practice No. 63,* Task Committee on Design of the Structural Plastics Research Council, ASCE, New York p. 1176.

50. Truss, R. W., Duckett, R. A. and Ward, I. M. (1981), Effect of hydrostatic pressure on the yield and fracture of polyethylene in tension, *Journal of Material Science*, **16,** 439–450.
51. Petrov, R. J. and Rowe, R. K. (1997), Geosynthetic clay liner (GCL) – chemical compatibility by hydraulic conductivity testing and factors impacting its performance, *Canadian Geotechnical Journal,* **34**, 863–885.
52. Koerner, R. M. and Daniel, D. E. (1997), *Final Covers for Solid Waste Landfills and Abandoned Dumps*, Thomas Telford, London, U.K.
53. Melchior, S. (2002), Field studies and excavations of geosynthetic clay liners in landfill covers. In *Clay Geosynthetic Barriers*. Zanzinger, Koerner and Gartung (Eds.). A.A. Balkema Publishers. Lisse. ISBN 90 5809 380 8, pp. 321–330.
54. James, A., Fullerton, D. and Drake, R. (1997), Field performance of GCL under ion exchange conditions, *Journal of Geotechnical and Geoenvironmental Engineering, ASCE*, **123**(10), 897–901.
55. Shen, X. (1996), An Application of Needle Punched Nonwoves in Wet Press Casting of Concrete, PhD Thesis, The University of Leeds, November 1996, p. 24.
56. Smith, J. L., Bhatia, S. K., Ridgway, J. and Hawkins, W. (1999), 'Are porosity and O_{95} important in the retention and particulate clogging behaviour of geotextiles?'. *Geosynthetics'99, Conference Proceedings*, Vol. **2**, pp. 813–832.
57. Koerner, R. M. (1990), *Designing with Geotextiles*, 2nd Edition, Prentice Hall, New Jersey, USA, p. 652.
58. Smith, J. L., Bhatia, S. K., Ridgway, J. and Hawkins, W. (1999) 'Are porosity and O_{95} important in the retention and particulate clogging behaviour of geotextiles?', *Geosynthetics'99, Conference Proceedings*, Vol. **2**, pp. 813–832.
59. Calhoun, C. C. (1972), Development of Design Criteria and Acceptance of Specifications for Plastic Filter Cloth, US Army Corps of Engineers W.E.S, Vicksburg, MS Technical Report S-72-7, p. 83.
60. Christopher, B. R. and Holtz, R. D. (1985), '*Geotextile Engineering Manual*', U.S. Department of Transportation, Federal Highway Administration, Washington, DC, USA, Report No. FHWA-TS-86/203, March 1985, p. 1044.
61. Fischer, G. R., Holtz, R. D. and Christopher, B. R. (1996), 'Characteristics of Geotextile Pore Structure', *Recent Developments in Geotextile Filters and Prefabricated Drainage Geocomposites,* Bhatia, S. K. and Suits, L. D., Editors, ASTM Special Technical Publication 1281. *Proceedings of a Symposium Held in Denver*, Colorado, USA, June 1995, pp. 3–18.
62. Smith, J. L., Bhatia, S. K., Ridgway, J. and Hawkins, W. (1999), Are porosity and O_{95} important in the retention and particle clogging behviour of geotextiles?. *Conference Proceedings, Geosynthetics*, Vol. **2**, 28–30 April 1999, pp. 813–831.
63. Jeon, H.-Y. and Mlynarek, J. (2004), Assessments of Long-Term Drainage Performance of Geotextiles, *57th Canadian geotechnical conference 5th joint cgs/iah-cnc conference*, p. 27–32.
64. Zhao, A and Montanelli, F. (1999), 'Effect of soil presence on flow capacity of drainage geocomposites under high normal loads' *Geosynthetics'99, Conference Proceedings*, Vol. **2**, pp. 783–798.
65. Dixon, J. H. (1992), Membrane Protection Layers for Landfills, A summary of German Experience, Wastes Management Workshop in Association with the Institute of Wastes Management, Solihull, 21 May 1992.

66. Brummermann, K., Blumel, W. and Stoewahse, C. (1994), Protection layers for geomembranes: effectiveness and testing procedures, *5th International Conference on Geotextiles, Geomembranes and Related Products*, Singapore, 5–9 September 1994.
67. Luders, G. and Seeger, S. (1999), Geotextile Protective Layer systems for Geomembranes, Draft translation, Integrated Research Programme, Development of Advanced Landfill Liner System, Germany.
68. El-Jack, A. M. A. (2004), Structure-Property Relation of Non-woven Geotextiles, PhD Thesis, The university of Leeds, May 2004.
69. Hemmer, S., Muller, R. and Janssen, E. (1996), Influences of Fibre Geometry and Production Parameters on the Technical Characteristics of Non-wovens, ITB Non-wovens, Industrial Textiles, April 1996.
70. Bernhard, L.(1996), Amoco Fabrics Europe, Gronau/Germany, Mineral filled geotextiles, The Ultimate Protection System for Geosynthetic Liners in Safe Waste Management, *World Textile Congress on Polypropylene in Textiles*, University of Huddersfield, UK, July 1996.
71. Richardson, G. N. and Bove, J. A. (1987), Testing and monitoring of high strength geosynthetics, *Geotextiles and Geomembranes,* **6**, 157–172.
72. John, N. (1987), Paulson geosynthetic material and physical properties relevant to soil reinforcement applications, *Geotextiles and Geomembranes,* **6**, 211–212.
73. Van Harten, K. (1986), The relation between specifications of geotextiles and their essential properties, *Geotextiles and Geomembranes*, **3**, 1.
74. Bonaparte, R., Holtz, R. D. and Giroud, J. P. (1987), Soil reinforcement design using geotextiles and geogrids. In: *Geotextile Testing and the Design Engineer,* J. E. Fluet, Jr (Ed.), ASTM STP-952, ASTM 1987.
75. Bonaparte, R. and Berg, R. (1987), Long term allowable tension for geosynthetic reinforcement, *Geosynthetics '87*, New Orleans, Louisiana, p. 1.
76. Jewell, R. A. (1986), Material properties for design of geotextile reinforced slopes, *Geotextiles and Geomembranes,* **7**(4), 2.
77. Leshchinsky, D. and Perry, E. B. (1987), A design procedure for geotextile reinforced walls, *Geosynthetics '87*, New Orleans, Louisiana, p. 11.
78. Fowler, J. and Koerner, R. M. (1987), Stabilization of very soft soils using geosynthetics, *Geosynthetics '87*, New Orleans, Louisiana, p. 21.
79. McGown, A., Andrawes, K. Z. and Kabir, M. H. (1982), Load-extension testing of geotextiles confined in-soil. *Proceedings of the Second International Conference on Geotextiles*, IFIA, Las Vegas August 1982.
80. Jewell, R. A., MiUigan, G. W. E., Sarsby, R. W. and DuBois, D. D. (1984), Interaction between soil and geogrids. *Proceedings of Symposium on Polymer Grid Reinforcement in Civil Engineering*, SERC/Netlon, London 1984.
81. Jewell, R. A., Milligan, G. W. E., Sarsby, R. W. and Dubois, D. (1984), Interaction between soil and geogrids. In *Proceedings of Symposium on Polymer Grid Reinforcement.* Thomas Telford Ltd, London, pp. 18–29.
82. Chang, J. C., Hannon, J. B. and Forsyth, R. A. (1977), Pull resistance and interaction of earthwork reinforcement and soil, *Transportation Research Record*, **640**, pp. 1–7.
83. Bergado, D. T., Chai, J. C., Abiera, H. O., Alfaro, M. C. and Balasubramaniam, A. S. (1993), Effect of particle size on the mechanical behaviour of sand-geotextile composites, *Geotextiles and Geomembranes*, **12**, 327, 349.

84. Ling, H. I. and Tatsuoka, F. (1994), Effect of particle size on the mechanical behaviour of sand-geotextile composites, *Journal of Geotechnical Engineering ASCE*, **120**(7), 1166–1184.
85. Porbaha, A. and Goodings, D. J. (1996), Effect of particle size on the mechanical behaviour of sand-geotextile composites, *Journal of Geotechnical Engineering, ASCE*, **122**(10), 840–848.
86. Zornberg, J. G. and Mitchell, J. K. (1994), Effect of particle size on the mechanical behaviour of sand-geotextile composites, *Geosynthetics International*, **1**(2), 103–148.
87. Mitchell, J. K. and Zornberg, J. G. (1995), Effect of particle size on the mechanical behaviour of sand-geotextile composites, *Geosynthetics International*, **2**(1), 265–307.
88. Swan, R. H. (1998), The influence of fabric geometry on soil/geotextile shear strength, *Geotextiles and Geomembranes*, **6**, pp. 81–87.
89. Jewell, R. A. (1985), Material properties for the design of geotextile reinforced slopes, *Geotextiles and Geomembranes*, **2**, 83–109.
90. Murray, R. T., Carder, D. R. and Krawczyk, J. V. (1979), Pull-out tests on reinforcements embedded in uniformly graded sand subject to vibrations. *Proceedings of the 7th European Conference Soil Mechanics and Foundation Engineering, Brighton*, **3**, pp. 115–120.
91. Haliburton, T. A., Anglin, C. C. and Lawmaster, J. D. (1978), Testing of geotechnical fabric for use as reinforcement. *Geotechnical Testing Journal*, GTJODJ **1**, 203–212.
92. Collios, A., Delmas, P., Gourc, J. P. and Giroud, J. P. (1980), Experiments on soil reinforcement with geotextiles. *Proceedings of Syrup Use of Geotextiles for Soil Improvement*, ASCE, Portland pp. 53–73.
93. Myles, B. (1982), Assessment of soil fabric friction by means of shear. *Proceedings of 2nd International Conference on Geotextiles*, Las Vegas, **3**, pp. 787–91.
94. Fourie, A. B. and Fabian, K. J. (1987), Effect of particle size on the mechanical behaviour of sand-geotextile composites, *Geotextiles and Geomembranes*, **6**, 275–294.
95. Richards, E. A. and Scott, J. D. (1985), Soil geotextile friction properties, *Second Canadian Symposium on Geotextiles and Geomembranes*, Canadian Geotechnical Society, Edmonton, Alberta September.
96. Fourie, A. B. and Fabian, K. J. (1987), Effect of particle size on the mechanical behaviour of sand-geotextile composites, *Geotextiles and Geomembranes*, **6**, 275–294.
97. Bouazza, A. and Djafer-Khodja, S. (1994), Effect of particle size on the mechanical behaviour of sand-geotextile composites, *Geotextiles and Geomembranes*, **13**, 807–812.
98. Nataraj, M. S., Maganti, R. S. and McManis, K. L. (1995), Interface frictional characteristics of geosynthetics. In *Proceedings of Geosynthetics '95*, Nashville, Tennessee, Vol. 3, pp. 1057–1069.
99. Ling, H. I. and Tatsuoka, F. (1994), Effect of particle size on the mechanical behavior of sand-geotextile composites, *Journal of Geotechnical Engineering ASCE*, **120**(7), 1166–1184.
100. Alwahab, R. M. and Al-ourna, H. H. (1995), Fiber reinforced cohesive soils for application in compacted earth structures. In *Proceedings of Geosynthetics '95*, Nashville, Tennessee, Vol. **2**, pp. 433–446.

101. Farrag, K. A. and Griffin, P. (1993), Pull-out testing of geogrids in cohesive soils. In *Geosynthetic Soil Reinforcement Testing Procedures*, ASTM STP 1190, ed. S. C. J. Cheng, Philadelphia, pp. 76–89.
102. Athanasopoulos, G. A. (1996), Results of direct shear tests on geotextile reinforced cohesive soil, *Geotextiles and Geomembranes,* **14**, 619–644.
103. Cedegreen, H. (1994), *Drainage of Highway and Airfield Pavements*, John Wiley and Sons., New York, NY.
104. Alawaji, H. A. (2001), Settlement and bearing capacity of geogrid-reinforced sand over collapsible soil, *Geotextiles and Geomembranes,* **19**, pp. 75–88.
105. Al-Qadi, I. L., Brandon, T. L., Valentine, R. J., Lacina, B. A. and Smith, T. E. (1997), How do geosynthetics improve pavement performance, *Proceedings of Infrastructure: New Materials and Methods for Repair*, ASCE, San Deigo, pp. 606–616.
106. Marienfield, M. L. and Baker, T. L. (1998), Paving Fabrics Interlayer System as a Pavement Moisture Barrier, Transportation Research Board.
107. Lawrence, C. A., Tebay, R. N., Caddick, V. R. and Warwick, R.G. (2002), Geosynthetic Protection of Ballast and Sub-grade in Railway Foundation Structures', *International Geosynthetics Society Annual Conference, Geosynthetics: State of the Art and Recent Developments*, NICE, 27–30 September 2002, 901–904.
108. Bonaparte, R., Holtz, R. D. and Giroud, J. P. (1987), 'Soil Reinforcement Design Using Geotextiles and Geogrids', In *Geotextile Testing and the Design Engineer, ASTM, Special Technical Publication 952*, J. E. Fluet, editor, American Society for Testing and Materials, pp. 69–116.
109. Humphrey, D. N. and Rowe, R. K. (1991), Design of Reinforced Embankments – Recent Developments in the State-of-the-Art, State of the Art Paper, Session 8A on Earth Reinforcement, Geotechnical Engineering Congress 1991, Geotechnical Special Publication No. 27, ASCE, Vol. **II**, 1991, pp. 1006–1020.
110. Leroueil, S. and Rowe, R. K. (2001), Embankments over Soft Soil and Peat, Chapter 16 in *Geotechnical and Geo-environmental Handbook*, R. K. Rowe, Editor, Kluwer Academic Publishers, Norwell, U.S.A. pp. 463–499.
111. Krieger, J. and Thamm, B. R. (1991), Studies of failure mechanisms and design methods for geotextile-reinforced soil walls, *Geotextiles and Geomembranes,* **10**, 53–63.
112. Jewell, R. A. (1985), Material properties for the design of geotextile reinforced slopes, *Geotextiles and Geomembranes,* **2**, 83–109.
113. Dabke C.S., Rawes B.C. and Lengart S.S. (1992), Role of geotextiles in marine construction, *Geotextiles and Geomembranes,* **11**, 223–233.
114. Tatsuoka, F., Uchimura, T., Tateyama, M. and Koseki, J. (1996), Geosynthetic-reinforced soil retaining walls as important permanent structures. Geosynthetics: applications, design and construction. *Proceedings of the First European Geosynthetics Conference*, Maastricht, pp. 3–24.
115. Kojima, K., Tateyama, M., Sakamoto, N. and Maruyama, O. (1996), Geosynthetic reinforced soil retaining wall using clay on a very soft ground for Hokuriku bullet train yard in Nagano. *Proceedings of the International Symposium on Earth Reinforcement*, Kyushu, Japan. Vol. **1**, pp. 227–232.
116. Kamon, M., Akai, T., Fukuda, M. and Nanbu, Y. (1996), *In-situ* failure test of high water content soft clay embankments reinforced by GHDs. *Proceedings*

of the International Symposium on Earth Reinforcement, Kyushu, Japan, Vol. **1**, pp. 215–220.

117. Montanelli F. and Recalcati, P. (2003), The Design of Reinforced Soil Retaining Walls using TENAX geogrids, Desgin Manual, Technical Document TDR002-04/03.
118. Pradhan, T. B. S., Shiwakoti, D. R. and Imai, G. (1996), Effect of normal pressure and width of geosynthetic horizontal drain in pullout behaviour using saturated clay. *Proceedings of the International Symposium on Earth Reinforcement*, Kyushu, Japan, Vol. **1**, pp. 133–138.
119. Murata, O., Tateyama, M. and Tatsuoka, F. (1992), 'Shaking Table Tests on a Large Geosynthetic-Reinforced Soil Retaining Wall Model', Recent Case Histories of Permanent Geosynthetic-Reinforced Soil Retaining Walls (Tatsuoka, F. and Leshchinsky, D., Editors), Balkema. *Proceedings of Seiken Symposium* No. 11, Tokyo, Japan, November, pp. 259–270.
120. Tatsuoka, F., Tateyama, M. and Koseki, J. (1995), Performance of Geogrid-Reinforced Soil Retaining Walls During the Great Hanshin-Awaji Earthquake, Earthquake Geotechnical Engineering, Ishihara, K., Editor, Balkema, Vol. **1**. *Proceedings of the First International Conference on Earthquake Geotechnical Engineering*, IS-Tokyo '95, Tokyo, Japan, November 1995, pp. 55–62.
121. Budhu, M. and Halloum, H. (1994), Seismic External Stability of Geotextile Reinforced Walls, *Fifth International Conference on Geotextiles, Geomembranes and Related Products*, Vol. **1**, Singapore, September 1994, pp. 529–532.
122. Adams, M. T. (2000), Reinforced soil technology at FHWA. Geosynthetic Fabric Report, Vol. **18**, No. 6, August 2000, pp. 34–37.
123. Kevin, Z. Z., Lee, J. and Wu, T. H. (2004), A synthesis of case histories on grs GRS bridge supporting structures with flexible facing, *Geotextiles and Geomembranes*, **22**, 181–204.
124. Abu-Hejleh, N., Wang, T. and Zornberg, J. G. (2000), Performance of Geosynthetic-Reinforced walls Supporting Bridge and Approaching Roadway Structures. *ASCE Geotechnical Special Publication* No. 103, *Proceedings*, Geo-Denver, pp. 218–243.
125. Ketchart, K. and Wu, J. T. H. (1997), In: Wu, Barrett. (Eds.), Performance of GRS Bridge Pier and Abutment in Denver, CO, USA. Special Presentation, Mechanically Stabilized Backfill. A.A. Balkema Publisher, Rotterdam, pp. 101–116.
126. Lauritzen, C. V. and Perterson, H. (1953), Butyl Fabrics as Canal Lining Materials, Utah Agricultural Experiment Station Bulletin No. 363.
127. Lauritzen, C. V., Frank, V. and Humpherys, A. S. (1955), Plastic Film for Controlling Seepage Losses in Farm Reservoirs, Utah Agricultural Experiment Station Bulletin No. 391.
128. Ng, H. B. (2009), HDPE Lined Water Reservoir for Power Generating Stations, Geosynthetics in Civil and Environmental Engineering, Part 11, pp. 769–774.
129. Mackey, R. E and Koerner, G. R. (1999), Biological clogging of geotextile filters: a five year study, *Geosynthetics '99, Conference Proceedings*, Vol. **2**, pp. 783–798.
130. Breitenbach, A. J. (2004), Improvement in slope stability performance of lined heap leach pads from design to operation and closure, *GFR Engineering Solutions*, **22**(1).

131. Martin, J. P., Koerner, R. M. and Whitty, J. E. (1984), Experimental friction evaluation of slippage between geomembranes, geotextiles and soils. In *Proceedings of International Conference on Geomembranes, Fabrics Association International*, St Paul, Minnesota, Vol. **II**, **4**, pp. 191–196.
132. Williams, N. D. and Houlihan, M. F. (1987), Evaluation of interface friction properties between geosynthetics and soils. *Proceedings of the Geosynthetics '87 Conference*, Industrial Fabrics Association International, St Paul, Minnesota, pp. 616–627.
133. Carroll, R.G. and Chouery-Curtis, V. (1991), Geogrid reinforcement in landfill closure, *Geotextiles and Geomembranes*, **10**, pp. 471–486.
134. Carroll, R. G., Rodencal J. and Collin, J. G. (1992), Geosynthetics in erosion control – the principles, *Geotextiles and Geomembranes*, **1**, pp. 523–534.
135. Simmonds, G. E., Bomberger J. D. and Bryner M. A. (2007), Designing nonwovens to meet pore size specifications, *Journal of Engineered Fibers and Fabrics*, **2**(1), pp. 1–15.
136. Jeon, H.-Y. and Mlynarek, J. (2008), Assessment of Long-Term Drainage Performance of Geotextiles, *57th Canadian Geotechnical Conference*, Session 4D, pp. 27–32.
137. Hamir, R. B., Jones, C. J. F. P. and Clarke, B. G. (2001), Electrically conductive geosynthetic for consolidation and reinforced soil, *Geotextiles and Geomembranes*, **19**, 455–482.
138. Activated Carbon and carbon photo-catalysis for anti-order cover, TUT (2005), No. 55, p. 15.

9
High performance fibers and fabrics for civil engineering applications

Y. GOWAYED, Auburn University, USA

DOI: 10.1533/9780857099075.351

Abstract: In fiber reinforced plastics (FRP), as a special type of polymer matrix composite, fibers provide the stiffness and strength while the surrounding plastic matrix transfers the stress between fibers and protects them. In this chapter, the role of fibers in FRP is delineated, their types and properties are discussed and the fabric forms in which they can be formed and used to reinforce FRP are presented. A special focus is given to the effect of the chemical structure of fibers on the stability and the level of anisotropy of their mechanical response. Furthermore, the effect of assembling these fibers into yarns and fabrics on the response of the FRP is presented as basis for further readings.

Key words: matrix, strength, stiffness, composite, carbon fibers, Kevlar® fibers, glass fibers, unidirectional fabrics, woven fabrics, stitched fabrics.

Note: This chapter is an adapted version of Chapter 1 'Types of fiber and fiber arrangement in fiber-reinforced polymer (FRP) composites' by Y. Gowayed, originally published in *Developments in fiber-reinforced polymer (FRP) composites for civil engineering*, ed. Nasim Uddin, Woodhead Publishing Limited, 2013, ISBN: 978-0-85709-234-2.

9.1 Introduction

Fiber-reinforced polymer (FRP) composites are made of a mixture of two solid materials, a material with high strength and stiffness surrounded by a homogeneous material that protects it and keeps it in place. The stiff material or reinforcement is typically made of a directional component such as fibers, rods, or sheets, while the surrounding material is typically isotropic and is called the matrix. This chapter will focus on the reinforcement.

Fibers are long solid objects with an extremely high aspect ratio and a unique set of directional properties. They are also flexible with a very low bending stiffness. Despite such combination of directionality, high aspect ratio, and flexibility, it is almost impossible to find a durable, high strength application for fibers without protecting them from the surrounding environment. They can easily be damaged by friction or even simple handling.

Coating a group of fibers with a polymeric film or any other material will not only protect them but also help transfer the stresses between them, making them behave as if they are a single body, rather than a loose group of fibers.

The role of fibers in FRP composites is typically limited to providing strength and stiffness while the matrix keeps the fibers in their intended location, protects them, transfers stresses between the fibers, and provides the rigidity for the FRP composite as a whole. This means the properties of the interface between the fibers and the matrix play a major role in how the stress is transferred between them, affecting the way the FRP behaves under sustained loads. A strong interface ensures that the entire load is transferred causing the composite to have high stiffness and strength. On the other hand, a weak interface may not allow such full load transfer, lowering the strength and stiffness of the composite.

The toughness of the FRP composite is also affected by the interfacial strength between the fibers and the matrix. Defining toughness as the energy needed to break the composite, it is important that a fracture-causing crack be forced to take a long and tortuous path consuming as much of its energy as possible to achieve a tough composite. This can only be achieved if the crack path is faced by different material phases with varying levels of strength and stiffness. A weak interface next to a stiff fiber would help deviate the crack path and accordingly enhance the toughness of the composite.

FRP composites are usually anisotropic with different material properties in different directions. Such anisotropy is caused by the alignment of the fibers in the composite. For example, if the fibers are only aligned in one direction, the composite is called unidirectional with properties along the direction of the fibers different than those in any other directions. If the fibers are aligned in two perpendicular directions, then the properties in those two directions are different than the properties in all other directions. Depending on the volume of fibers in each direction, the properties in both directions may not be the same.

The concept of anisotropy and, in particular, the ability to design for anisotropy has made composite materials what they are today; not only is it easy to achieve an optimum design, but they are also an inexpensive answer to critical loading scenarios. From an optimum design perspective, an anisotropic material design can provide certain properties only in required directions. Since a property such as stiffness can be qualitatively equated to cost, such optimum design will also reduce the cost. For isotropic materials, stiffness is provided in all directions of the material causing additional unnecessary cost, but anisotropic design strategies can meet target stiffness and strength in required directions without waste.

Fibrous composites are typically manufactured using either the process of stacking lamina or fabrics. In some cases, these fiber layers are pre-impregnated (typically called prepregs) with a resin and kept at a sub-zero temperature to slow down the chemical reaction for at least 6 months or even a year. These layers are typically placed in the expected load directions of the target product for an optimum design. After stacking, if the fibers are not 'prepreged', resin is introduced to the composite and cured, or, in the case of prepregs, heat is applied to initiate the curing process.

The ensuing sections introduce some of the fibers typically used in composite materials and their formation into fabric layers as well as other types of FRP. As the discussion goes into different material configurations, the reader has to keep in mind the importance of anisotropy in the selection of fibers and the formation of fabrics.

9.2 Fibers

High strength and modulus fibers that are commonly used in composite materials can be categorized based on their molecular conformation in three groups:

- polymeric fibers
- carbon fibers
- other inorganic fibers.

The first group has one-dimensional primary bonds that are somewhat aligned with the longitudinal axis of the fibers. This one-dimensional bond can be as simple as in polyethylene with a linear chain of carbon atoms attached to side groups of a single hydrogen atom or a much more complex chain configuration with rings of atoms and more complex side groups. Most of these fibers are either anisotropic or transversely isotropic with the plane of isotropy perpendicular to the fiber longitudinal axis.

Carbon fibers have two-dimensional (2D) graphite sheets in a hexagonal planar network of primary bonds that are aligned parallel to the fiber axis with secondary bonds connecting the sheets in the radial direction of the fiber. Further connection between the sheets is due to some disorder and imperfection in the alignment of the atoms in the sheets. Graphite/carbon fibers are expected to be transversely isotropic. The last group of fibers has a three-dimensional (3D) network of primary bonds that can provide stiffness and strength and, additionally, a good thermal stability at higher temperatures. These networks can be random or crystalline and typically exist in ceramic fibers such as glass. Most of the fibers in this type are expected to be isotropic or pseudo-isotropic. The following sections will provide some examples of the most common of the above-mentioned types of fibers. This by no means represents all that the market has of these types of fibers; to the contrary, there are many other fibers that are used in making composite materials and there are new fibers that are being developed.

9.2.1 Para-aramid fibers such as Kevlar® and Twaron®

The generic term 'aramid' designates a long chain synthetic polyamide molecule (-CO-NH-) and para-aramids, such as Kevlar®, that are wholly aromatic polyamides. DuPont started commercialization of Kevlar® as the first high strength aromatic polyimide fiber in 1971. Another fiber with a similar chemical structure is Twaron® produced by Akzo Nobel since the late 1970s.

Kevlar® fibers are manufactured by solution spinning of 1,4-phenylene diamine and terephthaloyl chloride in a condensation reaction yielding hydrochloric acid as a by-product. The spinning solution is extruded through the holes of a spinneret and subjected to mechanical stretch to elongate and orient the molecules in the axial direction of the fibers. The higher the level of mechanical stretch, the higher the strength and modulus of the fiber and the lower is its strain to failure. The resulting polymer (p-phenylene terephthalamide), shown in Fig. 9.1, exhibits a liquid-crystalline behavior in solution. The long chain molecules act as if they are rigid rods and when the solution is sheared their crystalline regions tend to reorient themselves in the direction of the flow, providing high modulus and strength in the axial direction of the fiber.

Kevlar® fibers are known for their ultraviolet (UV) degradation in the presence of oxygen. Upon degradation, the fiber changes in color from a lustrous golden hue to dark yellow or brown. Zhang *et al.* studied the effects of simulated solar UV irradiation on the mechanical properties of Twaron®. They showed a decrease in mechanical properties and reported that UV irradiation deteriorated the surface and defect areas of the fiber severely by photo-induced chain scission, while the crystalline structure remained almost unchanged (Zhang *et al.*, 2006). Wang *et al.* reported a reduced compressive strength and modulus of Kevlae/epoxy fabric composites as compared to its tensile strength (Wang *et al.*, 1995). In support of this finding, Fidan *et al.* reported that the compressive strength of Kevlar® fibers was reduced from their tensile strength due to the kink band formation in the

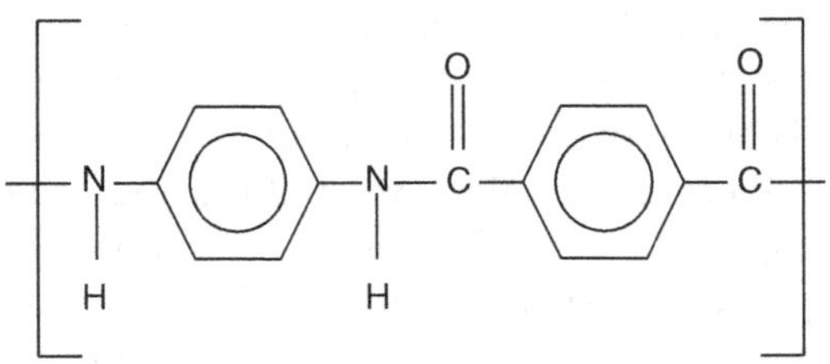

9.1 The chemical structure of Kevlar

Table 9.1 Typical properties of some Kevlar® fibers (manufacturer data)

	Kevlar® 29	Kevlar® 49
Tensile modulus (GPa)	70.5	112.4
Poisson's ratio	–	0.36
Strength (GPa)	2.92	3
Strain to failure (%)	3.6	2.4
Density (g/crre)	1.44	1.44
Thermal conductivity (W/m.°K)	0.04	0.04
Coefficient of thermal expansion (x10-6/°K)	–4	–4.9

fiber from microbuckling of separated microfibrils due to elastic instability in the fibers (Fidan *et al.*, 1993). Furthermore, the weak bonding between the chains caused the fibers to split into smaller fibrils and micro-fibrils. This phenomenon is typically witnessed during the failure of Kevlar® reinforced composites, especially those subjected to compressive stresses (Hull and Clyne, 1996). Different types of Kevlar® fibers are available in the market, covering a large range of strengths and moduli. Table 9.1 lists the properties of some of the most popular types of Kevlar® fibers used in composite reinforcement.

Kevlar® fibers are transversely isotropic with the tensile modulus in the axial direction around 52 times the transverse compressive modulus of Kevlar® 49 and 38 times for Kevlar® 29. Additionally, the longitudinal tensile strength is about 42 times the transverse compressive strength of Kevlar® 49 and 54 times for Kevlar® 29 (Kawabata, 1990).

9.2.2 Carbon fibers

Carbon fibers have been known since their development by Thomas Edison in 1870 and have been under continuous development for the past 60 years (Hearle, 2001; Peebles, 1995). The current interest in carbon fibers stems from their excellent mechanical properties and thermal stability. Carbon can be found in nature in many forms. Diamond, graphite, and ash are all made from pure carbon atoms with different atomic arrangements. Diamonds are made from covalent 3D bonds between carbon atoms. Graphite and carbon fibers are made from sheets of covalent-bonded carbon atoms. These sheets are connected to each other via a weak secondary bond.

Carbon fibers can be made by pyrolysis of a hydrocarbon precursor. Rayon was one of the first precursors used to make carbon fibers. During the processing of Rayon fibers into carbon fibers, only 25% of the fiber mass is retained. This made carbon fibers manufactured from Rayon precursors very expensive. Another precursor that has proved to be economical is the polyacrylontrile

(PAN) fiber with a conversion yield of around 50–55%. Carbon fibers made from a PAN precursor generally have higher strength than fibers made from other precursors. This is due to the lack of surface defects, which act as stress concentrators, and hence reduce tensile strength. Another commonly used precursor is the Pitch precursor which is a byproduct of petroleum refining. Pitches are relatively low in cost and high in carbon yield. Their most significant drawbacks are the irregular surfaces of the fiber, which reduces the fiber tensile and compressive strengths (Hearle, 2001). Currently, fibers are manufactured from either a PAN or a Pitch precursor. Rayon precursors are used in less than 1% of the production of carbon fibers (Department of Defense MIL-HDBK-17-5, 2002). Physical properties of some carbon fibers used by the industry are listed in Table 9.2 (Hearle, 2001).

There are several processes that a PAN precursor has to go through to be converted to a carbon fiber. Typically, the precursor-to-carbon-fiber conversion process follows the following sequence: stabilization, carbonization, graphitization, surface treatment, and application of sizing and spooling. Stabilization is carried out at temperatures < 400°C in various atmospheres. The fiber is held under tension during this stage to enhance molecular orientation, which increases fiber modulus and strength. Carbonization is accomplished at temperatures from 800 to 1200°C in an inert atmosphere. Fiber tensioning is still maintained during this process. Graphitization is an additional process at a temperature of > 2000°C in an inert environment. This process reduces the level of impurities and stimulates crystal growth. Various materials can be applied to the surface of carbonized/graphitized fiber during the surface treatment process. These materials will help control the interaction between the fibers and the matrix materials in a composite. Sizing is applied to fiber tows (yarns) to enhance their handling characteristics in further textile forming operations (e.g., weaving and braiding). After these processes, the fiber is spooled on a carrier tube to form a stable package. For

Table 9.2 Mechanical properties of some carbon fibers

Precursor type	Product name	Young's modulus (GPa)	Tensile strength (GPa)	Strain to failure (%)
PAN	T300	230	3.53	1.5
	T1000	294	7.06	2.0
	M55J	540	3.92	0.7
	IM7	276	5.30	1.8
Pitch	KCF200	42	0.85	2.1
	Thornel P25	140	1.40	1.0
	Thornel P75	500	2.00	0.4
	Thornel P120	820	2.20	0.2

Source: Hearle, 2001.

a pitch precursor, the petroleum mixture is heated above 350°C to allow a condensation reaction to occur, which produces large flat molecules, and then the mixture is extruded through a spinneret to produce a 'green fiber' aligning the molecules in the fiber axial direction. The fibers are further heated without tension at around 2000°C to produce a carbon fiber.

Most carbon fibers are transversely isotropic with the value of anisotropy scattered for both PAN and Pitch fibers. Kawabata reported that the ratio of the longitudinal tensile modulus divided by the transverse compressive modulus is 51.1 for T300 which is a PAN fiber, 19.1 for Thornel P25, 108.9 for Thornel P75, and 247.4 for Thornel P120 with all the Thornel fibers manufactured from a Pitch base. On the other hand, the strength tells a different story. The longitudinal tensile strength of a PAN-based fiber divided by its transverse compressive strength is around 1.1, while for a Pitch-based fiber like Thornel P25 it is 3.8, for Thornel P75 it is 21.5, and for Thornel P120 it is 43.4 (Kawabata, 1990).

9.2.3 Glass fibers

Drawing glass into fibers is an ancient art. These early fibers were used to reinforce clay vessels as well as clay statues. Recent development for glass fibers started in the early years of the twentieth century with the work by Griffith (1920), in which he used glass fibers as a model for his theories on fracture mechanics. Glass fibers are manufactured by melting various raw materials including silica and other salts followed by extruding the melt to form glass fibers. The primary component of glass fiber is sand, but it also includes varying quantities of limestone, soda ash, borax, sodium sulfate, boric acid, etc. Raw materials are heated in a furnace to temperatures ranging from 1500 to 1700°C and are refined and transformed through a sequence of chemical reactions to molten glass. The molten glass is forced through a heated platinum spinneret containing very small holes. The fibers emerging from the spinneret are immediately coated with a water-soluble sizing or a coupling agent. The coat protects the fiber from dust which can scratch its surface and reduce the fiber strength, while the coupling agent can enhance the interfacial adhesion of the fiber to matrix material in a composite. The coat also improves the yarn handlability and protects it during further fabric forming processes.

Fiber glass typically used to reinforce composite materials include E-glass (E for electrical) which is an alumino-borosilicate glass, C-glass (C for corrosion) which has a close chemical structure to that of E-glass but with a better resistance to corrosion, and S-glass (S for strength) which is an alumino silicate glass with a high tensile strength. Some of the typical fiber properties are listed in Table 9.3.

The basic building block of the glass fiber, as shown in Fig. 9.2, is a tetrahedral structural unit of silica (SiO_4), where the Si atom shows tetrahedral

Table 9.3 Mechanical properties of some glass fibers

	Tensile modulus (GPa)	Tensile strength (GPa)	Poisson's ratio	Density (g/cm³)	Strain to failure (%)
E-glass	72	3.45	0.22	2.55	1.8–3.2
C-glass	69	3.3	—	2.49	—
S-glass	87	3.5	0.23	2.5	4

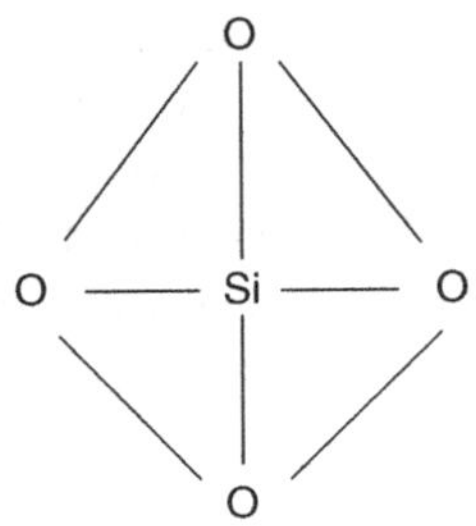

9.2 The chemical structure of glass fibers.

coordination, with four oxygen atoms surrounding a central Si atom. The presence of Ca, Na, and K breaks up the silica network lowering the stiffness and strength of the fiber but improving the formability of the melt to form the fibers. The structure of the fiber is typically amorphous, although crystalline regions may form after prolonged heating at high temperatures. The bonds are randomly oriented causing the fiber to be isotropic or pseudo-isotropic.

9.3 Fabrics

Polymer matrix composite materials are made by mixing fibers with a resin material (matrix). These fibers can either have a non-specific arrangement like in the case of chopped fiber (non-wovens) or long fibers grouped together, called tows or yarns, and assembled into fabrics such as wovens, braids, or knits. Since the fibers are the main source for the stiffness and strength of polymer matrix composites, it could be correctly inferred that the direction of these fiber will play a major role in the mechanical and thermal behavior of the composite. The more the fibers are aligned in the load direction(s), the higher the stiffness(s) and strength(s) of the composite in these direction(s). Hence, a chopped non-woven glass composite with fibers randomly oriented in many directions will have stiffness inferior to that of a woven fabric composite with fibers aligned in the load direction, and so on.

A logical conclusion from the above can be to align the yarns exactly in the load direction without involving any weaving or braiding which typically moves the yarns away from the load direction to achieve fabric integrity. This is typically done by using layers of unidirectional yarns in which each group of yarns is aligned in one of the load directions. This alignment of the yarns will definitely achieve the highest possible level of stiffness and strength in the direction of the aligned yarns.

The yarn arrangement utilized in unidirectional composites comes at the expense of the toughness of the composite: its ability to resist failure and crack propagation. Unidirectional layers of yarns have a clean matrix interface between planes of yarns. If a crack initiates in this matrix area, it will have a clear path to follow and break the composites. Besides having high moduli and strengths in the plane of the yarns, composites made from unidirectional lamina are known to have low toughness, low interlaminar shear strength as well as low out-of-plane modulus and strength. On the other hand, composites made from non-woven fibers, and woven, braided, or knitted fabrics have the advantage of fibers and yarns moving from one plane into another, This complex arrangement increases the energy required for the crack to propagate, increasing the toughness of the composite. So they are known to have in-plane moduli and strength lower than those of composites made from unidirectional layers, but have higher toughness, interlaminar shear strength, and out-of-plane moduli and strengths.

The above-mentioned yarn and fiber forms can come either in a 'dry' form or impregnated with a resin. Prepreg(ing) is a process where fibers and yarns, regardless of their form, are impregnated with a resin with a known fiber volume and yarn distribution in the form of a sheet. The chemical reaction of the resin is halted by placing the prepregs in a freezer for a specific period of time. The user typically takes these sheets of prepregs, aligns them in the proposed directions, stacks them, and then heats and presses them to initiate the chemical reaction and cure the composite. The following sections give some details of some typical fiber arrangements.

9.3.1 Unidirectional laminates

Most unidirectional composites are manufactured from prepreged layers of yarns. Each layer is called a lamina and a group of lamina is called a laminate. Each lamina in the laminate can have a specific direction based on the design requirements starting from an arbitrary direction. The only important matter to remember is that these laminates must have a mirror symmetry around their neutral axis; otherwise they will warp immediately after manufacture and cannot be straightened. This happens due to the difference in the thermal expansion coefficients between the fibers and the matrix and between different lamina causing one side to permanently expand more than

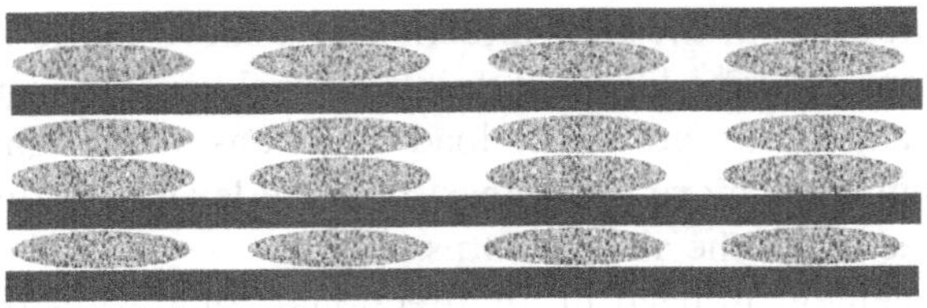

9.3 A (0,90,0,90)s laminate composite.

the other side and eventually warp. Figure 9.3 shows a schematic of a symmetrical laminate. Note the mirror symmetry in the drawing. The nomenclature of these composites follows the stacking procedure. For example, the laminate in Fig. 9.3 is called (0,90,0,90), where the subscript (s) entails the symmetry. Any other angle beside the 0 and 90 is possible.

9.3.2 Non-wovens (chopped fibers)

Non-woven fabrics are formed directly from short fibers that are assembled in sheets. The fibers are connected together to enhance the handlability of such sheet of otherwise loose fibers. Rows of needles are sometimes used to punch through the sheet of fibers and reorient some of the fibers to achieve mechanical adhesion between the fibers (needle punched fabrics). This process, although very successful with polymer fibers such as para-aramids, has less success with brittle ceramic fibers such as glass and carbon because fibers can easily break. Another method used to connect the short fibers is by using an adhesive to glue the short fibers to one another. In the case of short fiber composites, this glue will typically have a chemical affinity to the resins.

9.3.3 2D woven fabrics

2D woven fabrics are made by interlacing yarns in a weaving loom. Yarns are divided into two components: one called the warp, running along the length of the loom, and the other is the weft, running in the cross direction. The warp yarns are also divided into multiple parts and each part is passed through a group of eyelets held in a harness frame. Each of the frames is moved up or down and the weft yarns are inserted at each step following a defined pattern of movement, creating a specific fabric structure. Warp and weft yarns cross at cross-over points and sometimes they bend (called crimp) in order to pass over or under one another. A beat-up comb is used to stack the yarns and pack them into the fabric. The angle between the warp and the weft yarns is typically 90°. The fabric structure is characterized

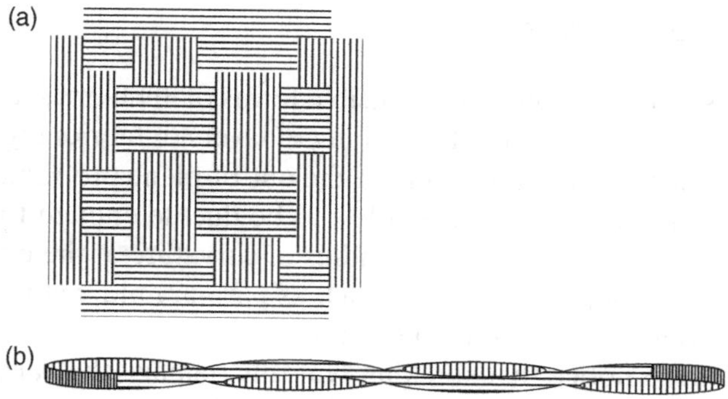

9.4 Drawing of a plain weave plan (a) and side views (b) as created by pcGINA© software (Gowayed and Barrowski, 2004).

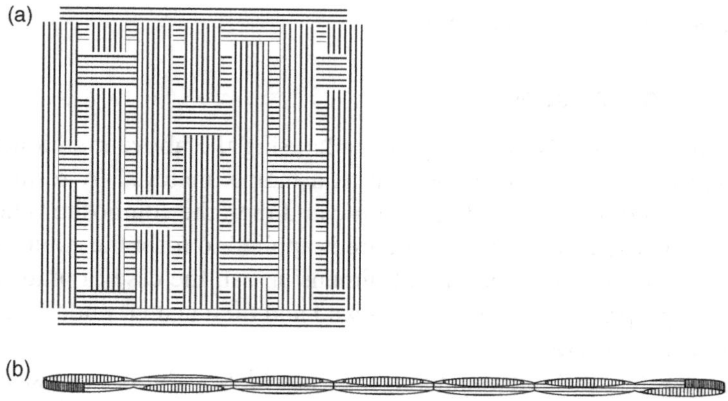

9.5 Drawing of a 5-HS weave plan (a) and side view (b) as created by pcGINA© software (Gowayed and Barrowski, 2004).

by the movement of its yarns as they relate to one another. For example, the plain weave shown in Fig. 9.4, has a 1/1 arrangement, where a warp yarn alternates being under or over a weft yarn at each crossover point.

A 5-harness satin (5-HS) weave will have its warp yarns running over four weft yarns and under one weft yarn as shown in Fig. 9.5. Other n-HS weaves used in composite materials include 8-HS, where the warp yarn passes over seven weft yarns and under one, and 12-HS where the warp yarn passes over eleven weft yarns and under one. It can easily be understood that the more yarn passes over the other yarn, the straighter and less crimped will be the yarns present in the fabric. The straighter the segments, the more the fabric composite behaves as a lamina in a laminated composite, increasing the in-plane properties at the expense of out-of-plane properties and interlaminar shear strength – and of course vice versa.

9.3.4 Stitched fabrics

Stitched fabrics represent one of the possible forms of 3D fabrics. These dry fabrics are manufactured by simple stitching of layers of fabrics together to allow for better handlability, and enhance the out-of-plane properties of the composite. The yarn used for stitching has to be ductile enough to sustain the stitching forces and the very small radius of curvature at the end of the stitch path. Typically a polymeric yarn such as para-aramid, nylon, or high density polyethylene is used to stitch fabrics.

It is important to notice that stitching typically damage the yarns of the fabric, especially if they are brittle fibers such as glass or carbon. The needle going from one side of the stack of fabrics to the other side encounters and damages various yarns and fibers. Accordingly, the density of the stitch repeat has to be low enough to limit the damage but high enough to achieve some advantage to the composite.

9.4 Composites

Composite materials are typically made using the above-mentioned fibers and fabrics, or other types, as well as a matrix material. A resin, such as epoxy, is used to impregnate the fabric and then the composite is heated to cure and harden. Sometimes pressure is applied to form the composite into a specific shape with exact target dimensions. In the case of unidirectional or fabric prepregs, the layers are stacked and then heated under pressure to form the composite.

Composite can be manufactured *in situ* like in the case of strengthening a column or a beam by wrapping the structural element with layers of the fabric, then impregnating them with a resin using hand lay-up or a vacuum bag. In other scenarios, the composite is manufactured in a separate facility then shipped to the construction site where it is fastened to the structural member, or adhesive is used to integrate it with the structural element. Composites manufactured outside the site come in different shapes. The most common type is in the form of a sheet of material with specific dimensions. These sheets can be manufactured in a non-continuous form such as using hand lay-up, vacuum bagging, or compression molding. They can also be manufactured in a continuous form using pultrusion, where the fibers, continuous or chopped, are fed through a resin bath and then pulled through a fixed orifice, with the final cross-section shape of the part, then immediately cured.

Many other forms can be manufactured and used in different civil engineering applications. One of these forms is rebars that are used to replace steel in order to prolong the life of the structure. Some of these rebars are made from unidirectional fibers at the core, mostly glass or para-aramid fibers, which are wrapped by a helical wrap or a braid and then infused with resin. These rebars are used to reinforce the concrete and must have a high resistance to alkaline environments (Abbasi and Hogg, 2005).

9.5 Future trends

The composite industry has achieved great strides in making a product that is light, stiff, has excellent strength, and is environmentally stable. As it continues to press forward with a product that has such unique characteristics, the industry realizes the need to advance in different directions, such as those of the fibers, the fabric geometry, and the composite application. There is a strong need for a high strength and environmentally stable polymer fiber that can be stacked and stitched without the fear of breaking like the case of ceramic fibers. The recent work on PBO fibers such as Zylon (Toyobo, 2001), which is a thermoset liquid crystalline polyoxazole, shows promise of such development, although reports suggest that these fibers are susceptible to strength degradation in high humidity and direct sunlight. Ultra high molecular polyethylene fibers such as Spectra and Dyneema are environmentally stable, have very low density, and excellent stiffness and strength. Further work is needed to enhance the interfacial strength of these fibers with the matrix material. Recent developments in M5 fibers by Akzo Nobel prepared by condensation polymerization of tetraaminopyridine and dihydroxyterephthalic acid using diphosphourus pentoxide as a dehydrating agent show great promise (Sikkema, 1998). It has properties similar to those of carbon fibers.

The composite fabric manufacturing industry has focused its efforts on producing 2D fabrics that have consistent and stable dimensions and with minimum damage. Work on 3D fabric structures has produced orthogonal fabrics (Bogdanovich and Mohamed, 2009) and angle interlock fabrics as

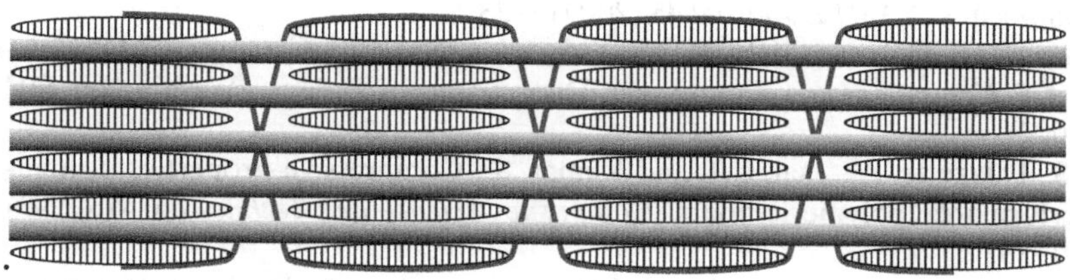

9.6 Orthogonal weave structure.

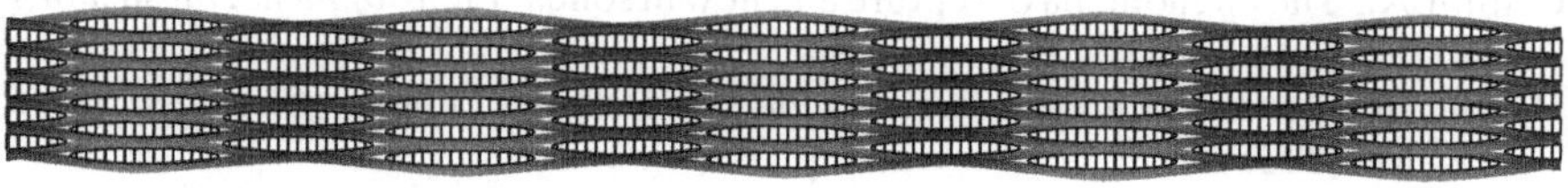

9.7 Angle interlock weave.

shown in Figs 9.6 and 9.7, as well as other new fabric structures to limit the need for stitched layers and the associated damage to the yarns.

9.6 Sources of further information and advice

The reader is encouraged to read more on the below mentioned subjects. On the subject of fibers, the book by J.W.S. Hearle (ed.), *High performance fibers* (CRC Press, Boca Raton, 2001) provides an excellent reference. Peter R. Lord and Mansour H. Mohamed wrote *Weaving: Conversion of Yarn to Fabric* (Merrow Publishing Co., Darlington, UK, 1976), which is considered one of the most important books in the field. Two- and three-dimensional fabrics are covered in the book by A. Miravete, *3-D Textile Reinforcements in Composite Materials* (Woodhead Publishing Limited, Boca Raton, FL, 1999). Two of the most concise composite mechanics of composite material books are S. Tsai and H. Hahn, *Introduction to Composite Materials* (Technomic Publishers, Lancaster, PA, 1989) and Robert M. Jones, *Mechanics of Composite Materials* (McGraw-Hill, New York, 1975). A. Bogdanovich and C. Pastore wrote a treatise on the mechanics of fabric composites in their *Mechanics of Textile and Laminated Composites* (Chapman and Hall, London, 1996). Finally on composite manufacturing, Timothy G. Gutowski edited the book *Advanced Composites Manufacturing* (Wiley-Interscience, New York, 1997).

9.7 References

Abbasi, A. and Hogg, P.J., 'Temperature and environmental effects on glass fibre rebar: modulus, strength and interfacial bond strength with concrete,' *Composites: Part B* **36** (2005) 394–404.

Bogdanovich, A.E. and Mohamed, M.H., 'Three-dimensional reinforcements for composites,' *SAMPE Journal* **45** (2009) 8–28.

Department of Defense MIL-HDBK-17-5, *'Handbook of Composite Materials,'* Vol. **5**, *Ceramic Matrix Composites,* Department of Defense, Washington, DC (2002).

Fidan, S., Palazotto, A., Tsai, C.T. and Kumar, S., 'Compressive properties of high-performance polymeric fibers,' *Composites Science and Technology* **49** (1993) 291–297.

Gowayed, Y. and Barrowski, L., 'pcGINA©, A pc-based Graphical Integrated Numerical Analysis of composite materials to calculate the properties of fabric based composites', version 10.2004.

Griffith, A., 'The phenomena of rupture and flow in solids,' *Philosophical Transactions of the Royal Society, London* A**221** (1920) 163.

Hearle, J.W.S. (ed), *High Performance Fibers,* Woodhead Publishing Limited, Boca Raton (2001).

Hull, D. and Clyne, T., *An Introduction to Composite Materials,* 2nd edn, Cambridge University Press, Cambridge (1996).

Kawabata, S., 'Measurement of the transverse mechanical properties of high performance fibers,' *Journal of the Textile Institute* **81** (1990) 432–447.

Peebles, L., *Carbon Fibers: Formation, Structure and Properties*, Woodhead Publishing Limited, Boca Raton (1995).

Sikkema, D.J., 'Design, synthesis and properties of a novel rigid rod polymer, PIPD or "M5": high modulus and tenacity fibres with substantial compressive strength,' *Polymer* **39** (1998) 5981–5986.

Toyobo Technical information, PBO Fiber Zylon, Toyobo Co. Ltd (revised 2001.9).

Wang, Y., Li, J. and Zhao, D., 'Mechanical properties of fiber glass and Kevlar woven fabric reinforced composites,' *Composites Engineering* **5** (1995) 1159–1175.

Zhang, H., Zhang, J., Chen, J., Hao, X., Wang, S., Feng, X. and Guo, Y., 'Effects of solar UV irradiation on the tensile properties and structure of PPTA fiber,' *Polymer Degradation and Stability* **91** (2006) 2761–2767.

10
High performance fibre-reinforced composites for sustainable energy applications

L. C. HOLLAWAY, formerly of the University of Surrey, UK

DOI: 10.1533/9780857099075.366

Abstract: This chapter will discuss advances in the properties, production and manufacturing techniques of the advanced synthetic fibre/polymer composite materials that are utilised in the manufacture of machines that produce sustainable energy. Furthermore, it will suggest methods for the repair, maintenance and recycling of advanced polymer composite wind turbine blades.

Key words: thermoplastic thermosetting polymers, carbon- aramid- glass-fibre and nano-fibre composites, land- sea- space-environments, composite manufacturing technologies, wind- tidal- wave- space- generators.

Note: This chapter is an adapted version of Chapter 20 'Advanced fibre-reinforced polymer (FRP) composites for sustainable energy technologies' by L. C. Hollaway, originally published in *Advanced fibre-reinforced polymer (FRP) composites for structural applications*, ed. Jiping Bai, Woodhead Publishing Limited, 2013, ISBN: 978-0-85709-418–6.

10.1 Introduction: current use of composite materials in sustainable energy technology

This chapter discusses the role that synthetic fibre/matrix composites have, and will have, in the manufacture of structural component parts used in the developed and developing technologies for sustainable power. It will discuss the types of in-service and mechanical properties of the synthetic fibres, the synthetic matrix and the composite materials that are required in the harsh environments of Earth, sea and space in which the material will operate; the methods of manufacture of the composites used will be presented. Furthermore, it will suggest synthetic fibre/matrix composite systems that could be used in the emerging tidal- and wave-power technology systems. It will be realised that with the technologies being developed to produce power, confidentiality prevents full discussion of the fibre/matrix composites which will be used. Finally, the future of fibre/polymer composites used in, and the trends, of sustainable energy will be discussed. Due to the rapid advances in the topic of sustainability, this chapter, which was completed in January 2013, will require updating in few years' time.

10.1.1 Advanced fibre-reinforced polymer (FRP) composites

Composites are made up of individual materials; these are referred to as constituent materials. The purpose of a composite is to create a material that combines its constituent parts in some beneficial way. The two main categories of constituent materials are the matrix and the reinforcement. The synthetic matrix materials are either thermoplastic or thermosetting resins. These polymers bind the reinforcement together and determine the physical in-service properties of the composite material. Polymers can also act as reinforcing material in composites; Kevlar for instance, is a synthetic polymer fibre that is very strong and imparts toughness to a composite.

Synthetic thermosetting polymers used in the construction industry are polyester, vinylester and epoxies; these materials are generally used to manufacture parts of the machines that produce sustainable energy generators. In addition, thermoplastic resins, such as polyetheretherketone (PEEK), polyethersulphone (PES) and various liquid crystal polymers (LCP) are also used. The latter high performance polymers also meet stringent out-gassing (relevant to space environments) and flammability requirements.

In any composite, synthetic fibres (such as glass, carbon and Kevlar fibres) carrying the load, and their type, volume fraction, orientation, and straightness determine their effectiveness; they are the dominant contributors to the mechanical properties of the composite. Glass fibre, the generic name given to this class of material, is used for applications where toughness, electrical non-conductivity, or abrasion resistance is required. From this it will be clear that there are a number of different types of glass fibre, all with specific mechanical and physical properties, Hollaway and Head (2001). Carbon fibre is used for applications requiring high strength and stiffness; likewise, there are a number of specific carbon fibres which may be selected having required properties, Smith (2000). The resin transfers loads between fibres, protects them and holds them in the correct location and orientation in the composite. Moreover, the type of resin used in the composite determines the resistance of the composite to water and chemical absorption and sensitivity, mechanical properties at elevated temperatures, and compressive strengths and stiffness. In addition, the resin type determines the method of fabrication of the final structural component and its cost relative to alternate resin types and fabrication methods.

10.1.2 Recently developed polymers

There are several firms that specialise in producing composite materials for machinery to generate sustainable energy; these include Advanced Composite Group (ACG, now Cytec), Gurit and Hexcel.

ACG/Cytec has developed several prepreg materials, for instance, the resin Variable Temperature Moulding (VTM) systems forming the resin VTM®260 series prepreg (which was used on the SeaGen Generator blades) and in the initial construction of the blades of the 'Quietrevolution' but later superseded by the resin MTM®57 systems. Based on this latter, superior system, ACG/Cytec have developed two new 80–120°C curing variants; these are the MTM®57-2 and the MTM®57-3. These polymer systems, used with heavyweight unidirectional (UD) carbon reinforcement prepregs, could in future form an integrated spar of the very large wind turbine blade. The skins of the future blades could be manufactured from the resin MTM®57-2 on a 1200 gsm glass ZPREG® rapid lay-up format, which would be combined with an in-mould surface primer film, MTF246.

For the repair of wind turbine blades, Gurit has developed the prepreg SPRINT™, SparPreg™, which requires no de-bulking. They have also developed the RENUVO™ blade-repair system for the repair and maintenance of wind turbine blades. SparPreg™ material is said to provide the following benefits in spar manufacture:

- fast material deposition rates;
- single debulk vacuum processing;
- 95°C curing and low exotherm;
- very low void content and material wrinkling in thick sections (e.g. 80 mm);
- wide range of industrial grade fibre choice.

GFPMS is an European composites distributor, and has established a strong portfolio of advanced composites products, including ACG/Cytec's prepregs and Sigmatex's carbon fibre materials; they do not manufacture polymers.

Hexcel (2005) have published an article on their prepreg technology. Since then Hexcel have developed the UD carbon fibre Hexcel HexPly M9G[R], which is a standard cure prepreg product. The UD carbon fibre prepreg HexPly M19G cures 15–20% faster than the HexPly M9G, whilst having the same handling and mechanical properties. Both prepregs have been certified by Germanischer Lloyd (GL) for use in the manufacture of wind turbine blades. Certification of wind turbines or components to harmonised requirements is a necessity, and therefore it is important for manufacturers, banks and insurers of wind turbines and components to know the different certification processes and guidelines as well as the keystones of their development. Consequently, GL, Hamburg, Germany has developed a new standard and innovations in certification of wind turbines, Woebbeking (2007), Woebbeking (2010). Hexcel (2005) prepregs are suitable for shells, spars and the root ends of wind turbine blades. Hexcel have also developed a new surfacing prepreg for wind energy applications. HexPly® XF2P provides a ready-to-paint surface; they claim it is tough and durable, without the need for a gel coat.

Renewergen Ltd is a tidal energy device developer but, as an offshoot to the firm's activity, undertakes blade repair using Gurit and ACG/Cytec materials; their application method is by squeegees. The repair area is ground with an angle grinder and scarified to an angle of between 25:1 and 40:1. The peel-ply is laid on the repair area and is covered with a perforated release film and breather absorber. The whole repair area is vacuum bagged, and a vacuum is applied by means of a vacuum pump to consolidate the repair whilst the peel-ply is cured.

Scott Bader Co. Ltd, Wollaston, England have recently launched Crystic Permabright, a high performance gel coat designed to provide long-term ultraviolet (UV) weathering performance by providing strong colour stability. This gel coat is designed for marine, construction, wind energy and transportation applications.

10.2 The use of nanoparticles in composites

As with advanced polymer composite, a nanocomposite is formed from the combination of two or more materials; however, one of the materials has nanoscale (< 100 nm) dimensions. Nanoparticles can be classified into three categories depending on the number of their nanoscale dimensions: (i) nano-spheres; (ii) nano-fibres; and (iii) nano-plates, having three, two and one nanoscale dimension, respectively, Thostenson *et al.* (2005). Paul and Robeson (2008) have given a comprehensive review of nanoparticles. Only nano-fibres and nano-plates will be mentioned in this chapter, as these are relevant to possible structures concerned with sustainable energy.

10.2.1 Nano-fibres

The paper of Iijima (1991) has generated unprecedented interest in carbon nanostructures, and has fuelled intense research in the area of nanotechnology.

Carbon nanotubes (CNT) consist of molecular cylinders of pure, hexagonally-arranged carbon atoms with a diameter of a few nanometres and a length of many microns. They occur in two main forms: the single-wall carbon nanotube (SWNT) composed of a single lattice cylinder of carbon, and the multi-wall nanotube (MWNT) version, consisting of concentric lattice cylinders of carbon; they resemble one cylinder within another. The ends of the tubes are usually closed off by a carbon end-cap, also in lattice form.

The strength of the sp^2 carbon–carbon bonds (the sp^2 hybridisation is the mixing of one s and two p atomic orbitals, which involves the promotion of one electron in the s orbital to one of the 2p atomic orbitals – the combination of these atomic orbitals creates three new hybrid orbitals equal in energy-level (Housecroft and Sharpe, 2008)) provides carbon nanotubes with extraordinary mechanical properties; the nature of the bonding of a nanotube is described by applied quantum chemistry, specifically orbital hybridisation, and the chemical bonding of nanotubes is composed entirely of sp^2 bonds, similar to those of graphite. The scale size, aspect ratio and properties of CNT provide advantages in a variety of applications, including electro-statically dissipative materials, advanced materials with combined stiffness, strength and impact for aerospace, space, composite mirrors and components with enhanced mechanical properties. Consequently, CNTs are a promising new material for blending with polymers and having the potential to obtain low-weight nanocomposites with exceptionally good mechanical, electrical, thermal and multifunctional properties ideal for use in the manufacture of structural components to produce sustainable energy.

The mechanical properties of various types of nanotubes have been extensively studied both by experimental and computational means, Thostenson *et al.* (2001), Qian *et al.* (2002), Ruoff *et al.* (2003). Their tensile strength, can vary between 100–600 GPa, which is about two orders of magnitude higher than that of current high strength carbon fibres, Sennett *et al.* (2003), Koziol *et al.* (2007), and their density is around 1.3 g/cm^3, lower than the density of commercial carbon fibres (1.8–1.9 g/cm^3). The enhancement in strength implies that, for the same performance, replacing commercial carbon fibres with CNTs will lead to significant reduction in the density and volume of the structural composite parts, Breuer and Sundararaj (2004). The compressive strength of CNTs is approximately two orders of magnitude higher than that of any known fibre, Louriel *et al.* (1998), Salvetat *et al.* (1999). CNTs are also one of the stiffest material structures ever made; compared to carbon fibres, which typically have a modulus of elasticity of up to 750 GPa, the modulus of elasticity of nanotubes ranges between 1 and 5 TPa, Iijima (1991), Sennett *et al.* (2003).

Loos (2011) reported that investigators from Bayer Material Science LLC, USA and Moulded Fibre Glass, Cleveland, USA have developed a prototype wind turbine blade 0.74 m long manufactured from polyurethane reinforced with carbon nanotubes (CNT PU). The researchers claim that the advanced material has a specific tensile strength five times and 60 times that of carbon fibre composite and aluminium, respectively, and is tougher than carbon fibre-reinforced polymer (CFRP); but the excellent properties of these materials come with a price penalty.

10.2.2 Nano-plates

Nano-plates are generally naturally occurring layered materials such as layered silicates (montmorillonite plates, a type of clay) which is dispersed within polymers for nanocomposite formation, Hackman and Hollaway (2006), Tran *et al.* (2006), Hackman (2007); initially, man-made materials, such as silicate acids, were used, Wang *et al.* (1996). Only a limited number of layered silicates are suitable for use as nano-plates. The objective in a nanocomposite produced from plate-like fillers is to disperse the latter in a polymer to take advantage of the large surface area of the plate and thus modify the polymer properties. Nano-plates were first produced by a group of Toyota researchers when they dispersed clay within Nylon-6, Kojima *et al.* (1993a); the nano-plate composites possessed considerable mechanical and permeability advantages, Kojima *et al.* (1993b). A number of extensive reviews have been published covering various areas of nanocomposite processing, behaviour and properties with respect to numerous types of polymer, Le Baron *et al.* (1999), Ray and Okamoto (2003), Utraki (2004).

10.2.3 Advantages of incorporating nano-plates into FRP composites

Barrier properties. At a high aspect ratio, which can be achieved in nanocomposites, (with exfoliated clay) some degree of decrease in water permeability can be predicted, and observed in practice, Hollaway and Hackman (2004), Hackman and Hollaway (2006), (who used layered silicates montmorillonite plates-EPON 828), Liu *et al.* (2005) (who tested a TGDDM-DDS (diaminodiphenylsulphone) system). However, Wang *et al.* (2006) investigated highly exfoliated randomly dispersed platelets in a DGEBA-Ethacure 100 system and found that this system slightly increased the uptake of moisture. It is therefore to be noted that the moisture barrier properties of polymers can be altered by inclusion of inorganic platelets with sufficient aspect ratio to alter the diffusion path of the penetrant molecules; it is clear that there is considerably more research needed to be undertaken in this area.

Flammability resistance. An increased flammability resistance has been noticed as an important property enhancement involving nano-platelets incorporated into polymers involving exfoliated clay; this involves the formation of a stable carbon/nano-platelet or nanofibre surface. This surface exhibits analogous characteristics to intumescent coatings, whereby the resultant 'char' provides protection to the interior of the specimen by preventing continual surface regeneration of available fuel to continue the combustion process. The primary advantage noted with nano-filler incorporation is the reduction in the maximum heat release rate.

Hydrogen is a promising alternative to fossil fuels, due to its cleaner combustion; the only combustion by-product of hydrogen is water. Compared to petrol, hydrogen is lightweight, can provide a higher energy density and is readily available. However, to replace petrol as a fuel, hydrogen must be safely and densely stored, but easily accessed; storage of hydrogen is a potential problem. Scientists at the US Department of Energy, Lawrence Berkeley National Laboratory have designed a composite material for hydrogen storage consisting of nanoparticles of magnesium metal sprinkled through a matrix of polymethyl methacrylate, a polymer related to Plexiglas. This nanocomposite rapidly absorbs and releases hydrogen at modest temperatures without oxidising the metal after cycling; this technology has been designed specifically for hydrogen storage, batteries and fuel cells. This work shows an ability to design composite nanoscale materials that can overcome fundamental thermodynamic and kinetic barriers to realise a materials combination. The unique properties of both the polymer and the nanoparticle to form a composite material may be applicable to related problems in other areas of energy research.

An area where nanocomposites could achieve a dramatic commercial prominence is in advanced polymer composites. CFRP composites have limited achievable properties, particularly in cross-ply composites, due to the low modulus and strength of the matrix phase. Modification of the matrix with CNT, at the lower scale of dimensions and carbon nano-fibres, at a higher dimensional scale, would allow for significant increase in the modulus and strength contribution of the matrix to the overall composite properties. Whilst this would offer some improvement in UD composites, it could be dramatic in the case of cross-ply composites, which are the major type of composite structure utilised in some advanced composite applications.

Whilst there have been some components manufactured using CNT and nano-plates in development, the nanoparticle markets have been constrained by three main issues, namely:

- A lack of commercially available material of consistently high quality.
- The nanoparticle technologies are at an early stage of industrial development. Whilst they are expensive compared to a fully commoditised product, such as carbon fibre, the price will fall as demand increases. Currently, they are probably too expensive for the field of sustainable energy, but their potential cannot be ignored.
- The incredibly small scale of the material poses some interesting challenges for advanced material and coatings science.

Such teething problems in developing new materials are not unfamiliar; carbon fibre, for instance, took many years to be widely accepted in the materials world from both cost and performance aspect.

10.3 In-service requirements of advanced fibre-reinforced polymer (FRP) composites for sustainable energy applications

10.3.1 Land and sea environments

The properties of synthetic fibres, matrices and fibre arrangement within the composite and the fibre volume fraction govern the final strength and stiffness value properties of the composites. These parameters have been illustrated in references, Smith and Yeomans (2002), Hollaway and Teng (2008). The matrices for the composites that could be, and are, used in the manufacture of wind turbine blades are polyester, vinylester and epoxies. Glass, Kevlar and carbon fibres can be used with any of the polymers mentioned but, as the rotor blades become larger, a hybrid construction of glass and carbon is used. The hybrid concept is often a compromise between the improved performance of carbon fibres and their high cost. Few rotor blades have been made completely of carbon fibre composites.

The wind turbine rotor blades, both for on-shore and off-shore wind farms, are exposed to various hostile conditions such as extreme temperatures, humidity, rain, hail impact, snow, ice, solar radiation, lightning and salinity. In order to withstand these external conditions without diminishing safety, a sound knowledge of the fatigue behaviour of the material and structural properties is needed.

The cyclic loading of the structure of a wind turbine could cause failure if some critical level of damage is exceeded. Once initiated, the damage will grow with the load cycling until failure occurs. The failure process would occur because of one of the following:

- The net section stress, accounting for the loss of section caused by the damage, exceeds the ultimate strength of the material.
- A critical crack forms due to accumulation of damage.

Forces relevant to fatigue. The S-N curves provide an indication of the relevant fatigue properties; they do not take into account the complex effects of the large number of different cyclic forces which act on a wind turbine blade during operation. These forces arise due to the self-mass of the blade and the force of the wind acting upon it, and include:

- The gravitational force, which leads to compression and tension through each cycle.
- The centrifugal force, due to the rotation of the blade.
- The wind thrust which is a force that is perpendicular to the plane of the rotor blade; it varies relatively slowly.
- Other rapidly varying forces arise from the wind turbulence, which increases as the stall conditions are approached.

Research workers have concluded that the relatively low-frequency, high-amplitude and wind-thrust forces primarily contribute to fatigue damage.

Environmental factors affecting fatigue. Environmental attack can rapidly reduce the material's fatigue strength. This could happen in two ways:

- Blade topography – the topography of the surface of the blade may be modified by minute erosive and corrosive pits from sand or rain impingement. These would act as stress concentrations during cyclic loading, causing localised cracking to be initiated. The erosion attack would occur near the blade tips, where the rotational velocities can reach the equivalent of 100 m/s.
- Bulk material properties – the bulk material properties may be altered, thus reducing fatigue strength through the blade wall thickness or through the surface layers. To overcome this problem protective coatings are applied, initially at the manufacturing stage and, if necessary, during maintenance procedures. The leading edge of wind turbine blades will require a special finish.

The fatigue properties of composite materials depend on the inherent strength and stiffness of their component materials, as well as on their structure. Experimental full scale simulations on glass fibre-reinforced polymer (GFRP) turbine blades have been conducted and the results indicate a satisfactory service life under normal conditions. However, laboratory test data show a steadily decreasing S-N curve indicating a finite service life; it is advisable to monitor rotor blades under operational conditions. Composite materials containing higher modulus or stiffer fibres possess better fatigue properties, if the cyclic stress is applied parallel to the fibre orientation. CFRP composites exhibit excellent fatigue performance when compared to GFRP composites, particularly when subjected to tension fatigue in the fibre direction.

10.3.2 Seawater environment

Fibre/polymer composites have a long history of use in marine vessels, piping, corrosion equipment and underground storage tanks, and anecdotal evidence and limiting testing shows that they can be successfully engineered to have long service life in contact with moisture and aqueous solutions, Helbling and Karbhari (2007). Therefore, the use of composite materials for tidal energy power generators and similar technologies looks promising.

Durability and dynamic failure properties are critical parameters for fibre/polymer composites in seawater. In general, all carbon fibre/epoxy engineering composite structures are subject to a multiaxial stress state; furthermore, carbon/epoxy composites are susceptible to environmental degradation from long-term saltwater exposure. It is important that the long-term effects of saltwater on the multiaxial fatigue behaviour of these composites are understood, but research results of impact resistance of composite materials when immersed in seawater are sparse and not well documented, Monaghan and Wang (2004). Chiou and Bradley (1997) investigated the fatigue and static edge delamination on $[45/0/-45/90]_s$ CFRP laminates, in both the dry state and in pre-soaked seawater specimens, to examine any effect moisture absorption had on fatigue crack development; the results were monitored by optical microscopy and ultrasonic C-scan. It was found that seawater changed the dominant edge-cracking mode from the −45/90 interlaminar delamination in the un-aged specimen to intralaminar cracking in 90° plies in the aged saturated seawater specimens. The edge-crack growth rates in both un-aged and aged specimens were similar. Monaghan and Wang (2004) studied the long-term saltwater effects on multiaxial fatigue degradation, failure modes and stress-life relationship. A series of tension–torsion fatigue experiments were conducted on hoop-wound carbon/epoxy tubes in a variety of saltwater environments. It was concluded that the composite elastic properties degrade slightly during cycling. Long-term soaking in saltwater provides the most significant reduction in fatigue life, modulus and monotonic strength. Xu (2010) tested more than 50 specimens for 21 months under saturated seawater exposure; he showed that after impact the compression strengths of the wet specimens reduced less than 9% compared to the dry baseline specimens. From these results he concluded that the durability of composites exposed to seawater were much better than had previously been realised.

10.3.3 Space environment

Space-based solar power (SBSP) (see Appendix 10.12.2) geostationary structural systems would be exposed to harsh environments, and therefore the physical properties of the composite materials must resist such loadings.

The necessary basic criteria for a wholly space resistant, space tailored composite are:

- Atomic oxygen (AO) and atomic nitrogen (AN) resistance.
- Radiation resistance.
- Low out-gassing – it is important that the out-gassing is low as the volatile compounds can migrate from the polymers and condense onto nearby surfaces.
- Fatigue resistance.
- Vacuum stability.
- High specific strength and stiffness.
- The thermal differential stresses mentioned above.

Hollaway (2011) has discussed the various space environments.

The type of environment encountered in low earth orbit (LEO) (approximately 250 km above the Earth's surface) to which space systems would be exposed contains:

- High levels of magnetic flux and trapped solar radiation. This environment may degrade a composite quite quickly, with the polymer cross-linking embrittlement occurring as a consequence of the high ionisation levels that exist in this orbit.
- Residual atmospheric drag effects and chemical attack from highly energetic atoms, principally from AO and AN radicals, Leger *et al.* (1986), Leger *et al.* (1987).
- Incessant and rapidly varying thermal fluctuations as the systems travel through the Earth's umbra, Annandale (1986).

The LEO environment would be experienced for a period of time, albeit a short period during the space structure's fabrication and deployment (see Section 10.3.3).

At the higher altitude of geostationary orbit (GEO), at 36 000 km above the Earth, the space structure is exposed to a different loading regime. The environment to be encountered in GEO to which space systems would be exposed features:

- The solar wind, which is hydrogen plasma of varying intensity and is dependent on solar flare activity; its velocity is approximately 400 km/s. The effect of this plasma can be expected to lead to substantial degradation of the space structure's material thermal properties over its lifetime.
- The essentially beneficial warming influence of planetary infra-red (IR) radiation.

- The greatly diminished Earth albedo at these high altitudes.
- The space-system not being attacked by AO at these heights.

10.4 Manufacture of FRP composite materials for sustainable energy technologies

The following processing methods of synthetic fibre/matrix composite materials have been used in sustainable energy technologies and are referred to where necessary.

10.4.1 Wet lay-up

This process is one of the open mould processes for the manufacture of fibre/polymer composite. Initially, the mould cavity is coated with either polyvinyl alcohol or a non-silicon wax to aid component release. If a resin-rich smooth surface to the final composite is required for environmental protection purposes, a gel coat is applied to the mould surface. Furthermore, for improved surface finish and corrosion resistance of the composite a surface veil is used, which is applied with an embedded fabric for reinforcement. Each layer of the designed fibres is positioned on the mould. An accelerator and a catalyst are mixed in with the resin, and the whole mixture is applied to the fibres, by brush or poured onto the fibres and rolled to ensure complete wetting of the fibres; during this process, air bubbles are removed. For large mouldings, the cold cured resin is polymerised at room temperature for a period of 16 h at 40°C, but these figures depend upon the resin system used; if the temperature is lower than 40°C, the composite must be post-cured for a further time, the length of time being dependent upon the post-cure temperature. If a hot-cured resin is used, an accelerator would not be employed. For health and safety considerations, care must be taken at the time of manufacture of the composite due to styrene emissions. This method is not suitable, in terms of strength or weight critical primary structure, as the fibre orientation and local resin content cannot be well controlled.

To improve the quality of the final composite by reducing the voids caused by air becoming trapped in the laminate and improving the consolidation at the wet lay-up stage, a vacuum-assisted technique can be applied. The wet composite is sealed within a bagging material and a vacuum is applied to the wet composite using a vacuum pump to extract the enclosed air; the polymer is then cured.

10.4.2 Resin infusion technology

Resin infusion technology is a process using a dry fibre laminate preform. A dry stack of fabric material is placed into a mould tool; the fabric is sometimes pre-stressed to the shape of the mould and held together by a binder. A second mould is then clamped over the first, and the resin component is added by an infusion methodology; alternatively, a flexible film membrane can be placed over the composite material. The thermosetting polymer enters the space between and around the dry fibre laminate preform through feeder pipes and a runner system; the composite is then cured in place. A high quality, dimensionally accurate polymer composite is formed with a high quality surface finish. Typical polymers used are epoxies, vinylester, polyester or phenolic with glass or carbon fibre preform reinforcement which is stitched.

The process of resin infusion is often divided into two distinct categories, based on the manner in which the resin is infused into the preform and mould cavity; these divisions are not universally accepted:

- Resin transfer moulding (RTM), or the recently invented light resin transfer moulding (LRTM), in which the resin is introduced into the mould and the preform at greater than atmospheric pressure.
- Vacuum infusion process (VIP), generally referred to as vacuum-assisted resin transfer moulding (VARTM), in which the resin is introduced into the mould and the preform at less than atmospheric pressure (i.e. a vacuum).

Process innovations have led to a number of advances in this field which have improved the process, or have developed a process to meet structural or design needs. Examples of these are:

- The Seemann Composites Resin Infusion Molding Process (SCRIMP™) system, which was developed to infuse very large and complex structures more quickly than other RTM methods by taking advantage of an increased rate of flow through the porous media over large surface areas such that infusion travelled the minimum distance (e.g. through the thickness).
- Resin film infusion (RFI), where dry fibres are laid up interleaved with layers of semi-solid resin film supplied on a release paper. The lay-up is vacuum bagged to remove the air through the dry fabrics and then heated to allow the resin to first melt and flow into the air-free fabrics, and then to cure.
- Autoclave pressure, temperature–time sequencing, and vacuum draw similar to conventional prepreg procedures are utilised to cause the resin to flow in the minimum thickness direction.

10.4.3 Prepreg technology

A pre-impregnated (prepreg) laminate consists of a specially formulated pre-catalysed resin system, using machinery designed to ensure consistency; the resin is combined with fibres or fabrics reinforcement.

One of the most common methods of preparing a composite prepreg is to draw a sheet of woven fibre through a bath, or to impregnate it through gravity from a container, containing the pre-catalysed resin system. A heating chamber removes the solvent through evaporation and partially cures the resin. Ideally, the resin and cross-linking agent have reacted to form short polymer chains, producing a viscoelastic solid, but which has not cross-linked or vitrified it into a glassy solid. The prepreg should be slightly tacky, so that layers do not slip over each other during lay-up, and it should be sufficiently pliable such that it forms (drapes) to the mould plate. Controlling the extent-of-cure is extremely important. The prepregs are then covered by a flexible backing paper, and can be readily handled and remain pliable for a certain time period (out-life) at room temperature. Curing of prepreg laminates is achieved by heating at elevated temperatures under pressure according to the manufacturer's specifications. Conventional prepregs are formulated for autoclave curing, whilst low-temperature prepregs can be fully cured by using vacuum-bag pressure alone at much lower temperatures. Before use, prepregs are typically stored at sub-freezing temperatures (generally –20°C) to prolong their useable life. The prepreg technology, including the manufacturing method for the production of the prepreg, is discussed in European Pultrusion Technology Association (EPTA) Publication (2008).

10.4.4 SPRINT® technology

When using prepreg materials in thick laminates (greater than 3 mm) it becomes difficult to remove air entrapped between plies, and around laminate details such as ply overlaps, during the curing procedure. To overcome this problem, in traditional prepreg materials a number of warm debulking stages are introduced during lay-up to remove the trapped air and thus significantly reduce the manufacturing time. Gurit has patented a number of modified prepreg products that enable the manufacture of high quality (low void content) thick laminates in one processing step; two of these are:

- SPRINT® materials, consisting of a resin film sandwiched between two dry fibre layers. Once placed in the mould, a vacuum is applied to extract all of the air in the laminate (monolithic and/or sandwich) before heat is applied to allow the resin to soften and impregnate the dry fibre layers and then cure.
- SparPreg™, a UD prepreg also for use with thick laminate sections. The excellent 'breathability' of the material produces laminates of exceptional quality with low void content, without the need for debulking or additional dry fabric reinforcement to aid air removal during the application of vacuum and the subsequent curing process.

The main advantages of the SparPreg™ technique are:

- High fibre volumes can be accurately achieved with low void content for very thick laminates (100 mm).
- High resin mechanical properties, due to the solid state of initial polymer material and elevated temperature cure.
- It enables the use of lower cost heavyweight materials (e.g. 1600 gsm), and allows fast deposition rates reducing component manufacturing costs.
- It is a very robust and repeatable process, as the resin content is accurately controlled and the complexity of the infusion process is very low.
- It is environmentally safe, as the resin has already been mixed, so there is no emission of volatile gases typically present when mixing two component resin systems.

The main disadvantages are:

- Materials cost is higher than for non-pre-impregnated fabrics.
- Tooling needs to be able to withstand higher temperatures compared to the infusion processes (typically 80–140°C).

10.4.5 Film-stacking technology under elevated temperature and pressure

Prepreg sheets of thermoplastic aromatic polymers (a PES or PEEK) can be moulded by a hot press using vacuum or compressed air. The equipment consists of a fixed frame that can be heated and a two-step heating system for providing a uniform temperature distribution to the surface and cross-section of the sheet. The heater is designed so that the temperature will reach 270° to 280°C within 40–50 s. To form tubes, a film-stacking and compression moulding technique is used. It consists of steel moulds and an expandable mandrel (a suitable material would be PTFE) of the correct size for the finished internal diameter of the rods; the initial cure temperature of a carbon fibre – polyethersulphone (CF-PES) would be 290°, and a pressure of 3 MPa would be applied to the mould. To fabricate the tubes the prepreg films are wrapped on to the PTFE mandrel at the correct fibre orientation and stacking sequence. The cure temperature must remain at the constant value (i.e. 290°C) for half an hour; it is then reduced to room temperature. The PTFE material has a coefficient of thermal expansion much greater than that of the steel moulds, and this leads to high compaction of the polymer against the metal mould and a low void ratio.

10.4.6 Pultrusion

The pultrusion technique is a closed mould system, and is a fully automated continuous process. Continuous fibrous reinforcement rovings and strand mat, or other designed reinforcements, are pulled through a reservoir of resin and a heated die. Alternatively, the fibres can be impregnated with the resin by injecting it through port holes in the heated die as the fibres pass through. The fibre placement, resin formulations, catalyst level, die temperature and pull speed are all critical process variables. These variables must be established during the design of the product and during manufacture; continuous monitoring must be undertaken to assure that the finished pultruded unit has the correct appearance and specific physical and chemical properties. The pultrusion technology has been discussed in EPTA Publication (2008).

10.5 Composite materials/fabrication techniques for on-shore and off-shore wind turbines

10.5.1 Introduction

A wind turbine transforms the kinetic energy of the wind into mechanical energy and then into electrical energy and/or heat. There are three components forming the turbine, viz. the rotor (an aerodynamic device and a rotor blade with an aerodynamic shape for rotation), the nacelle (containing the gearbox and generator), the yaw mechanism (sensors to monitor and rotate the turbine directly into the wind in order to generate maximum power), and the tower. The blades rotate around a horizontal hub, which is connected to a gearbox and generator located inside the nacelle. The latter houses the electrical components, and is mounted at the top of the tower (column of the turbine). This type of turbine is referred to as a 'horizontal axis' machine and is the commonest of all turbines.

Synthetic fibre/polymer composites are the materials of choice for rotor blades in wind turbines, as these have the best balance between stiffness and density for the application, Brøndsted *et al.* (2005), although there are other composite materials which could be used for wind turbine blades, Griffin and Ashwill (2003). As mentioned in the previous sections, the most common design is a thermosetting matrix (polyester, vinylester or carbon) with glass fibre. Although glass reinforcements are not as stiff and strong as carbon fibres, they provide similar reinforcements at a reduced cost, albeit at an increased volume content. Different volume fractions of fibre, matrix and fibre arrays can be employed to provide different mechanical strengths and stiffness. Furthermore, carbon-fibre-reinforced load-bearing spars have been identified as a cost-effective means for reducing weight and increasing stiffness.

Typical large wind turbine blades consist of outer skins supported by a main spar and stiffeners. The blades are generally constructed using synthetic fibre/polymeric matrix composites, and may have a sandwich construction,

Allen (1969), with low density polymer foam or balsa wood core. The epoxy-based composites are of greatest interest to wind turbine manufacturers, because they deliver a combination of environmental, production and cost advantages over other resin systems. Epoxies also improve wind turbine composite blade manufacture, by allowing shorter cure cycles, increased durability and improved surface finish. The utilisation of epoxy infusion or prepreg manufacture (see Sections 10.4.2 and 10.4.3) further improves cost-effective operations by reducing processing cycles, and therefore manufacturing time, compared with the former method, wet lay-up procedure. However, vinylester, phenolic and high temperature polymers such as polyimides, cyanate esters and bismaleimides (BMI) have also been used. The primary motivation for the use of carbon fibre is its lower density, higher stiffness and tensile strength relative to glass fibre. With the increase in the very large rotor diameters currently being developed, it could be expected that there will be an increase in the utilisation of carbon fibre in the spars and selected regions of the aero-shell.

Figure 10.1 illustrates a typical section of fibre lay-up through a Gurit manufactured wind turbine blade. Further information on the manufacture of turbine blades, the materials used, their description and properties may be obtained from the Gurit *Handbook – Materials for Wind Turbine Blades*.

10.5.2 Fabrication techniques for wind turbine blade moulds

Metal moulds, on which the blades are formed, are still used by some manufacturers of smaller blades, but polymer composites are invariably used for the multi-megawatt (MW) wind turbines; the following description of the manufacture of FRP composite moulds has been derived from Marsh (2007). The moulds are generally manufactured using epoxy (although vinylester or polyester polymers can be used), reinforced with glass fibres and/or carbon fabric, woven rather than stitched. The blade is manufactured by moulding two half shells into a pair of composite moulds. These mould materials are much lighter than all-steel ones, and have less thermal inertia and their coefficients of thermal expansion are then matched with those of the wind turbine blade materials; a mismatch between two unlike materials could create difficulties during in-mould thermal cycling. The composite moulds are supported on mild steel support frames. The complete tool pair and for, say, a 40 m wind turbine blade, can weigh about 16–18 tonnes; the polymer composite in the tool would weigh about 4 tonnes and the rigid steel backing frame would weigh about 12 tonnes, Marsh (2007). It is possible to achieve a smooth FRP mould surface containing a high vacuum integrity with the infusion technique, which is rapidly becoming the process

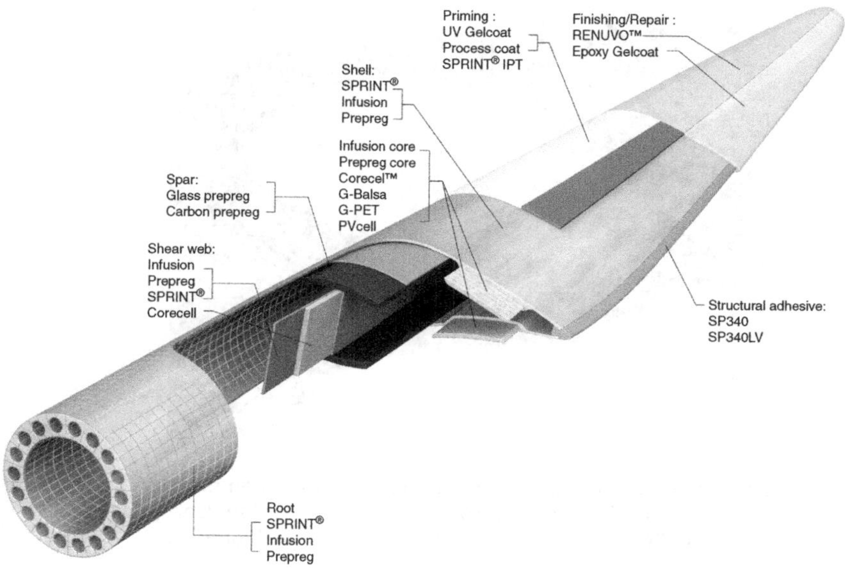

10.1 A typical section through a Gurit wind turbine blade (image courtesy of Gurit).

of choice of moulds for significant series production; it is also possible to use prepreg material. Polymer composite mould tools are generally rated to produce 300–400 wind turbine blade sets, but they rarely achieve this value before the blade becomes obsolete.

A set of tools is required for the manufacture of a turbine blade; these include the two half shells in a pair of moulds, the web stiffeners and the spar cap. Further moulds may be associated with blade edges and root ends.

A spar cap, web stiffeners and other details are installed in one of the halves of the blade mould, and adhesive is applied to the exposed bonding edges of these units. The second half of the mould, containing the half shell, is then turned over and lowered onto the first; the adhesive is then allowed to cure, thus joining the two halves of the wind turbine blade together. The steel support frames are hinged together at one side to enable the mould to open clamshell fashion. A gel coat can be used on the inner mould surface to obtain a blade surface free of voids that can be present in mouldings produced by wet lay-up or from prepregs. Figure 10.2 shows an image of an open composite mould.

To obtain the configuration of one of the outer surfaces of the turbine blade that will be formed from one of the inner surfaces of the mould, a polystyrene plug is first shaped to the exact profile of the outer blade surface; the plug is generally about 6 m long and can be readily joined to and separated from its neighbour as required – this provides the flexibility to

10.2 An image of the composite mould in the open position for the manufacture of the wind turbine (courtesy of Solent Composite Systems (SCS); the system uses the SCS SmartMould™).

change individual modules as modifications are made to the blade design. The plug is then reversed and the other face is machined to the profile of the second face of the turbine blade; the two faces of the plug are shaped to fine tolerances and are then ready for the application of the release system and subsequent lamination of the mould.

10.5.3 Examples of on-shore and off-shore wind turbines

This section will describe one unique on-shore and one unique off-shore wind turbine.

Quietrevolution on-shore wind turbine

The 'Quietrevolution' wind turbine was conceived by Quietrevolution, which was designed and developed by XCO2 in conjunction with Aviation Enterprises Ltd. The rotating section is 5 m high and 3.1 m in diameter, and is designed to withstand a maximum wind speed of 120 mph. It is vital that this wind turbine's weight is kept to a minimum, due to the centrifugal forces developed of up to 200 G. XCO2 states that the operation of the turbine is more efficient compared with the traditional turbine as it does not have to change direction constantly with the wind; this advantage reduces the vibrations and the aerodynamic noise. A further advantage is that the cylindrical structure with the vertical 'S' blades occupies less space than the conventional turbine, and is able to integrate into the urban environment; it is very pleasing aesthetically. The above information has been based upon the company's web site: www.quietrevolution.com. Figure 10.3 shows an image of the 'Quietrevolution' wind turbines – these are placed upon the roof of City House Building in Croydon.

The blades of the Quietrevolution wind turbine were originally manufactured from ACG/Cytec's resin VTM system, but were changed to ACG/Cytec's resin MTM®57 systems to reduce costs. The blades are now manufactured using a prepreg system which is similar to MTM®57 system.

Aerogenerator off-shore wind turbine

Many of the most innovative wind energy systems proposed in recent years target the off-shore market. Off-shore is the ideal position for wind turbines,

10.3 'Quietrevolution' wind turbines on the roof of City House Building in Croydon.

10.4 The 'Aerogenerator'.

but the environment can be hostile. The technology for this type of turbine is based on the same principles as that for on-shore turbines; the off-shore winds tend to flow at higher speeds than those on-shore, thus allowing turbines to produce power more economically.

As an example of the rapid development of off-shore wind turbine technology, the 10 MW 'Aerogenerator' is a revolutionary design for a vertical axis off-shore wind turbine, conceived by Wind Power Ltd. in conjunction with architects Grimshaw, Cranfield University, Rolls Royce, Arup, BP and Shell; the 'Aerogenerator' is shown in Fig. 10.4. The first stage has been funded by the Energy Technologies Institute. In its presently conceived form, it is twice the size and power of any conventional wind turbine and because of its economies of scale its capacity could transform the global energy market. It is believed the first turbines will be built in 2013–2014, followed by 2 years of testing, The Engineer (2011); this is probably an optimistic forecast. It is different from the typical wind turbine, in that it features a set of blades mounted on a vertical axis with a blade span of just over 270 m.

The two arms and the two sail (blades) design comprises a steel hub and the lower arm sections. The two upper arms are 140 m long and are manufactured from carbon fibre fabric skin, as are the spar and rib box structures. The glass fabric composite leading and trailing edge fairings are bonded to this. The sails use similar materials selection to the structures connected by steel pinned fittings. The composite parts of the structure have a total mass of about 150 tonnes.

The manufacturing features a novel automated fabric laminating system for the one-piece skin and spar components. For ease of construction, the *in situ* resin infusion technique and the curing of the composite assembly is undertaken in a factory, ideally located at the dockside where the fabrication of the machine is being assembled.

The size of the structure presents some structural and manufacturing challenges, in particular the strut and cable stay joints, the structural box adhesive bonding, the carbon fibre tow supply, and the mass of the components and structure during assembly.

The next generation of this type of system is the Aerogenerator X, which is a vast machine and has been described as a system with more power and less weight, and the only real alternative solution available to help deliver the UK's off-shore wind strategy in a reliable and cost-effective manner. The original Aerogenerator had a span of 275 m and could produce ten megawatts of electricity. All systems float, and their weight is concentrated at their base, thus giving a low level centre of gravity. The systems could, therefore, reduce the costs of deep water off-shore wind energy. Both systems mimic a spinning sycamore leaf and use techniques developed for semi-submersible oil platforms. This new design is half the height of an equivalent horizontal axis turbine.

10.5.4 Repair and maintenance of wind turbine blades

Maintenance of wind turbine blades is a critical element in the lifecycle cost of turbines and the efficient running of the system; poor maintenance will lead to erosion and an inefficient wind turbine. Damage to the blades may occur during transportation to the final site, erection of the complete nacelle onto the tower or during operation in hostile environments such as rain, wind erosion and lightning strike. Wind turbulence and changes in wind direction will have been taken into account during the design, but these loads will have a severe impact on the blade fatigue, and can lead to early wear and tear damage. In the past, the repair systems used were specific to the original blade substrate construction (polyester, vinlyester or epoxy) and were fabricated by a wet process and thus limited to use in a narrow temperature window during curing. The utilisation of this process when maintaining blades in their fixed position is difficult, particularly in high wind speed and low temperature.

Gurit has developed and launched the RENUVO™ blade-repair system (see Section 10.1.2) which overcomes the above disadvantages, Cripps (2011). Depending on the thickness of the composite laminate, a high intensity UV cure of 60–180 s cures the material to its full value. The RENUVO™ materials are specially formulated, and exhibit excellent handling, adhesion and cured properties at temperatures as low as +5°C and at high humidity levels. It has been through the combination of the low-temperature operating window and fast UV cure system that the RENUVO™ blade repair has received certification by the GL (See Section 10.1.12 for definition).

Further information may be obtained from Marsh (2011).

10.5.5 Recycling wind turbine blades

Currently, there are three possible routes for disposing of thermosetting polymer/fibre wind turbine blades, which are: (i) to send them as waste to landfill; (ii) to incinerate them; or (iii) to recycle them. As countries are seeking to reduce the landfill option generally, and in particular GFRP due to its high organic content, the first disposable option is losing its credibility. The most popular option is to incinerate the material to create power (electricity). However, there are objections to this route. After incineration of the GFRP about 60% of residue remains and, due to the possible inorganic pollutant nature of this residue, two courses of action would be considered: (i) if it were a pollutant it would go to landfill; and (ii) if it were not a pollutant it could be recycled as a substitute construction material. Furthermore, the inorganic material emits hazardous flue gases, and the small glass fibres in the residue may cause problems to the flue gas cleaning stages. Before the blades are recycled they have to be crushed, which places a further strain on the environment in terms of energy used.

10.6 Composite materials/fabrication techniques for tidal energy power generators

10.6.1 Introduction

Tidal stream systems use the kinetic energy from the ebbing and surging tides, and wave power uses the oscillations of the waves created by the wind blowing across the surface of the water thus forming ripples – the stronger these winds become, the larger and stronger will be the waves. There are two types of tidal energy that are able to produce electricity:

- The tidal stream system uses the kinetic energy from the ebbing and surging tides.
- Tidal barrages are designed to utilise the potential energy from the difference in height of the tidal waves.

The former system is the one with which this section is concerned, and three tidal energy generators will be discussed in the following section.

Converting tidal energy current to electrical current poses a major polymer composite materials challenge. The hydrosphere is a particularly hostile environment as the water is approximately 830 times denser than air; the exact value will depend upon its depth. The flow is varying, often turbulent, and will reverse in direction with the ebb and flow of the tide. Moisture ingress is a constant hazard, exacerbated by device motion and equipment maintenance; the latter event has to be infrequent, as generally the access to submerged installations is limited. Furthermore, fatigue, which is not often a significant consideration with composites, becomes important when a typical service life of 20–25 years in a turbulent environment is proposed. It is clear that the UK has by far the highest potential in the European Union for converting tidal energy into electricity. The advantage of off-shore tidal generators is that tidal currents are sub-surface, so tidal generators have minimum visual impact, unlike wind farms or estuary barrage schemes.

Three tidal wave systems will be discussed in the following sections, viz. SeaGen Generator, Atlantis Tidal Generator and Pulse Tidal Generator.

10.6.2 SeaGen tidal power generator

In 2010 SeaGen tidal stream generator was the largest tide-harnessing turbine in the world. It captures the energy of the tide by placing the generator into the path of flowing water and is currently known as the 1.2 MW SeaGen, which is installed in Strangford Narrows, Northern Ireland. Figure 10.5 shows an image of SeaGen. It was the first tidal turbine to produce energy for the National Grid on a commercial scale and in March 2011 it passed the British Marine Current Turbines, the UK Government's operating performance criteria for emerging tidal and wave energy technologies. It has the capacity to deliver about 10 MWh per tide, which is equivalent to 6000 MWh per year, Fraenkel (2011). It consists of twin axial-flow rotors of 16 m in diameter, each driving a generator through a gearbox similar to the wind turbine; each twin rotor sweeps over 200 square metres of flow. These generators have a patented feature by which the rotor blades can be pitched through 180°, allowing them to operate in both flow directions (on the ebb and the flood tides). The power units of each system are mounted on arm-like extensions on either side of a tubular steel monopile some 3 m in diameter; the arms and the power units can be raised above the surface for maintenance access; the inset image of Fig. 10.5 shows the blades acting under water. Each blade of the SeaGen rotor comprises a hollow carbon fibre composite box spar as the main load-bearing member, along with carbon ribs, and a glass fibre composite envelope bonded to this skeleton. The tubular steel monopile system is situated on piles drilled into the bedrock of the seabed; SeaGen weighs 300 tonnes, Rush (2008). Sea Generation Ltd. is

10.5 The SeaGen tidal power generator.

the project company, which is a wholly owned subsidiary of Marine Current Turbines (MCT) Ltd. based in Bristol, and SeaGen was developed and manufactured by MCTs. Many pre-production tidal stream devices are now in operation, or are currently being installed in several locations around the British Isles. SeaGeneration (Wales) Ltd, a development company, has been set up by MCT and the British utility company RWE nPower Renewables to develop a 10.5 MW tidal energy farm in 'The Skerries' off the coast of Anglesey; it is anticipated that construction will commence in 2012.

Aviation Enterprises Ltd., Lambourn, UK, was responsible for manufacturing the blade assemblies of SeaGen. ACG/Cytec's Variable Temperature Moulding, VTM®266 carbon fibre/epoxy polymer prepreg, was selected for the manufacture of the 65 mm thick structural spars which run down the full length of the inside of the blade; the spars are the critical elements that add strength and rigidity to the 8 m long blades. The spar and blades were manufactured separately. The blade skins were manufactured in two halves using the VTM®260 prepregs, and the spar and two half blade skins were bonded in one operation. The spar was bonded to a steel root fitting to transfer the loads from the composite spar into the hub. High strength carbon fibre, with a tensile strength greater than 4000 MPa and a modulus of elasticity value of 230 GPa, was used; the polymer was a modified bisphenol A/epoxy cured with an accelerated DICY system. The 8 m long spars had to be made using a material that possessed the requisite design qualities, performance, and high strength and stiffness characteristics with minimum mass. From the processing standpoint, the chemistry of the VTM®266 provided a safe, low-temperature cure processing route, thus making these prepregs ideally suited for this particular application where the overall composite thickness could result in severe exothermic reaction. The out-of-autoclave vacuum-bag system and a large zone controlled oven were used for curing and processing the prepreg material; the vacuum-only route was particularly suitable for this size of component.

10.6.3 Atlantis tidal generator

The Atlantis tidal generator (AK1000™ tidal turbine) is currently the world's largest tidal turbine, and was installed on the sea bed and connected to the grid at a dedicated berth at the European Marine Energy Centre in Orkney, Scotland during late summer 2011; it was developed by Atlantis Resources Corporation, a developer of electricity-generating tidal current turbines. AK1000™ is a horizontal axis turbine designed for open ocean deployment in the harshest environments on Earth. It is a series turbine, featuring a unique twin rotor set with fixed pitch blades, eliminating the requirement for subsea nacelle rotation to improve operational reliability; it has a height of 22.5 m off the seabed with an 18 m rotor diameter and weighs 1300 tonnes. The two sets of blades are manufactured from GFRP and generate power from both ebb and flood tides. Figure 10.6 shows an image of the AK1000™ tidal turbine on the deck of Aker Wayfarer before it was lowered onto its foundation subsea; it was installed by the Aker Wayfarer, which is as an off-shore construction vessel designed for ultra-deepwater with state-of-the-art equipment.

The blades were manufactured from non-woven E glass multiaxial fabric and epoxy resin composites using the vacuum infusion technique; this method presented a number of challenges and the blades for the next generator will

10.6 The Atlantis tidal generator (AK1000™) on the deck of Aker Wayfarer just before it entered the sea.

10.7 The manufacturing procedure of the Atlantis tidal generator blade with the composite material under vacuum during curing (courtesy of Atlantis Resources Corporation).

most probably be made using the prepreg technique. Figure 10.7 shows the manufacturing procedure of the blade with the composite material under vacuum during curing; the top half of the blade is manufactured separately from the bottom one, and the two halves are then bonded together.

The fibres of the AK1000™ blade are orientated along its length to provide an axial stiffness at its root of 24 and 15 GPa in the chord. At the junction between the main hydrodynamic profile and the root, the fibres are aligned to give an axial laminate modulus of 29 and 13 GPa in the chord.

Due to time constraints and the availability of materials, only E glass fibre was considered, and the laminates were constructed from multiple layers of the commonly available 600 gsm UD fibres and + 45 fabrics.

On completion of the blades' manufacturing procedures, static uniformly distributed load tests of 40 tonnes were applied to them.

10.6.4 Pulse tidal generator

One of the first Pulse Stream generators was the 100 kW test rig situated in the Humber estuary, which currently feeds power into a chemicals company on the banks of the river. The shells of the blades of one of these Pulse Stream 100 tidal generators were manufactured in glass fibre/epoxy polymer composites; the spar was fabricated in carbon fibre/epoxy polymer composites. The spar, which is thick compared with the shell, was manufactured by the infusion technique, but for subsequent generators the prepreg technique will be used to make the spars. Figure 10.8 shows the Pulse Stream 100 tidal generator.

Pulse tidal machines operate in tidal currents which move horizontal blades up and down to drive a generator; during operation, the system sits

10.8 Pulse Stream 100 tidal generator.

on the sea bed and is fully submerged even in shallow water. Pulse Tidal Company, the Sheffield-based tidal power provider, is now deploying a 1.2 MW tidal power-generating machine which can operate in a mean water level of 9 m, with a 4m tidal range on either side of the 9 m; this machine will, in 2014, be deployed if environmental studies and permission in an area of seabed near Lynmouth in Devon is sanctioned. Figure 10.9 shows the Pulse Tidal generator.

10.6.5 Wave energy

Wave energy is a technology that uses the motion of ocean surface waves to create electrical energy. There have been several proposed generator systems relying on floating buoys that rise and fall with passing waves. The

10.9 Pulse tidal generator.

Pelamis Energy Converter became the world's first off-shore wave machine to generate electricity into the National Grid in 2004.

The converter is a wave energy machine, and is made of five metal tube sections linked by universal joints which allow flexing in two directions. At the moment the converter is made of steel, but as steel corrodes in sea water, the ideal manufacturing material would be FRP composites; tests on composites are being conducting currently. The machine floats semi-submerged on the surface of the water and inherently faces into the direction of the waves. As waves pass down the length of the machine, and the sections bend in the water, the movement is converted into electricity via hydraulic power take-off systems housed inside each joint of the machine tubes, and power is transmitted to shore using standard subsea cables and equipment. The machine is rated at 750 kW with a target capacity factor of 25–40%, depending upon the conditions at the chosen project site. Figure 10.10 shows diagrammatic views of the operation of the Pelamis converter, one semi-submerged and the other completely submerged.

10.7 Composite materials/fabrication techniques for solar energy applications

10.7.1 Introduction to space-based solar power (SBSP)

It is advisable to refer to the first paragraph of the Appendix (Section 10.12) before reading the following sections. The following discussion refers only to Satellite 1 (Little, 2011) positioned at GEO as discussed in the Appendix. The support structures for the solar collectors and

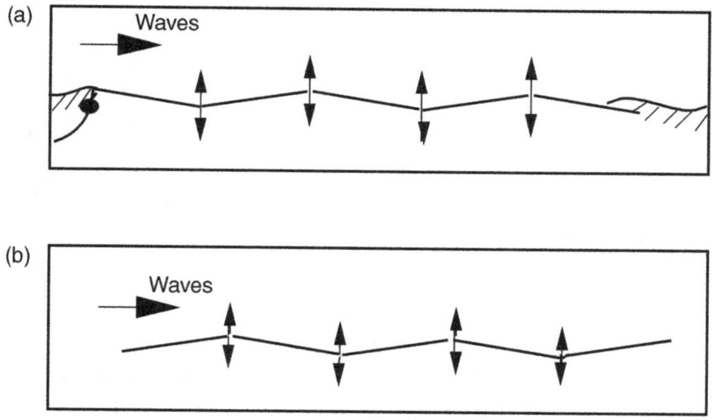

10.10 The Pelamis machine in wave situations: (a) on surface of ocean and (b) under surface of ocean.

equipment in Satellite 1 that would be required for the SBSP system are shown in Fig. 10.11.

The two types of support structures for the solar collectors and equipment that will be considered are the rigid deployable skeletal (RDS) structure and the rigidised inflatable flexible continuum (RIFC) structure; both types of structure need to be lightweight and rigid and, in the case of the skeletal structure, high modulus composite materials would be used. In addition, the structures must be able to be folded to a minimum volume to be placed in the cargo bay of the Launch Vehicle, launched and deployed at LEO, and then taken to GEO by space tug, Wingo (2004).

The environment of space to which SBSP support structural systems would be exposed is harsh. Full account must be taken of the environmental loadings to which the composite materials are to be subjected when in orbit. Primary among these loading situations are thermal-differential-induced stresses, which can be either gradually varying, due to member self-shadowing in a planet-orientated orbit, or abruptly periodic, such as during passage through the Earth's shadow, Hollaway and O'Neill (1991). Hence, an accurate understanding of the thermal response characteristics of candidate composite materials is important before any complete assessment of their suitability for orbital applications can be made.

The space frame would be permanently placed at GEO at an altitude of 36 000 km Earth, but would remain in LEO for a short length of time whilst the support structure was deployed and the collectors and equipment were fixed to it by extra-vehicular activity (EVA). At GEO, the whole structure is exposed to a different loading regime from that of LEO. Although safe from AO attack at GEO, the structures would be exposed to the effects of solar

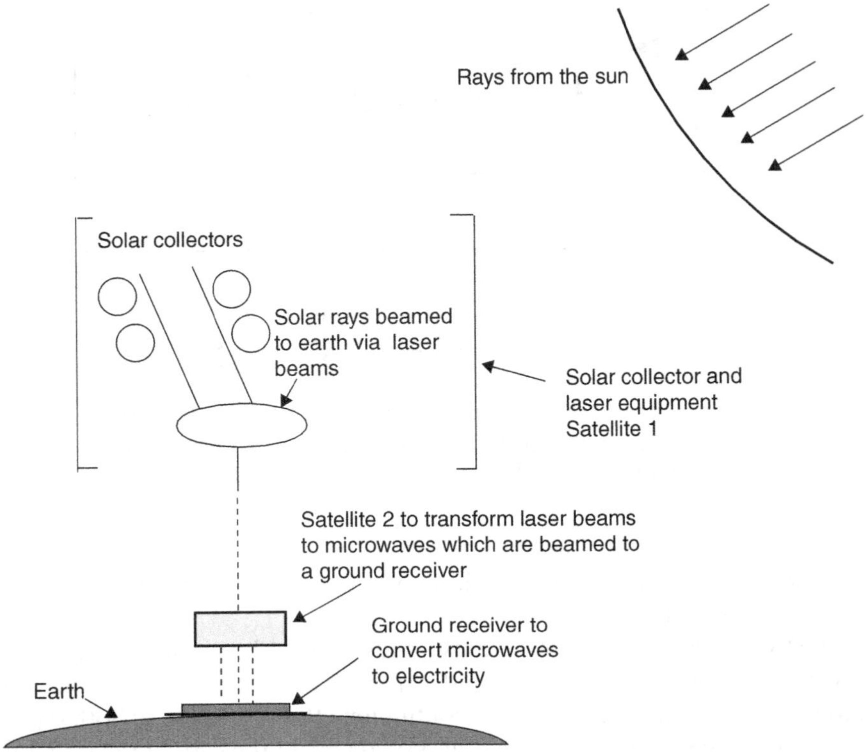

10.11 Support structures for the SBSP collectors and equipment.

wind. This wind is hydrogen plasma of varying intensity, and is dependent on solar flare activity; its velocity is approximately 400 km/s. The effect of this plasma can be expected to lead to degradation of the space structure's material thermal properties over its lifetime; overcoming this problem would require a protective gel coat to be applied to the polymer surface. However, the essentially beneficial warming influence of planetary IR radiation and Earth albedo is greatly diminished at these high altitudes. Annandale (1986) has discussed the environment in GEO, the thermal shocks during the vernal and autumnal equinoxes and at these two periods when the structure will be shadowed by the Earth for up to 72 min per day for around 50 days. The environment in GEO is not nearly as hostile as that at LEO.

10.7.2 Carbon fibre-reinforced thermoplastic composites

The polymer composites which are utilised in space must be lightweight, oxidatively and thermally stable, have good mechanical properties and be resistant to the hostile environments of space, Hollaway (2011). Currently, the matrix materials in conjunction with carbon fibres to form a FRP composite material which might be considered for use to construct a backing skeletal space structure are:

- *The bismaleimide (BMI) resins* (BMI) matrix systems, which are polyimides used in high performance structural composites require superior toughness and high temperature resistance. These resins have processing characteristics similar to those of the epoxy resins, and are used as laminating resins and in prepregs. Advanced Composite Group, (ACG/Cytec), Derbyshire has developed HTM®556, which is a toughened MBI matrix system with improved handling characteristics and, after post-cure, thermal cycling capability up to 250°C. It has been specifically used for structural applications where continuous service temperatures exceed 200°C. In addition, HTM®556 offers improved toughness and resistance to microcracking, which has long been a short-fall of bismaleimide resins.
- *The cyanate ester resins* are high performance thermosetting resins, characterised by their T_g value, which can be up to 400°C, and have excellent dielectric and mechanical properties. Cyanate esters are based on a bisphenol or novolac derivative, in which the hydrogen atom of the phenolic OH group is substituted by a cyanide group. The resulting product is named a cyanate ester. The cyanate ester matrix prepreg system (HTM®143) developed by ACG/Cytec,, Derbyshire for use in space applications is cured at 180°C, is inherently flame-retarded and, after a suitable post-cure, can achieve a T_g of 250°C. The manufacturers state that the matrix has:

    1. Very low moisture absorption.
    2. Very high temperature resistance.
    3. Very low out-gassing.
    4. Substantial weight savings compared to a metal matrix composite (MMC).

- *The high technology thermoplastic aromatic polymers*. During the mid 1980s, ICI developed a new generation of thermoplastic aromatic polymers, in particular the amorphous polymer PES and the semi-crystalline polymer PEEK, Cogswell (1989); they also developed the pre-impregnation of continuous fibres with thermoplastic polymer melts without the need for solvents in the pre-impregnation process. These thermoplastic polymers offered significant advantages over equivalent matrices available at that time. These advantages included high temperature resistance, increased resistance to embrittlement at low temperatures in orbit and the potential for large-scale thermoforming production, as well as improved toughness and higher damage tolerance, Barnes and Cogswell

(1989). Another aromatic polymer within this family of thermoplastic polymers was the semi-crystalline polymer polyphenylene sulphide (PPS), which offered better impact toughness and had a greater resistance to vacuum and thermal cycling than its thermoplastic and thermosetting counterparts, Cogswell (1992).

10.7.3 Rigid deployable skeleton support structure (RDSS) for solar collectors and equipment

During the late 1980s and through the 1990s, research work was undertaken in the USA and in the UK to ascertain whether a skeletal backing frame made from the high technology thermoplastic aromatic polymer composite would be suitable to be placed in the environment of space to support large reflectors for telecommunication satellites. The outstanding in-service and structural properties of the aromatic polymer composite materials naturally led them to be investigated as structural systems for space.

The polymer chosen in the UK for preliminary investigations was the PES, due to its ease of manufacture, its low out-gassing in space and light weight, Hollaway and Thorne (1987). Prepreg sheets of the polyethersulphone/carbon fibre (PES/CF) composite were moulded by the technique of film-stacking technology.

It is suggested that the technology proposed for the backing frame to support the large telecommunication reflectors in space could be used to support the solar collectors and transmission equipment to beam the laser energy to Satellite 2. The backing frame used in the 1990s consisted of carbon fibres/PES tubes fabricated in the form of a double-layer skeletal structure, Thorne and Hollaway (1990). Each tube of the skeletal structure was 1 m in length and 25 mm diameter with a wall of thickness 1.25 mm; they were manufactured from five layers of the high strength Toray 'Torayca' T-300 carbon fibre with a 90/10 weave prepreg interspersed and a total of twelve layers of neat grade 4800 PES film. Ninety per cent of the longitudinal fibres were 10^0 off-axis, with a stacking sequence of 10/-10/0/10/-10; the remaining 10% of the fibres were placed at right angles to the main reinforcement. The fibre lay-up of the tubes resulted in a high specific stiffness and strength, coupled with dimensional stability over the wide range of operating temperatures to which they would be exposed in orbit. The size of the final unit building block of the backing structure was dictated by the size of the 3.5 m Solar Simulation Chamber at Royal Aircraft Establishment (RAE, now QinetiQ), Farnborough in which the structure was tested under near-space environments. The chamber's set of three Xenon arc lamps supplied up to 1.4 solar equivalents in a highly evacuated environment maintained by cryogenic pumping of the chamber down to a pressure of 10^{-7} bar. The cold heat-sink effect of deep space was simulated by circulating liquid nitrogen in the shrouds lining the full length of the chamber, and the shrouds were maintained at a constant temperature of 90 K; this compares with the typically expected temperatures at LEO and GEO of 15 K and 4 K, respectively, Goodbody and Kimber (1996). Further information may be found in Hollaway *et al.* (1990), Thorne and Hollaway (1990), Hollaway and Sparry (1991). The manufacturing method for the tubes was developed by Complas Ltd., Studham, Buckinghamshire and the University of Surrey, by adapting a film-stacking and compression moulding technique which used steel moulds and an expandable PTFE mandrel of the correct size for the finished internal diameter of the rods; the initial cure temperature of the CF-PES was 290°C, with a pressure of 3 MPa applied to the mould. To form the composite tubes, the prepreg films were wrapped on to the PTFE mandrel at the correct fibre orientation and stacking sequence. The cure temperature was maintained at a constant value for half an hour and then reduced to room temperature. The PTFE material of the mandrel had a coefficient of thermal expansion much greater than that of the steel moulds, and this led to a satisfactory compaction of the polymer against the metal mould and a low void ratio; Hollaway (2011) has discussed further the material used.

The structural form of the system was fabricated into a building block in the form of a three-way double-layer grid (tetrahedral truss) with 21 tubular members connected at nine cluster node points, the (9/21) system. The particular self-deployment method used in that study incorporates energy-loaded joints at all nodal points and also at the centre of the top and bottom members of the skeletal structure. To prove the deployment of an RDS structure, a 5 m system was manufactured in CFRP composite; it deployed satisfactorily Fanning (1993). Figure 10.12 shows the unit building block backing frame which was proposed for the telecommunications system and is now being proposed for a possible support system for space solar collectors (Satellite 1); the backing frame may be extended by connecting several unit building blocks together at LEO. The joints in this structure are fixed. Figure 10.13 shows a type of energy-loaded folding joint which would be suitable for positioning at nodal joints of the structure for the deployment of the unit building block; these were manufactured in steel to prove the system. Figure 10.14 shows the joint manufactured in CFRP composite in the closed position, the springs providing the energy for deployment.

10.7.4 Rigidised inflatable flexible continuum (RIFC) support structure for solar collectors and equipment

Rigidised inflatable structures are those that are fabricated from shape-memory polymers (SMP)/carbon fibre composite materials; the methods

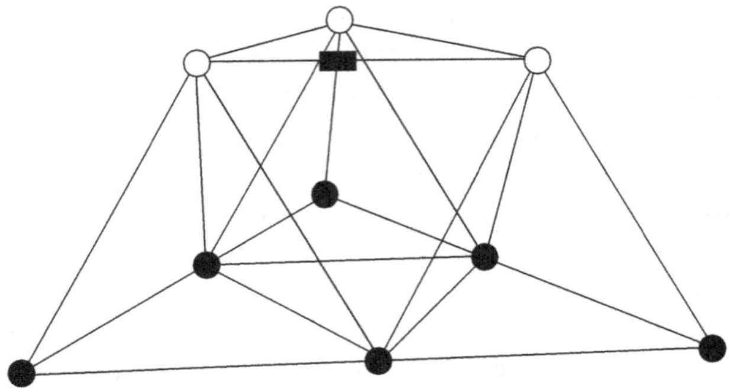

10.12 A unit building block backing frame for supporting the SBSP system; the nodal joints are fixed.

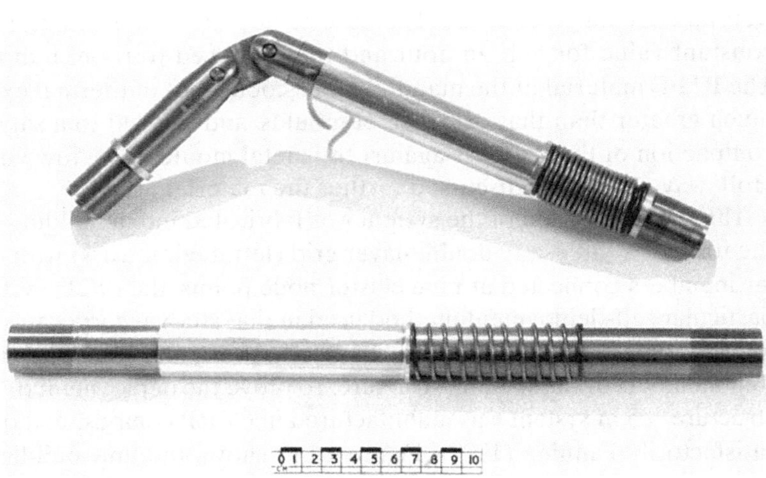

10.13 A type of energy-loaded folding joints suitable for the deployment of the backing frame.

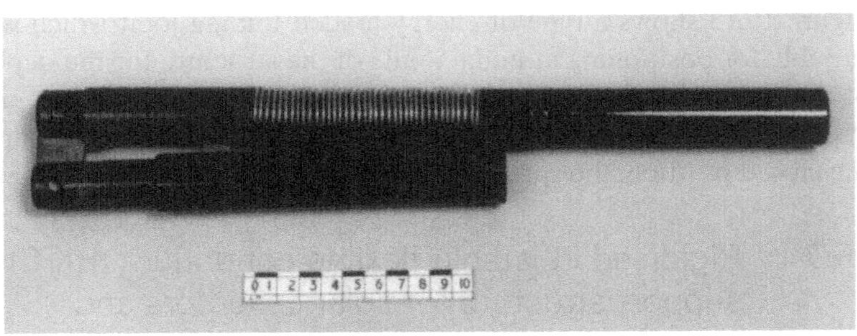

10.14 The joint manufactured in CFRP composite in the closed position.

utilised are similar to those used for traditional thermosetting composite structures, except that the epoxy that is used is a lightly cross-linked one. The final deployed shape of the SMP composite structure is established during its initial cure cycle. Once the SMP material is completely cured, it can be heated to the folding temperature (typically 20°C above the glass transition temperature (T_g)), where it becomes flexible and can be tightly packed into a minimum volume. The flexibility of the SMP composite material at the folding temperature is highly dependent on both the resin and fibre properties. Once the structure is packed, it is constrained in that position until cooled to approximately 15°C, or lower, below its T_g value, at which point the SMP composite structure will remain frozen in the packed position until it is again heated above its T_g value. When the structure is heated, internal strain energy will return it to its initial cured shape. The speed and the accuracy of the original shape are a function of the shape-memory recovery force of the composite. Lin *et al.* (2006) have discussed the shape-memory rigidised inflatable materials technology and their utilisation in structural space systems. Figure 10.15 illustrates the shape-memory cycle.

The SMP polymer materials which have been considered or used by researchers for rigidised inflatables have not been completely defined in the literature, and the suitability of their physical and mechanical properties for space applications at GEO have not been discussed in detail, to this author's knowledge; some trade names of the polymers used for rigidised inflatables are:

- A PAN-based carbon fibre tow coated with a novel, low-cure-temperature thermosetting resin, Hydrosize U-Nyte® Set 201 binder.

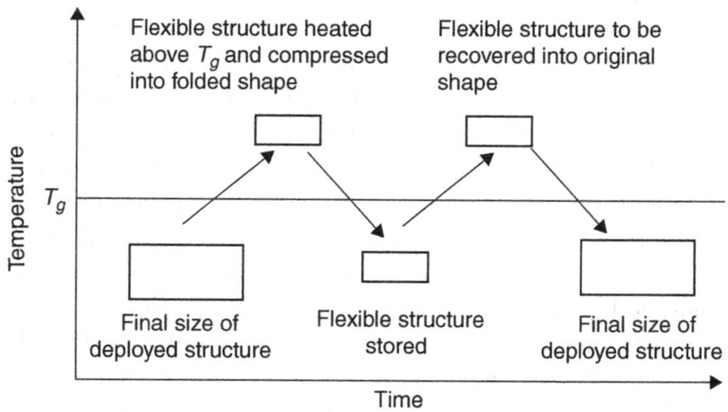

10.15 The shape-memory cycle.

- TEMBO® Elastic Memory Composites are stated, by the manufacturers, to be a family of thermoset (epoxy) SMP developed by Composite Technology Development Inc., 2600 Campus Drive, Lafayette Co., 80026–3359, USA.
- Scarborough and Cadogan (2006), Lan *et al.* (2009) used materials that were an 11.5 × 11.5 count, 203 g/m^2, 5 harness satin (5HS) weave fabric made from 6K IM7 carbon tows and TP407, a thermosetting polyurethane SMP resin with a T_g of approximately 55°C.
- Lan *et al.* (2009) in their study used a fibre-reinforced SMP composite in a deployable hinge as discussed, but little information on the space environment properties has been given.

Heat is the stimulus that is currently used for SMP. However, their recovery can be triggered, by various other external stimuli, Everhart *et al.* (2006), Liu *et al.* (2007):

- *electricity*, Schmidt (2006), Lv *et al.* (2008a),
- *magnetic field*, Buckley *et al.* (2006),
- *light*, Hy *et al.* (2006),
- *moisture*, Yang *et al.* (2006), Lv (2008b).

It is possible that these effects could distort the final configuration of the rigid skeletal deployable structure, as they are all within the environment of space. The effect of these influences might be negated by incorporating various blends to the polymer but, in so doing, a new polymer formulation would be formed; this would require a completely new environmental testing programme.

The low heat that is required to provide the internal strain energy to return the structure to its initial cured shape from its folded state is the main concern in the utilisation of SMP polymers for rigidised inflatable structures. From the cold state the polymer/fibre system is heated above its T_g to return the structure to its original cured shape. Once the structural system is deployed and then cooled, at least to 15°C below T_g, it will again be structurally rigid; this maximum temperature of 65°C for a T_g value of 80°C (maximum of any lightly cross-linked thermosetting polymer quoted by manufacturers) is low for the temperature to which the structure could be exposed in space. If the temperature rises above 65°C, the structure will again become distorted; consequently, it is advisable that the T_g value of the polymer that is to be used should be higher. It is vitally necessary that all space structures should resist the environmental loadings of space; consequently, the RDS Structures should show satisfactory thermal test results – if they do not, there may be a possibility that the structure will not deploy as required at LEO.

Weight and volume restrictions are issues that affect the cost of transporting space structural systems, from the point of view of the number of trips to LEO the launch vehicle has to make to complete the fabrication of the backing frame. As the requirement for the number of structures in space grows, their weight and volume become important issues. Therefore, inflatable structures with respect to their storage capacity in the cargo bay of the space launch vehicle offer the possibility of reducing the overall structural weight, typically, by 50% of that required by the FRP composite skeletal structure. In addition, inflatable systems can typically be packaged to a volume less than 25% of that necessary for the standard skeletal structure; moreover, they can be packaged to any shape depending on the particular mission requirements.

The advantages and disadvantages of the RDSS support structure and the rigidised inflatable skeletal structure (RISS) are:

- The RISS for the continuum backing frame can typically be packaged to a volume less than 25% of that necessary for standard mechanical structures, and can be packaged to, essentially, any shape depending on the particular mission requirements.
- The RISS is able to be folded into a compact volume for launch; consequently, a greater volume of stored structural units is able to be stowed in the cargo bay of the space transporter to LEO, thus offering the possibility of reducing the overall structural weight, typically, by 50% of that required by the RDSS. Thus, there will be fewer flights, but the weight per launch of the rigidised inflatables will be greater than that of the RDSS. The lesser number of flights will influence the cost of placing satellites into LEO; the volume and therefore weight of the structure will be a major consideration in the transportation costs.
- The quoted values of the T_g by the manufacturers of the SMP material is of the order of 50–80°C; these are low values for composite materials in a space environment.
- Only limited analysis associated with the long-term durability of the two types of materials when they are in the harsh environment of space has been undertaken; further detailed work must be done before these two systems are able to be used in space.

The National Space Society (2007) has suggested that the cost of space solar-power development should be compared to the cost of not undertaking this development work.

10.7.5 Composite materials/fabrication techniques for deployable skeletal support systems for earth-based solar panel (EBSP) generators

This section is concerned with the EBSP generation as part of a solar-power plant; solar panels as part of a building design for dwellings are outside of the scope of this chapter.

Fibre/polymer composite could have been used for the backing frames instead of the steel system used by BrightSource Energy Company to support the photovoltaic (PV) panel equipment to collect solar rays from the EBSP generators. The manufacturing technique to produce these fibre/polymer composite frames could be either the pultrusion or the prepreg technique; these FRP members would probably be fabricated into a tubular double-layer skeletal frame. If the FRP material of the frame required special shielding from any hostile environments on Earth, it is likely that a solution is already available for protection; the only requirement for the above scheme would be a large acreage of land, which might be a limiting factor in the development of EBSP energy. However, one ideal area would be the Sahara Desert, and the southern parts of countries bordering the North African coast, where there are high energy concentrations from the sun and a large area of vacant land suitable for erecting EBSP generators. The great advantage in utilising the desert would be to revitalise the land by using the energy collected from the solar panels to pump water to irrigate and thus fertilise the land to feed the people and animals of that area, so reversing the migration of the people. Power from this area could also be piped across the Mediterranean to Southern European countries. A great political will would be required to undertake such a big project but, if successful, it would solve many problems. This project would be ideal to utilise composites for the manufacture of the backing frame for the solar panels.

10.8 Conclusions and future trends

Market figures show the current wind energy growth rate is set to continue for the foreseeable future. Enforcement of the Kyoto protocol, which was initially adopted on 11 December 1997 in Kyoto, Japan, and became active on 16 February 2005, ensures that renewable, sustainable energy sources are a priority for national governments. Development of materials, manufacturing processes and designs will continue to cater for the demand for larger, better performing turbines to produce sustainable power.

Off-shore wind farms are developing quickly, and massive potential is seen in this area. They are particularly attractive because water has a very low level of roughness (i.e. the water 'interferes' with the wind less than the land does) and wind speeds in general are higher and less turbulent off-shore. Wind turbines off-shore can therefore be much larger than is the case with those on-shore, and they remove the noise and visual pollution associated with inland farms. The disadvantages of off-shore fields are the problems of assembly on to foundations that extend to the sea bed, and the connection to a grid – although the floating wind turbines can be assembled on-shore in dry-dock and floated out to their moorings. Off-shore fields are currently more expensive than on-shore ones. The current estimates, based partly on European experience since 1991, indicate off-shore wind energy costs less than 6 cents per kWh. Capital costs are around 30–50% higher than on-shore, due to larger machine size and the costs of transporting and installing at sea. This is partially offset by higher energy yields, as much as 30%.

However, these prices are expected to drop as technology improves and the more experience that is gained as happened with the on-shore systems, Ocean Energy Council (2011). The European Environmental Agency (EEA) (2011) has estimated that off-shore wind energy capacity in Europe will increase 17-fold between 2010 and 2020. The EU has the target of generating 20% of its energy consumption from renewable sources by 2020. Member States each have different individual targets to allow this overall target to be met; they are obliged to provide detailed roadmaps describing how they will meet their legally binding target. These figures will drop even more in future, as a number of companies are working on commercial deep-water wind turbine technologies. The American off-shore rig builder Principle Power has developed off-shore wind turbines which are manufactured in shipyards on-shore, mounted on semi-submersible platforms and towed into place. A great advantage in siting wind farms farther from shore in deep water is that the exploitable wind resource is superior to that nearer to shore.

Hydrokinetic power drawn from the Earth's oceans and rivers is about 15 years behind the progress of the wind power technology, but it is an interesting and an exciting form of sustainable energy; Northern Europe, and in particular the UK, is well placed to take full advantage of the tides around countries with sea-boarders. SBSP beamed to Earth via lasers or microwaves is probably the most exciting of all sustainable power technology. However, there is a down-side to the harnessing of power from wind, waves or solar – they are expensive technologies; but all are at the development or early production stage and therefore will appear to be expensive compared with the established technologies. Nevertheless, the EEA (2011) has also estimated that the newer renewable technologies, such as concentrated solar power and wave/tidal power, will also increase more than 11-fold. European countries are also expected to significantly boost solar PV power, on-shore wind, and other renewable technologies over the next decade. Centrica, which plans to invest £1.5 billion a year in new power-generating equipment and gas storage until 2020, is building a 75 turbine wind farm off the Lincolnshire coast at a cost of £725 million, The Times (2010). In the same Times article, Citigroup has estimated that the cost of installing one megawatt of off-shore wind is about £3.5 million and this is roughly five times the cost of building a gas-fired power station with the same capacity. These projections can only boost the utilisation of polymer/fibre composites in the future.

The improved performance of wind turbine blades, and therefore the capacity of the generator systems, will be dependent upon enhancing the reliability of the component materials of the blades and increasing the energy capture. Consequently, future designs will utilise larger rotors with longer blades fabricated from advanced composite materials with high strength and stiffness-to-weight ratios; a thorough knowledge of composite materials and safety factors will be necessary. In particular, a thorough knowledge of the effects of damage and failure mechanisms and the effects and interpretation of stochastic loadings, multiple stress states, environmental effects, size effects and thickness effects must be understood. Turbine materials need to exhibit excellent fatigue and corrosion-resistance properties to ensure durability and hence cost-effectiveness of projects. Wang and Ren (2010) have written an excellent report on an innovative technique for evaluating the integrity and durability of wind turbine blades made from composites, by quantifying the fracture behaviour of composite structures under mixed mode loading conditions, particularly under combined Mode 1 (flexural or normal tensile stress) and Mode III (torsional shear stress) loading; they have proposed a new testing technique based on the spiral notch torsion test.

The manufacturing techniques of fibre/matrix composite materials to form parts of sustainable energy systems have utilised the thermosetting polymer material, but it is important to consider future materials for the manufacture of wind turbines. One material would be to manufacture wind turbine blades using thermoplastic polymer composites such as the polyamide 6 (PA-6). The thermoplastic materials generally have a higher resistance to abrasion, fatigue and higher toughness than most thermosetting materials. van Rijswijk and Bersee (2006) undertook tests on the vacuum infusion manufacturing process (the reactively processed PA-6 outperformed the melt processed PA-6 in all temperatures and humidities that were tested) and found that the manufacturing technique is about six times faster than that for the thermosetting polymer, as the former material does not undergo the irreversible curing process of the thermosetting materials. Therefore, the thermoplastic material is readily able to melt and remould, allowing it to be reused. There are some disadvantages with the thermoplastic polymers; they have a lower tensile strength when compared with thermosetting materials. A glass/epoxy thermosetting composite has a tensile strength of the order of 1166 MPa, whereas a PA6-thermoplastic composite has a tensile strength of the order of 869 MPa; the actual values for these two composites will depend upon the type of glass fibre and its array in the composite. Moreover, they also require a much higher processing temperature, which in turn requires more energy and an increase in cost. Prabhakaran *et al.* (2011a) have given an overview of suitable thermoplastic material systems for the VIP and, as an illustrative example, a resin selection for a vacuum infused wind turbine blade was shown to demonstrate the intricacies involved in the proposed methodology for resin selection. In a state-of-the art study of thermoplastic polymer matrix materials for fibre composites, Prabhakaran *et al.* (2011b) has identified the polyamide 6 (PA6) as a potential candidate thermoplastic polymer relevant for manufacturing large composite structures such as wind turbine blades. The mechanical properties of PA6 are highly sensitive to moisture, and, if PA6 is used as matrix material in a fibre composite, the properties of the fibre composite will depend on the moisture content of the material. The matrix dominated properties, such as the shear stiffness, the shear strength and the stiffness and strength across the fibre direction, are the ones which are most affected by the moisture content in the material. Naderil *et al.* (2011) have investigated the effect of nanoclay and poly-epichlorohydrin-co-ethylene oxide (ECO) content on the microstructure and mechanical properties of PA6/ECO thermoplastic elastomers (TPEs). In thermoplastic elastomer nanocomposites, the tensile modulus samples increased from 30% to 80% depending on the ECO content and the nanoclay used. The exfoliated structure, in which the layers of the clay have been completely separated and the individual layers are distributed throughout the organic matrix, resulted in a reduction of the degree of crystallinity, because the dispersed clay silicates acted as nucleating agents.

10.8.1 Observations

This chapter has introduced the progress being made, and the synthetic composite materials being used, in the field of sustainable energy supplies, and the research and development which is being undertaken currently to develop power that is not reliant upon fossil fuels. During future decades more sophisticated structural systems will be introduced into wind-power, hydro-power and solar-power systems, which will require lightweight materials that are resistant to the hostile environment of Earth, sea and space; these materials will involve new technologies and manufacture. Some of these structural areas have been discussed, and the fibre/polymer composite materials which might be suitable for their construction have been given. Market figures show the current wind energy growth rate is set to continue for the foreseeable future. Enforcement of the Kyoto protocol ensures that renewable, sustainable energy sources are a priority for national governments. Development of materials, manufacturing processes and designs will continue to cater for the demand of larger, better performing turbines. However, the success of these developments is subject to environmental and social appraisal that is currently taking place as new sites are found.

The economic viability of composite construction increases with the number of units produced, where the cost of design development and tooling can be amortised. For large renewable energy systems, the reduced weight of the composite components when compared to metallic construction can greatly reduce transportation and erection costs. Perhaps the biggest advantage of composites for large energy projects is reduced maintenance costs over an expected 20–30 year service life. For large, unmanned engineered structures, corrosion resistance will be paramount for long-term economic viability.

Finally, in an attempt to moderate the demand for coal for power generation, several countries conclude that nuclear energy must also grow significantly, but for instance, in contrast to coal, nuclear is one of the more difficult energy sources to expand quickly on a global scale. Building capacity for uranium mining and nuclear power station construction takes time; added to that, it is difficult to dispose of nuclear waste. In the future, this form of energy will probably produce power alongside the renewable forms of energy.

10.9 Sources of further information and advice

The following list of professional bodies associated with civil engineering FRP composites is based upon a network group for composites in construction, (ngcc) 11 May 2012:

1. *Composites UK* – The mission of Composites UK, as the representative body of the UK composites industry, promoting the use of composite materials to the widest market spectrum.
2. *British Composites Society* (BCS) – this is one of the technical arms of the Institute of Materials (IOM), Minerals and Mining. The British Composites Society provides a focus for the exchange of knowledge on all aspects of composite materials. It is a national contact point for communication with similar bodies on a worldwide basis.
3. *The Institute of Materials, Minerals and Mining* (IOM3) recognised by the UK's Privy Council on 26 June 2002. It was created from the merger of The IOM and The Institution of Mining and Metallurgy (IMM). The Institute is potentially the leading international professional body for the advancement of materials, minerals and mining to governments, industry, academia, the public and the professions.
4. *International Institute for FRP in construction* (IIFC) – the aim of the Institute is to advance the understanding and application of FRP composites in civil infrastructure, in the service of the engineering profession and society.
5. *Welsh Composites Consortium* (WCC) – the Consortium acts as a technology transfer network, consisting of a number of partner organisations

with a wide range of expertise in the field of composites, particularly to SMEs in Wales in the form of advisory visits.

6. *Construction Industry Research and Information Association* (CIRIA).
7. *The Italian Association for Composites in Construction* (AICO) – the AICO was formed in 1996. It is active in the field, with membership from industry and universities
8. *European Composites Industry Association* (EuCIA) – the primary goal of EuCIA is to unite the composites industry at European level into one single European association.
9. *COBRAE* – the objective of COBRAE is to promote research, development, standardisation and application of FRP composites in rehabilitation, upgrade and new bridge constructions and infrastructure applications.
10. *European Construction Technology Platform* (ECTP) – it is hoped that it will raise the sector to a higher world-beating level of performance and competitiveness. This will be achieved by analysing the major challenges that the sector faces in terms of society, sustainability and technological development. Research and innovation strategies will be developed to meet these challenges, engaging with and mobilising the wide range of leading skills, expertise and talent available to us within our industry over the coming decades, in order to meet the needs of society.
11. *Intelligent Sensing for Innovative Structures* (ISIS), Canada.
12. *Canadian Association for Composite Structures* (CACS) – CACS is a network of individuals and corporate members (suppliers, fabricators, equipment manufacturers, distributors, consultants, technologists, research centres, materials specialists, researchers, teachers and government employees) working to develop and enhance new and existing applications for composite structures and materials.

10.10 Acknowledgements

This chapter could not have been written without the help of manufacturing firms concerned with composite materials and structural components made from those materials which are associated with the field of sustainable energy. These firms include: ACG/Cytec, Derbyshire, UK; Gurit, Isle of Wight, UK; Solent Composite Systems, Isle of Wight, UK; Wind Power, Ltd., Bury St Edmunds, Suffolk, UK; Aviation Enterprise, Lambourn, UK; Pulse Tidal Ltd., Sheffield, UK; Marine Turbines, Bristol, UK; Exxonmobil, Leatherhead, UK; Grimshaw, Architects, London, UK; *the Reinforced Plastics* Journal, Elsevier Publisher, Oxford. The author would like to express his thanks to the members of staff of these firms for their invaluable help.

10.11 References

Allen, H.G. (1969), *Analysis and Design of Structural Sandwich Panels'*, Published by Pergamon Press, Oxford, London, Edinburgh, New York, Toronto, Sydney, Paris.

Annandale, R.W. (1986), 'Thermal and Structural Analyses of Large Space Antenna Reflectors', PhD Thesis, University of Surrey, Guildford Surrey, England.

Barnes, J.A. and Cogswell, F.N. (1989), 'Thermoplastics for Space' SAMPE Quarterly, Vol. **20**, No. 3.

Breuer, O. and Sundararaj, U. (2004), 'Big returns from small fibres: A review of polymer/carbon nanotube composites', *Polymer Composites*, Vol. **25**, Issue. 6, pp. 630–645.

Brøndsted, P., Lystrup, A. and Lilholt, H. (2005), 'Composite materials for wind power turbine blades', *Annual Review of Materials Research*, Vol. **35**, pp. 505–538.

Buckley, P.R., McKinley, G.H., Wilson, T.S., Small, W., Benett, W.J., Bearinger, J.P., McElfresh, M.W. and Maitland, D.J. (2006), 'Inductively heated shape memory polymer for the magnetic actuation of medical devices', *IEEE Transactions on Biomedical Engineering,* Vol. **53**, No. 10, pp. 2075–2083.

Chiou, P.-L. and Bradley, W.L. (1997), 'Effect of seawater on strength and durability of glass/epoxy filament wound types as revealed by acoustic emission analysis', *Journal of Composites Technology and Research*, Vol. **19**, No. 4, pp. 214–221.

Cogswell, N. (1989), 'Thermoplastics find their feet', *Advanced Composites Engineering*, January 1989.

Cogswell, N. (1992), '*A study of the Structure, Processing and Properties of Carbon Fibre Reinforced Polyetheretherketone and Related Materials*', Published by Woodhead Publishing Limited, Cambridge, England.

Cripps, D. (2011), 'The future of blade repair', *Reinforced Plastics*, Vol. **55**, Issue 1 January–February 2011, pp. 28–32.

EPTA Publication (2008), 'What is Pultrusion' *9th World Pultrusion Conference*, 'Profiting from Pultruded Profiles' 26–28 March 2008.

Everhart, M.C., Nickerson, D.M. and Hreha, R.D. (2006), 'High-temperature reusable shape memory polymer mandrels', *Procceeding of 11th International Symposia on Smart Structures and Materials*, Vol. **6171**, Editor, E.V. White, SPIE, San Diego.

Fanning, P. (1993), 'Development and Analysis of a Deployable Skeletal Reflector for Spacecraft Antennas', PhD Thesis, University of Surrey.

Fraenkel, P. (2011), 'Underwater Windmills – Harnessing the world's marine currents', Ingenia (on line) Issue 46, March 2011, (The Royal Academy of Engineering, quarterly magazine), ISSN 9768. (www.ingenia.org.uk).

Goodbody, C. and Kimber, R. (1996), 'The UoSAT solar cell experiment-over 4 years in orbit', *Proceedings of the Photovoltaic Specialists Conference, 1996*, Washington DC, 13–17 May 1996, pp. 235–238.

Griffin, D.A. and Ashwill, T.D. (2003), 'Alternative composite materials for megawatt-scale wind turbine blades: Design considerations and recommended testing', *Journal of Solar Energy Engineering*, Vol. **125**, p. 515.

Hackman, I. and Hollaway, L.C. (2006), 'Epoxy-layered silicate nanocomposites in civil engineering', *Composites, Part A*, Vol. **37**, pp. 1161–1170.

Hackman, I. (2007), 'Nanocomposites in civil engineering' PhD. thesis the University of Surrey, Guildford, Surrey.

Helbling, C. and Karbhari, M.K. (2007),'Durability of composites in aqueous environments' Chapter 3, *Durability of Composites for Civil Structural Applications*, editor V.M Karbhari, Published by Woodhead Publishing Limited, Cambridge, England.

Hexcel (2005), 'Prepreg Technology' Publication No. FGU 017b, March 2005, Published by Hexcel Corporation.

Hollaway, L.C. and Thorne, A. (1987), 'High technology thermoplastic polymers reinforced with fibres for space applications' MOD (PE) Contract No. A57A/1733, Final Report, for the Space Department RAE Farnborough, Hampshire.

Hollaway, L.C., Thorne, A. and Rankin, I. (1990), 'High Technology Thermoplastic Polymers reinforced with fibres for space structures application', Final Contract Report, Contract No. 4064/107/RAE/XR(F) BNSC, November, 1990.

Hollaway, L C. and Sparry, D. (1991), 'Design study and vibration analysis of large deployable space antennas', Contract No. 2064/109/RAE (F).

Hollaway, L.C. and O'Neill, M. (1991), 'Thermal and structural analyses of carbon fibre/polyethersulphone tubes for space applications', Contract Number D/ER/1/9/4/2064/129/SP(F), Final Report for the MoD.

Hollaway, L.C. and Head, P.R. (2001), '*Advanced Polymer Composites and Polymers in the Civil Infrastructure*, Published by Elsevier, Oxford.

Hollaway, L.C. and Hackman, I. (2004), 'Strengths and limitations of fibre reinforced polymers in the civil infrastructure, material advances and the influences on present and future developments', *Conferences proceedings FRP Composites in Civil Engineering* (CICE 2004), Edited by Dr R. Seracino, Published by A.A. Balkema Publishers, Leiden, London, New York, Philadelphia, Singapore.

Hollaway, L.C. and Teng, J.G. (eds) (2008), '*Strengthening and Rehabilitation of Civil Infrastructures Using Fibre-Reinforced Polymer (FRP) Composites*', Published by Woodhead Publishing Limited.

Hollaway, L.C. (2011), 'Thermoplastic/carbon fibre composites could aid solar based power generation – A case study of a possible support system for solar power satellites', *Journal of Composite for Construction*, Vol. **15**, No. 2, pp. 239–247.

Housecroft, C.E. and Sharpe, A.G. (2008), '*Inorganic Chemistry*,' 3rd Edn., Published by Pearson Education Limited.

Hy, J., Kelch, S. and Lendlein, A. (2006), 'Polymers move in response to light', *Advanced Materials*, Vol. **18**, No. 11, pp. 1471–1475.

Iijima, S. (1991), 'Helical microtubules of graphitic carbon' *Nature,* Vol. **354**, pp. 56–58.

Kojima, Y., Usuki, A., Kawasumi, M. and Okada, A. (1993a), 'Synthesis of Nylon-6 clay hybrid epoxy', *Journal of Materials Research*, Vol. **8**, No. 5, pp. 1179–1183.

Kojima, Y., Usuki, A., Kawasumi, M. and Okada, A. (1993b), 'Mechanical properties of Nylon-6 clay hybrid, *Journal of Materials Research*, Vol. **8**, No. 5, pp. 1185–1189.

Koziol, K. Vilatela, J., Moisala, A., Motta, M., Cunniff, P., Sennett, M. and Windle, A. (2007), 'High-performance carbon nanotube fiber', *Science,* 21 Vol. **318**, No. 5858, pp. 1892–1895.

Lan, X., Liu, Y., Lv, H., Wang, X., Leng, J. and Du, S. (2009), 'Fiber reinforced shape-memory polymer composite and its application in a deployable hinge', *Smart Materials and Structures*, Vol. **18** (6 pp).

Le Baron, P., Wang, Z. and Pinnavaia, T. (1999), 'Polymer-layered silicate nanocomposites: An overview', *Applied Clay Science*, Vol. **15**, pp. 11–29.

Leger, L., Visentine, J. and Santos-Mason, B. (1986), 'Selected material issues associated with space station', *18th International SAMPE Technical Conference,* held at Seattle, Washington, USA, 7–9 October, pp. 1015–1026.

Leger, L., Visentine, J. and Santos-Mason, B. (1987), 'Selected material issues associated with space station', *SAMPE Quarterly*, January 1987, pp. 48–54.

Lin, J.K, Knoll, C.F. and Willey, C.E. (2006), 'Shape-memory rigidised inflatable (RI) structures for large space systems applications', *47th AIAA/ASME/ASCE/AHS/ASC Structures, Structural Dynamics and Materials Conference*, 1–4 May 2006, Newport, Rhode Island, pp. 1–10.

Little, F.E. (2011), 'Meeting the challenges of implementing portable space-based solar power', *Proceedings of 30th General Assembly and Scientific Symposium of the International Union of Radio Science*, Istanbul, Turkey, 13–20 August 2011, Paper CHGBDJK.3. (No page numbers.)

Liu, C., Qin, H. and Mather, P.T. (2007), 'Review of progress in shape-memory polymers', *Journal of Materials Chemistry*, Vol. **17**, No. 16, pp. 1543–1558.

Liu, W., Hoa, S. and Pugh, M. (2005), 'Fracture toughness and water uptake of high performance epoxy nanocomposites', *Composites Science and Technology*, Vol. **66**, pp. 2116–2125.

Loos, M. (2011), 'Case Western Reserve University researchers build carbon nanotube reinforce wind turbine blade', netcomposites, 6 September 2011. (Accessed 8 September 2011).

Louriel, O., Cox, D.M. and Wagner, H.D. (1998), 'Buckling and collapse of embedded carbon nanotubes', *Physical Review Letters*, Vol. **18**, Issue 8, pp. 1638–1641.

Lv, H.B., Leng, J.S. and Du S.Y. (2008a), 'Electro-induced shape-memory polymer nanocomposite containing conductive particles and short fibers'. *Procceedings of 15th International Symposia on Smart Structures and Materials*, Vol. **6932**. San Diego: SPIE; 2008.

Lv, H.B., Leng, J.S. and Du S.Y. (2008b), 'Shape memory polymer in response to solution', *Advanced Engineering Materials,* Vol. **10**, No. 6, pp. 592–595.

Marsh, G. (2007), 'Tooling up for large wind turbine blades', *Reinforced Plastics*, Vol. **51**, Issue 9, pp. 38–43.

Marsh, G. (2011), 'The challenge of wind turbine blade repair', *Renewable Energy Focus*, Magazine, Elsevier, Ltd.

Monaghan, D.A. and Wang, S.S. (2004), 'Effect of Saltwater on Multiaxial Cyclic Fatigue of Carbon Fiber/Epoxy Composite Material', Prepared for: Research Partnership to Secure Energy for America, Subcontract No. R-511, RPSEA, Unconventional Gas Technology and ChevronTexaco Energy Research and Technology Company, Bellaire, TX 77401, Report CEAC-TR-04-0103, (September 2003–September 2004).

Naderil, G., Razavi-Nouril, M., Taghizadeh, E., Lafleur, P.G. and Dubois, C. (2011), 'Preparation of thermoplastic elastomer nanocomposites based on polyamide-6/polyepichlorohydrin-co-ethylene oxide', *Polymer Engineering & Science*, Vol. **51**, No. 2, pp. 278–284.

Ocean Energy Council, Inc. © (2011), 'Offshore Wind Energy', OEC, 11985 Southern Blvd., Suite 155, West Palm Beach, Florida 33411. (Accessed from internet on 6 September 2011).

Paul, D.R. and Robeson, L.M. (2008), 'Polymer nanotechnology: Nanocomposites', *Polymer*, Vol. **49**, Issue 15, pp. 3187–3204.

Prabhakaran, R.T.D., Andersen T.L. and Lystrup, A. (2011a), 'Attribute based selection of thermoplastic resin for vacuum infusion process: A decision making methodology', *International Journal of Manufacturing, Materials and Mechanical Engineering*, Vol. **1**, Issue 3, pp. 31–52.

Prabhakaran, R.T.D., Andersen T.L. and Lystrup, A. (2011b), 'Influence of moisture absorption on properties of fibre reinforced polyamide 6 composites', 26th *ASC Annual Technical Conference* (the 2nd Joint US-Canada Conference on Composites), Quebec, Montreal, 26–28 September 2011, Paper ID: 1065, pp. 1–11.

Qian, D., Wagner, G.J., Liu, W.K., Yu, M.F. and Ruoff, R.S. (2002), Mechanics of carbon nanotubes. Applied Mechanics Review, **55** (2), pp. 495–533.

Ray, S. and Okamoto, M. (2003), 'Polymer/layered silicate nanocomposites: A review from preparation to processing', *Progress in Polymer Science*, Vol. **28**, pp. 1539–1641.

Ruoff, R.S., Qian, D. and Liu, W.K. (2003), 'Mechanical properties of carbon nanotubes: Theoretical predictions and experimental measurements', *Comptes Rendus Physique*, **4**, pp. 993–1008.

Rush, J (2008), 'Power generation: The new wave'. Channel 4, 31 March 2008.

Salvetat, J.-P., Bonard, J.-M., Thomson, N.H., Kulik, A.J., Forró, L., Benoit, W. and Zuppiroli, L. (1999), 'Mechanical properties of carbon nanotubes', *Applied Physics A*, Vol. **69**, pp. 255–260.

Scarborough, S.E. and Cadogan, D. (2006), 'Applications of Inflatable Rigidisable Structures' Technical Paper Published by ILC Dover, Frederica, DE, (USA), 16 pages.

Schmidt, A.M. (2006), 'Electromagnetic activation of shape memory polymer networks containing magnetic nanoparticles', *Macromolecular Rapid Communications*, Vol. **27**, No. 14, pp. 1168–1172.

Sennett, M., Welsh, E., Wright, J.B., Li, W.Z., Wen, J.G. and Ren, Z.F. (2003), 'Dispersion and alignment of carbon nanotubes in polycarbonate', *Applied Physics A*, Vol. **76**, pp. 111–113.

Smith, P.A. (2000), 'Properties of CFRP', in Comprehensive Composite Materials, editors-in-chief, A.Kelly and C.Zweben, vol. **2**, *Polymer Matrix Composites*, eds. R. Talreja and J.-A.E. Manson, pp. 107–150, Elsevier Science Ltd.

Smith, P.A. and Yeomans, J.A. (2002), 'Benefits of fiber and particulate reinforcement' Vol. **II**, Materials Science and Engineering, Knowledge for Sustainability – An Insight into the Encyclopaedia of Life Support Systems, (UNESCO Publishing-Eolss [e-books], Publishers, Paris, France, Oxford, UK). http://www.eolss.net/Eolss-sampleAllChapter.aspx.

The European Environmental Agency (EEA) (2011), 'Huge renewable energy growth this decade, if EU countries meet projections', Published by the European Environment Agency, 28 November 2011, in the National Renewable Energy Action Plans of the European Member States, (2011 update). (Accessed from internet on 20 November 2011).

The International Energy Agency (2008), 'Energy Technology Perspectives', OECD/IEA. Paris 2008.

The National Space Society (2007), 'Space solar power – limitless clean energy from space', Report to the Director, National Security Space Office, National Space Society, Washington, DC, USA.

The Times (2010), 'High cost could halt Brown's wind farm plan' 26 February 2010, Energy Editor, Robin Pagnamenta.

The Engineer (2011), 'Wind Power reveals 10 MW turbine design' 26 July 2010. http://www.theengineer.co.uk/news/wind-power-reveals-10mw-turbine-design/1003935.article#ixzz1ISWKqPKT. (Accessed 4 April 2211).

Thorne, A. and Hollaway, L.C. (1990), 'High-Technology carbon-fibre/polyether-sulphone composite for space applications', *Proceedings of ESA Symposuim: Space Applications of Advanced Structural Materials*, ESTEC, Noordwijk, NL, 21–23 March 1990, pp. 207–211.

Thostenson, E.T., Ren, Z.F. and Chou, T.-W. (2001), 'Advances in the science and technology of carbon nanotubes and their composites: A review', *Composites Science and Technology*, Vol. **61,** Issue 13, pp. 1899–1912.

Thostenson, E.T., Li, C. and Chou, T. (2005), 'Nanocomposites in review, *Composites Science and Technology*, Vol. **65**, pp. 491–516.

Tran, N.H., Wilson, M.A. Milev, A.S., Dennis, G.R., Kannangara, G.S.K. and Lamb, R.N. (2006), 'Dispersion of silicate nano-plates within poly(acrylic acid) and their interfacial interactions', *Science and Technology of Advanced Materials*, Vol. **7**, No. 8, pp. 786–791.

Utraki, T. (2004), 'Clay containing polymeric nanocomposites', Vol. **1**, *RAPRA Technology*, Vol. **44**, pp. 2441–2446.

van Rijswijk, K. and Bersee, H. (2006), 'Thermoplastic composite wind turbine blades', Dutch Wind Workshop – Results and future of wind energy research in the Netherlands, 11–12 October 2006.

Wang, Z., Lan. T. and Pinnavaia, T. (1996), 'Hybrid organic-inorganic nanocomposites formed from an epoxy polymer and a layered silicate acid (magadiite), *Chemistry of Materials*, Vol. **8**, pp. 2200–2204.

Wang, L., Wang, K., Chen, L., He, C., Wang, L. and Zhang, Y. (2006), 'Hydrothermal effects on the thermomechanical properties of high performance epoxy/clay nanocomposites', *Polymer Engineering and Science,* Vol. **46**, pp. 296–299.

Wang, J.J. and Ren, F. (2010), 'An innovative technique for evaluating the integrity and durability of wind turbine blade composites', Report FY2010, prepared by Oak Ridge National Laboratory, Tennessee, for the U.S. Department of Energy (Contract DE-AC05-00OR22725).

Wingo, D.R. (2004), 'Orbital recovery's responsive commercial space tug for life extension missions' *2nd Responsive Space Conference*, 19–24 April, 2004, Los Angeles, CA, USA, (9 pages).

Wiser, R., Barbose, G. and Peterman, C. (2009), 'Tracking the Sun – The installed cost of Photovoltaics in the US from 1998–2007' Pub. Environmental Energy Technologies Division, Lawrence Berkeley National Laborastory, USA.

Woebbeking, M. (2007), 'Development of a new standard and innovations in certification of Wind Turbines' IEC WT 01 vs. IEC 61400–22, pub. Germanischer Lloyd Industrial Services GmbH, Renewables Certification (GL), Hamburg, Germany.

Woebbeking, M. (2010), 'The new guideline for the certification of wind turbines', (Edition 2010), pub. Germanischer Lloyd Industrial Services GmbH, Renewables Certification (GL), Hamburg, Germany.

Xu, L.R. (2010), 'Intrinsic Impact and Fatigue Property Degradation of Composite Materials in Seawater', final technical report, published by Vanderbilt University, Nashville, Tennessee, USA, for the Office of Naval Research, $ 130,000. 10/22/2007–2/28/2010.

Yang, B., Huang, W. M., Li, C. and Li, L. (2006), 'Effects of moisture on the thermomechanical properties of a polyurethane shape memory polymer', *Polymer Engineering and Science*, Vol. **47**, No. 4, pp. 1348–1356.

10.12 Appendix: solar power

Solar power has great potential and it is the largest energy source available to mankind for consumption on Earth, and is limitless; if utilised, it could supply energy to mankind to meet many times the present demand, The International Energy Agency (2008). The basic idea of SBSP was first investigated in the 1970s, when solar panels (at low orbit) were positioned on a satellite to beam the collected energy from the sun to a receiver on Earth to be converted into electricity. When a satellite is placed in GEO, it can be exposed to sunlight for 24 h per day with no interruption due to cloud cover. With microwave transmission at frequencies up to about 10 GHz, they can move through Earth's thick atmosphere with little absorption, allowing most of the power collected to travel from the solar collector to the receiver on Earth. However, microwaves tend to spread out as they travel, so for great distances large receivers are required to capture the energy being beamed. Consequently, solar collectors at a GEO would require a microwave receiver on Earth to cover hundreds of square miles; this clearly is not a practicable option. Little (2011) has outlined a design for a space-based solar platform that first beams a laser from a solar-collecting satellite to another satellite positioned some 20 km above the surface of the Earth. This satellite would be equipped to transform the laser to microwaves and would then beam that energy to a ground receiver.

10.12.1 Earth-based solar power (EBSP) technology

EBSP is abundantly available, but it is variable and intermittent; it is less effective in overcast or cloudy conditions, and cannot generate electricity at night. Its conversion tends to be material-intensive, leading to high investment costs; these are decreasing as more experience is gained in this area, Wiser *et al.* (2009).

The two most frequently discussed solar technologies for the production of electricity are solar PV, which use semiconductor materials to convert sunlight into electricity, and concentrating solar power (CSP), which concentrates sunlight on a fluid to produce steam to drive a turbine, thus producing electricity.

The solar receiver systems concentrate the solar radiation for large-scale energy production including distribution. CSP systems use lenses or mirrors and tracking systems to focus a large area of sunlight into a small beam. The concentrated heat is then used as a heat source for a conventional power plant. One technology, and the most advanced, uses rows of parabolic troughs to focus the small beam onto a central-pipe receiver which runs above the troughs. Pressurised water and other fluids, generally molten salts, are heated in the pipe and are used to generate steam to drive a turbogenerator for electricity production or to provide industry with heat energy.

BrightSource Energy Inc., Oakland, California, manufacture power plants to generate power from solar thermal technology by creating high temperature steam to turn a turbine. Their solar thermal system uses proprietary software to control thousands of heliostats, each of which consists of two flat glass mirrors, supported by a lightweight steel support structure, that are mounted on a single pylon equipped with a computer-controlled drive system. Composites could readily be used as the support structure for the mirrors and would be an advantage in hostile environments due to their resistance to corrosion. The largest solar thermal power plant in the world is currently being built by BrightSource Energy Inc., at Ivanpal, California.

EBSP requires a considerable acreage of land for collection and production of electricity, therefore this technology competes with agriculture and forestry; consequently, the availability of land is a limiting factor in the development of EBSP energy.

10.12.2 Space-based solar power (SBSP) method

Considering an outline design as suggested by Little (2011), the collected electrical energy from solar collectors in Satellite 1 at GEO would be beamed to Earth by lasers onto Satellite 2 positioned some 20 km above the Earth. Satellite 2 would support the equipment to transform the laser to microwaves, which would be beamed to a ground receiver. Ideally, the structural system to support the collectors and equipment for Satellites 1 and 2 would be fabricated from a polymer/fibre composite skeletal structure; however, it is the structural support system for Satellite 1 at GEO which this chapter has discussed. The two satellites in relationship to the sun and Earth are shown in Fig. 10.11.

Currently, there are two techniques for placing large backing frames in space to support collectors and equipment; it is suggested that these are manufactured from polymer composite material systems as:

- A rigid deployable skeleton support structure.
- An inflatable and flexible continuum structure (RIFC).

The rigid deployable skeletal system

The unit building blocks of the RDSS and its deployment mechanisms would be manufactured on Earth and collapsed into a minimum volume as compact packages, and stowed in the cargo bay of the space transporter in their operating configurations for launch and deployment at LEO. One method for deployment of the space skeletal structure could be achieved by releasing potential energy which would be stored in the joints and in the centre of certain members of the skeletal structure in the stowed configuration of the structure; this is the system to be discussed here.

An alternative method could be by external means, such as electrically driven motors or inflatable devices; the former system will be discussed in this chapter. If necessary the unit building blocks after deployment at LEO would be joined in space by EVA and also any equipment that would be used at GEO would be attached at LEO; this would involve either astronauts or specifically designed robots. The completed structure could then be taken, if required, to GEO, by space tug, Wingo (2004), or under its own method of propulsion. The skeletal space structure positioned at 20 km above the Earth would be manufactured in a similar way. This system is described in the chapter.

The inflatable and flexible continuum structure

The unit elements of the RIFC structure would be stowed into a minimum volume in the cargo bay of the space transporter, launched and deployed at LEO. The various unit elements would be inflated and joined by EVA. Any equipment required would be connected at LEO and the whole would be transported to GEO. This system is described in the chapter.

Index

CPSIA information can be obtained at www.ICGtesting.com
Printed in the USA
LVOW04*2020140914
404040LV00001B/19/P

9 781845 691806